Fundamentals of Power System Resilience

IEEE Press
445 Hoes Lane
Piscataway, NJ 08854

Fundamentals of Power System Resilience

Disruptions by Natural Causes

Mathaios Panteli
University of Cyprus, Cyprus

Rodrigo Moreno
University of Chile, Chile

Dimitris Trakas
National Technical University of Athens, Greece

Magnus Jamieson
University of Strathclyde, Glasgow, United Kingdom

Pierluigi Mancarella
The University of Melbourne, Australia

Goran Strbac
Imperial College London, United Kingdom

Nikos Hatziargyriou
National Technical University of Athens, Greece

IEEEPress
WILEY

Published by John Wiley & Sons, Inc., Hoboken, New Jersey.

Library of Congress Cataloging-in-Publication Data Applied for:

Hardback ISBN: 9781119815990

Cover Design: Wiley
Cover Images: © Raymond Gehman/Getty Images, © Visual News Pakistan/Getty Images, © Mike Hardiman/Shutterstock, © Arctic-Images/Getty Images

Set in 9.5/12.5pt STIXTwoText by Straive, Chennai, India

Contents

About the Authors

Mathaios Panteli is an Assistant Professor in the Department of Electrical and Computer Engineering and the KIOS Research and Innovation Center of Excellence, University of Cyprus, since January 2021. He also holds an Honorary Lecturer position in the Department of Electrical and Electronic Engineering, Imperial College London, UK, and an Andre Jaumotte Chair Professor in Université Libre de Bruxelles (by the Royal Academy of Belgium). Prior to joining UCY, he was a Lecturer at the Power and Energy Division of The University of Manchester, serving as the Deputy Lead of the Sustainable Energy Systems research cluster. Through his academic career, Mathaios has successfully developed a multi-million research portfolio, attracting funding from research councils, industry, and European Commission. His main research interests include techno-economic reliability, resilience and flexibility assessment of future low-carbon energy systems, grid integration of renewable energy sources, and integrated modeling and analysis of co-dependent critical infrastructures. He has published over 130 research articles in high-impact peer-reviewed journals and conference proceedings. Mathaios is an IEEE Senior Member; an IET Chartered Engineer (CEng); the Chair of the CIGRE working group C4.47 "Power System Resilience" and the CIGRE Cyprus National Committee, an invited member of multiple IEEE, CIGRE, and CIRED working groups; and a Fellow of the Higher Education Academy (UK). He served as an Associate Editor for *IEEE Transactions on Sustainable Energy, IEEE Journal of Modern Power Systems and Clean Energy*, and *IET Renewable Power Generation*. He is a recipient of the prestigious 2018 Newton Prize and he led the team that received the 2022 Innovation Radar Prize from the European Commission.

Rodrigo Moreno is a faculty member in the Department of Electrical Engineering, Faculty of Physical and Mathematical Sciences, University of Chile. He coordinates the Dual PhD Program in Electrical Engineering between the University of Chile and the University of Manchester, UK, participates in the Instituto Sistemas Complejos de Ingeniería (ISCI) and leads the Enlace Energy Centre, where he develops strategic projects for governments and the private sector across Latin America and the Caribbean. With over 20 years of experience in power systems and electricity markets, his work has influenced public policy, regulation, and investment decisions, including the design of energy policies, capacity mechanisms, and network security standards. His research has been recognized by IEEE and INFORMS, as well as by the UK Government through the Newton Prize for his contributions to Power System

Resilience. He has published over 100 scientific papers and serves as an Associate Editor for *IEEE Transactions on Energy Markets, Policy and Regulation*.

Dimitris Trakas received his PhD in Power System Resilience Assessment and Enhancement from the National Technical University of Athens (NTUA) in 2018. From 2012 to 2021, he was a researcher with the SmartRue research group at NTUA, contributing to multiple EU- and industry-funded R&D projects. In 2018, he worked at the Research, Innovation, and Development Department of the Independent Power Transmission Operator (IPTO) of Greece, also served as a technical advisor to the CTO. Since 2021, he has been involved in developing software solutions for the virtualization of distribution and transmission systems, enabling more efficient and reliable monitoring, control, and operation.

Magnus Jamieson has spent his career researching what extreme weather and climate change are going to do to the power system, society, and the world more generally and currently works as a researcher at the University of Strathclyde. He maintains a broad interest in energy system security both within Scotland and across Europe and how the whole energy system interacts with society, the economy, and the environment.

He completed his bachelor's degree in Physics at Heriot-Watt University, Edinburgh, before studying for an MSc in Energy, Water, and Environmental Management at the University of Abertay, Dundee. His PhD research at the University of Strathclyde, Glasgow, and Imperial College London concerned how to qualify and quantify the effects of extreme weather on power system resilience and reliability.

Pierluigi Mancarella is Chair Professor of Electrical Power Systems at The University of Melbourne (Australia) and Professor of Smart Energy Systems at The University of Manchester (UK). He is the author of several books and book chapters, as well as over 400 research papers and reports. His key research interests include grid and market integration of renewables and distributed energy resources, energy infrastructure planning under uncertainty, techno-economic modeling and analysis of multi-energy systems, and the security, reliability, and resilience of low-carbon networks.

Pierluigi is a Fellow of the IEEE, an IEEE Power and Energy Society Distinguished Lecturer, a Senior Editor of the *IEEE Transactions on Power Systems*, a Senior Editor of the *Oxford Open Energy* journal, an Editor of the *IEEE Transactions on Energy Markets, Policy and Regulation*, the convenor of the CIGRE working group C6/C2.34 "Flexibility provision from distributed energy resources," the inaugural chair of the Energy Working Group of the IEEE European Public Policy Initiative, and the technical chair of the 2022 International Microgrid Symposium. He has been visiting researcher/professor in a number of international institutions, including NREL in Colorado, Tsinghua University in China, Ecole Centrale de Lille in France, the Universidad de Chile, and

DTU in Denmark as an Otto Mønsted Visiting Professor. Pierluigi is the Australian Director of the USA-UK-Australia Global Centre in Climate Change and Clean Energy, "Electric Power Innovation for a Carbon-Free Society (EPICS)."

Goran Strbac is a Professor of Energy Systems, with extensive experience in advanced modeling and analysis of operation, planning, security, and economics of multi-energy systems. He led the development of novel advanced analysis approaches and methodologies that have been extensively used to inform industry, governments, and regulatory bodies about the role and value of emerging new technologies and systems in supporting cost-effective evolution to smart and secure low carbon energy future. He is currently Director of the joint Imperial-Tsinghua Research Centre on Intelligent Power and Energy Systems, Member of OFGEM Climate Resilience Expert Panel, Member of the UK Smart System Forum, Member of the European Technology and Innovation Platform for Smart Networks for the Energy Transition, and Member of the Joint EU Programme in Energy Systems Integration of the European Energy Research Alliance. He co-authored 4 books and published over 500 technical papers.

Nikos Hatziargyriou has been with the Electrical and Computer Engineering School, National Technical University of Athens (NTUA), since 1984 as a Professor in Power Systems since 1995, and as Professor Emeritus since 2022. He is also a part-time Professor at the University of Vaasa, Finland. He has over 10 years of industrial experience as Chairman and CEO of the Hellenic Distribution Network Operator (HEDNO) (2015–2019) and as Executive Vice-Chair and Deputy CEO of the Public Power Corporation (PPC) (2007–2012), where he was responsible for the Transmission and Distribution and the Islands Divisions. He has authored more than 350 journal publications and 600 conference proceedings papers. He was included in the 2016, 2017, and 2019 Thomson Reuters lists of the top 1% most cited researchers and has been included for the past several years in the global list of top 2% Scientists released by Stanford University. He is the 2020 Globe Energy Prize laureate, recipient of the 2017 IEEE/PES Prabha S. Kundur Power System Dynamics and Control Award and the 2023 IEEE Herman Halperin Electric Transmission and Distribution Award. He was the Chair of the Power System Dynamic Performance Committee of IEEE and Editor-in-Chief of the IEEE Transactions on Power Systems. He is currently EiC-at-Large for IEEE PES Transactions. He is an honorary member of CIGRE and past Chair of CIGRE SC C6 "Distribution Systems and Distributed Generation." He has participated in more than 60 R&D projects funded by the EU Commission, electric utilities, and industry for fundamental research and practical applications.

Foreword by Professor Chongqing Kang

The resilience of our energy systems has never been more critical. Across the globe, societies are witnessing the mounting toll of climate-driven disasters and the technical complexities of integrating vast amounts of renewable energy. These forces are converging to expose the limitations of traditional reliability standards and to demand a more comprehensive, future-oriented perspective.

This book rises to that challenge. It stands out as one of the most thorough and forward-looking explorations of power system resilience to date, offering a body of knowledge that is both conceptually clear and practically applicable. Its contribution lies not only in defining resilience but in demonstrating how it can be measured, modeled, operationalized, and governed in the service of secure and sustainable energy futures.

One of the book's most significant achievements is the clarity it brings to a concept that is often misunderstood. Too often resilience has been conflated with reliability, creating confusion in both scholarship and practice. Here, resilience is carefully articulated as a dynamic capacity that involves anticipation, absorption, adaptation, and recovery. This framing provides readers with a vocabulary and perspective that can be applied consistently across diverse contexts, moving the discussion beyond simplistic metrics and into a more nuanced, multidimensional understanding.

Equally important is the book's methodological innovation. It showcases how resilience can be quantified and integrated into planning and operations using advanced approaches such as fragility modeling, stochastic optimization, and machine learning. By moving decisively beyond deterministic "N-1" criteria, these methods reflect the true realities of uncertainty, complexity, and extreme events. The result is a set of tools that not only capture risk more accurately but also guide operators and decision-makers toward more robust and adaptable solutions.

The strength of this work also lies in its practical orientation. Resilience here is not treated as an abstract or aspirational concept, but as a tangible set of measures that can inform day-to-day decision-making. It demonstrates how resilience thinking can shape investment strategies, operational planning, and the use of distributed resources such as storage and microgrids. The emphasis on preventive, emergency, and restorative responses illustrates how resilience is enacted over time, showing that it is not a one-off measure but a continuous process of preparation, response, and recovery.

What makes the book particularly compelling is its recognition that resilience must be understood at multiple levels. It is not confined to infrastructure alone, but extends to operations, organizations, markets, and governance. By addressing this multi-layered nature, the book reflects the reality that vulnerabilities in one domain can quickly cascade into others, and that true resilience requires coordination across technical, institutional, and societal spheres.

This breadth is further reinforced by the way the book bridges engineering with economics, regulation, and policy. Resilience is presented not only as an engineering challenge but as a socio-technical imperative, requiring market incentives, regulatory frameworks, and governance structures that enable and reward resilient outcomes. This interdisciplinary lens ensures that the insights are relevant to a wide range of stakeholders, from researchers and engineers to policymakers, regulators, and industry leaders.

Another valuable contribution is the grounding of its insights in real-world experience. Drawing lessons from practical disruptions, the book demonstrates how resilience strategies have been tested, refined, and proven under actual conditions. This grounding not only builds confidence in their feasibility but also provides concrete models that can be adapted and replicated in other contexts.

Above all, the book is forward-looking in its vision. It makes clear that resilience and decarbonization are not competing agendas but mutually reinforcing imperatives. By placing resilience at the center of the energy transition, it shows how societies can create power systems that are at once more robust, more sustainable, and more equitable. In highlighting the roles of digitalization, distributed resources, and new operational paradigms, it positions resilience as a cornerstone of the low-carbon, climate-resilient future we must build.

Taken together, these qualities make the book much more than a technical reference. It is at once a scholarly synthesis, a practical guide, and a policy roadmap. It provides the intellectual and practical tools needed to design energy systems that can withstand shocks, adapt under stress, and recover quickly from disruption. In doing so, it challenges us to think differently about risk, to embrace uncertainty as an inherent design condition, and to adopt resilience as a guiding principle for infrastructure, markets, and governance alike.

As the energy sector undergoes one of the most profound transformations in its history, the ideas advanced here will prove indispensable. For researchers, practitioners, and policymakers committed to safeguarding the infrastructures upon which modern society depends, this book is not only timely—it is essential.

Professor Chongqing Kang
Dean of Department of Electrical Engineering
Tsinghua University
Beijing, China

President of Tsinghua University
Energy Internet Research Institute

President of Tsinghua Sichuan
Energy Internet Research Institute

Foreword by Professor Ian Dobson

When we toggle a light switch, our expectation is that the light turns on. Occasionally, perhaps several times a year, there is no electric power due to a blackout, but power is often restored in minutes or a few hours. This is addressed as power system reliability by monitoring and investing in the electric infrastructure to minimize the most common smaller blackouts. But what happens when extreme weather causes extensive damage to the electric power system, and the power is out for days or weeks? Mitigating these large blackouts is addressed by power system resilience. Large blackouts are rare, but not so rare that they can be dismissed as outliers or "perfect storms." Indeed, large blackouts have high risk due to their huge impact on our daily life and work. Moreover, the frequency and severity of extreme weather will slowly increase as global warming continues. It costs too much to build an electrical infrastructure that can completely withstand all the weather extremes, so the practical problem is to analyze, plan, and invest to minimize the damage and to quickly restore power.

Electric utilities, regulators, and researchers around the world are continuing to work to improve resilience. However, resilience is challenging and multi-faceted, and there is no consensus on how to measure resilience and invest to improve resilience. Many exciting new approaches are emerging but in a scattered way. My international research colleagues Mathaios Panteli, Rodrigo Moreno, Pierluigi Mancarella, Goran Strbac, and Nikos Hatziargyriou are not only world leaders in the electrical engineering research on resilience, but have written this comprehensive textbook to define, consolidate, and clearly explain the state of the art. Fundamental questions about resilience are thoroughly addressed: How is resilience defined and how does it differ from reliability? What are the phases of resilience and how does one track performance in the phases with multiple metrics? How do resilience methods focus on, quantify, and reduce the risk of extreme events? How does one model the power system strength and simulate the extreme weather damage and its consequences? How does one distinguish infrastructure and operational resilience and plan investments in them? How can microgrids and distributed generation improve resilience? How can regulators and utilities reduce extreme event risk with coherent cost-effective policies, and, last but not least, who should provide and pay for resilience? This textbook gives coherent answers to these questions and is required reading for all those dedicated to keeping the lights on.

Ian Dobson
Iowa State University

Preface

As power systems globally undergo a profound transformation driven by decarbonisation, digitalisation, and the growing integration of renewable energy sources, the need to move beyond traditional reliability standards towards risk-aware, resilience-focused strategies is becoming increasingly urgent. Extreme weather, natural disasters, and other high-impact, low-probability events are exposing the vulnerabilities of modern power systems, urging for new approaches that combine robustness, adaptability, and rapid recovery. This book addresses explicitly these challenges by providing a comprehensive exploration of the fundamentals of power system resilience, from conceptual frameworks and metrics to advanced modeling approaches, operational measures, and the exploitation of distributed energy resources such as microgrids. It also examines the critical roles of investment strategies, policy, regulation, and market design in shaping resilient energy futures. Through theoretical insights, practical tools, and real-world case studies, the book equips researchers, practitioners, and policymakers with the knowledge needed to design and operate power systems capable of withstanding and adapting to an increasingly uncertain world.

This book brings together more than 15 years of close collaborative work among the co-authors, who have pioneered the topic of resilience in the context of power systems. Drawing on their experience in leading international research projects, developing advanced methodologies, and engaging with utilities, regulators, and policymakers, the authors bring a unique blend of academic rigor and real-world insight. The content reflects not only theoretical progress but also lessons learned from practical applications, case studies, and industry partnerships, making the book both a research reference and a guide for practitioners seeking to strengthen the resilience of modern power systems.

Chapter 1 introduces the scope of power systems resilience, and examines the transition from reliability-centered frameworks to resilience-focused strategies in power systems. It initiates the discussion by highlighting that conventional N-1 and N-2 reliability standards are proving insufficient against high-impact, low-probability (HILP) events such as extreme weather and natural disasters. Resilience, by contrast, stresses robustness, adaptability, and rapid recovery to ensure systems can endure and bounce back from such shocks. After reviewing historical reliability practices, both deterministic and probabilistic, the chapter highlights the need for risk-aware approaches. It then outlines the book's structure, including resilience metrics, data challenges, investment and operational measures, the role of distributed energy resources and microgrids, and the influence of policy and regulation.

Building on Chapter 1, Chapter 2 then explores the shocks and stresses affecting the critical power infrastructure, showing how their combined effects can undermine resilient performance. It reviews common definitions of reliability before introducing key definitions of power system resilience, clarifying the frequent confusion between the two. Resilience is presented as a multi-dimensional concept, with essential capacities for withstanding, absorbing, and rapidly recovering from high-impact, low-probability events. The chapter then introduces popular resilience frameworks, such as the resilience triangle and trapezoid, and links them to core resilience features. It concludes with a discussion of different resilience types, including infrastructure, operational, and organizational.

Laying on the foundations of Chapter 1 and 2, Chapter 3 introduces the need for resilience metrics in the power sector and presents a quantification framework based on the resilience trapezoid from Chapter 2, with a focus on the multi-temporal and multi-spatial resilience metric systems. A detailed example on a simplified UK transmission network demonstrates step-by-step system-level resilience assessment. The chapter then examines risk metrics to better capture tail risks and black swan events, before presenting a novel cascading model that simulates climate-induced outages and their propagation. Illustrative results show how the model integrates seamlessly with established resilience frameworks.

Chapter 4 then presents methodologies for integrating failure rate, network-branch, and weather data to quantify and visualise overhead line failure probabilities. These are linked with wind power projections to assess the interplay between power transport risks and generation availability during extreme wind events. A case study on the Northern Scottish transmission network demonstrates the approach through a simple Monte Carlo simulation, with exemplar results provided. The chapter also addresses data ambiguity – the uncertainty inherent in statistical models connecting natural hazards to component failure probabilities. While not a full system risk analysis, the examples lay the groundwork for risk calculation methods, supported by pseudocode and practical illustrations.

Chapter 5 examines strategies to strengthen power system resilience against natural hazards, incorporating high-impact, low-probability events into investment planning. It introduces mathematical models for resilient network design and highlights the need for balanced investment across grid expansion, infrastructure hardening, and smart technologies. Probabilistic tools and risk metrics such as Conditional Value at Risk (CVaR) are emphasized, alongside stochastic and robust optimization approaches. The chapter also distinguishes resilience from traditional reliability, proposes a holistic planning framework, and considers fairness in investment decisions. Finally, it explores the roles of distributed energy resources, non-wire solutions, long-term storage, and climate change impacts in shaping future resilient energy systems, offering guidance for planners and policymakers.

Chapter 6 focuses on operational measures that enhance power system resilience through "smart" control actions. While such measures cannot prevent physical damage from external threats, they can mitigate outages and help avoid widespread blackouts. The discussion covers resilience strategies across all phases of extreme events, modeling their impacts on system operation. It also examines machine learning applications and the challenges of operating low-inertia, renewable-rich grids under stress. Case studies, including preventive unit commitment and defensive islanding, illustrate the effectiveness of these approaches. Complementing Chapter 6, Chapter 7 examines the role of Distributed Energy Resources, with a focus on microgrids (MGs), as a key operational measure for strengthening power system resilience. MGs can operate either grid-connected or islanded, supporting resilience before, during, and after extreme events. Practical examples from past natural disasters are reviewed, along with a basic framework for MG formation as a preventive or corrective measure. The chapter also addresses the use of MGs in system restoration, the challenges of restoring MGs themselves, and optimal resource scheduling to enable feasible islanding with minimal consumer disruption.

Concluding the book, Chapter 8 examines the economic, market, regulatory, and policy dimensions of power system resilience, supported by real-world case studies. It highlights how the shift to low-carbon, weather-dependent energy systems can increase grid fragility, creating a need for new standards, techno-economic models, and decision-making tools to address security, reliability, and resilience. Beyond network resilience, the chapter considers resource adequacy and generation resilience, raising questions about the roles of technical, market, regulatory, and governmental instruments in achieving resilience objectives, shaped by stakeholder risk attitudes. In the context of global decarbonisation and more frequent climate-driven hazards, we also explore the interplay between decarbonisation policies and resilience, and the opportunities offered by distributed energy resources and grid digitalisation.

Next to the co-authors of the book, there are many researchers who have contributed to the material of this book by their knowledge, research efforts and fruitful collaborations over the years. We are indebted to all of them, but we feel obliged to refer to some of them individually and apologize in advance for the names we might forget. We would like to start by thanking Robin Preece, Matthias Noebels, Yitian Dai, Balaji Venkatasubramanian, Seyedsina Hashemi, Marios Shimillas and Georgios Paphitis, and the members of the CIGRE Working Group C4.47 "Power System Resilience" (particularly Andrea Pitto, Diego Cirio and Emanuele Ciapessoni, RSE Italy) for the fruitful discussions on defining and conceptualizing resilience. We would also like to acknowledge the contributions of Farshad Mohammadi and Mostafa Sahraei-Ardakani for their leading contributions to the research on applying machine learning techniques to enhance resilience, Ektor-Ioannis E Stasinos for his contribution to the research on the role of microgrids in power system resilience and Kaiyuan Pang for his fundamental contributions in Microgrids formation. We would also like to thank Jacob Kelly, Keith Bell and Simon Tindemans for their support and contributions on data and fragility analysis of power systems. We are also grateful to the researchers involved in and around the Newton-Picarte project, whose work ultimately led to the Newton Prize awarded by the UK Government, particularly Hugh Rudnick, Duncan Shaw, Juan Carlos De La Llera, Fernando Ordoñez, Tomas Lagos, Alejandro Navarro, Juan Carlos Araneda, Rafael Sacaan, Alexandre Street, Diego Alvarado, Alex Villamarín and Jenny Moreno, whose interdisciplinary insights and collaborative spirit were essential to advancing resilience research. Last but not least, we would like to thank Julian Eggleston and Christiaan Zuur for the insightful discussions on resilience-oriented regulatory developments at the Australian Energy Market Commission, as well as Farhad Billimoria and Francisco Muñoz for the enlightening conversations on the increasingly complex interplay between physics and economics in a fast-changing grid.

1

From Reliability to Resilience

1.1 Why Power System Resilience?

Extreme events – primarily natural disasters and climate-driven severe weather – have increasingly exposed the limitations of current, traditional reliability frameworks in power system design and operation. Historically, power system planners, regulators, and policymakers have not accounted for such events in network reliability standards. Instead, the design and operation of electric power infrastructure have centered on so-called credible (or "average") outages, such as single or double faults – commonly referred to in power system terminology as N–1 or N–2 contingencies [1]. While these approaches have ensured continuity of supply during minor, everyday disruptions, they fall short in addressing the challenges posed by extreme events.

Recent disasters have underscored the inadequacy of traditional methods. Bushfires in Australia, the Americas, and Greece; flooding events in the United Kingdom; storms in the Americas; and earthquakes in Pacific Ring of Fire countries have caused disruptions far beyond the scope of N–1 or N–2 outages. For example, Chile experiences more than 300 earthquakes exceeding 4.5 Mw annually, resulting in significant power interruptions each year. In the United Kingdom, severe storms and floods have led to prolonged outages, such as the 56-hour interruptions caused by flooding in North West England in 2016. Table 1.1 provides an overview of how natural events have triggered significant power outages. Globally, the World Meteorological Organization reports over 11 000 weather- and climate-related disasters between 1970 and 2019, resulting in 2 million deaths and $3.64 trillion in economic losses [2].

The need for change is urgent: power systems must evolve to address high-impact, low-probability (HILP) events, which are often underestimated or neglected by traditional reliability frameworks. This necessity is further amplified by aging infrastructure, increasing societal dependence on electricity, and emerging threats such as climate change. To meet these challenges, decision-makers must adopt resiliency-driven and risk-aware engineering practices that explicitly integrate resilience into operational and planning standards. This shift is critical to addressing both existing vulnerabilities and the growing challenges posed by climate change and other stressors.

The Intergovernmental Panel on Climate Change (IPCC) emphasizes the urgency of this transformation for all critical infrastructure. Projections indicate a rise in extreme heat, fire weather, heavy precipitation, flooding, and severe windstorms across Europe, alongside similar global challenges [3]. Therefore, resilience must become a cornerstone of future power system design and operation.

This book seeks to address pivotal questions: How can resilience-thinking be effectively incorporated into power system methodologies? What constitutes the optimal portfolio of measures for

Fundamentals of Power System Resilience: Disruptions by Natural Causes, First Edition.
Mathaios Panteli, Rodrigo Moreno, Dimitris Trakas, Magnus Jamieson, Pierluigi Mancarella, Goran Strbac, and Nikos Hatziargyriou.

Table 1.1 Significant blackouts caused by natural hazards.

Country	Date	Disaster name	Size of blackout	Effects on society
United States	August 25, 2005	Hurricane Katrina	2.7 million people without power, 10+ days	Major displacement, economic loss, and deaths due to lack of power for critical infrastructure like hospitals
Japan	March 11, 2011	Tōhoku Earthquake and tsunami	21 GW, several weeks	Nationwide economic disruption, rolling blackouts, and societal stress after the Fukushima nuclear disaster
India	July 30, 2012	2012 India blackouts (Monsoon flooding)	32 GW, affected 620 million people	Largest power outage ever, massive economic disruption, halted transportation, and widespread societal stress
Puerto Rico	September 20, 2017	Hurricane Maria	3.7 million people without power, ~11 months	Longest blackout in US history; major humanitarian crisis, public health emergencies, and infrastructural collapse
Philippines	November 08, 2013	Typhoon Haiyan	1 month	16 million people affected; loss of livelihoods and severe societal disruption
China	May 12, 2008	Sichuan Earthquake	4 million households without power, ~15 days	Infrastructure damage, agricultural and industrial losses, and disrupted supply chains
Chile	February 27, 2010	2010 Maule earthquake	4522 MW loss, 2 weeks	Severe disruption to infrastructure, widespread blackouts, and difficulty restoring electricity
Italy	September 28, 2003	2003 Italy blackout (storm-induced grid overload)	177 GWh of energy not supplied (ENS), 13 hours	Nationwide disruption, halted trains and airports, and emergency services stretched
Canada	January 05, 1998	North American ice storm of 1998	5 million people without power, up to 1 month	Severe economic impacts, fatalities due to cold, and reliance on emergency shelters
Australia	September 28, 2016	2016 South Australian blackout (severe storms)	~1.1 million people, 8 hours	South Australia-wide blackout; economic impact on manufacturing and agriculture
Mexico	September 19, 1985	Mexico City earthquake	1/3 of the city hit by power outages	Infrastructure collapse, humanitarian crisis, and significant rebuilding costs
United States	February 10, 2021	2021 Texas winter storm	~4.5 million people, ~7 days	Severe disruption in Texas, major fatalities (246+), loss of heating during extreme cold, and economic impact of ~$195 billion

enhancing power grid resilience? And how can we build power grids that are both robust and flexible enough to withstand unprecedented disruptions? By presenting a resilience-driven framework that bridges theory and practice, this book equips decision-makers with the tools to enhance power system resilience worldwide. It offers methodologies, regulations, theories, and real-world examples to inspire informed and practical solutions for the energy sector.

1.2 The Historical Approaches to Power System Reliability

1.2.1 The Historical N-k Network Security Criterion

The N-k (deterministic) network security criteria or standards require that networks withstand the loss of one or two circuits (i.e. N–1 or N–2) without causing overloads on other circuits, and that such outages do not compromise the integrity of system operations [1]. These standards are applied across both short-term operational and long-term planning time horizons. In short-term operations, network security is maintained by limiting power transfer across the network, often through generation re-dispatch. For long-term planning, security is ensured through investments in network reinforcements. Both approaches generally justify network redundancy (in the form of network congestion in the short term and network investments in the long term) for the provision of security.

Within this framework, outages beyond those deemed credible – and their associated consequences – are generally not analyzed. Network security is achieved through a combination of preventive and corrective control actions. These include managing network congestion and implementing post-contingency controls such as adjusting generation setpoints and utilizing flexible AC/DC network components like HVDC lines, line switches, series compensators, and phase-shifting transformers. However, deterministic security standards are typically biased toward preventive control measures, emphasizing network redundancy.

Furthermore, traditional deterministic standards assume that the system should remain stable under any credible contingency without requiring load shedding. While these rules were designed with transparency and cost-effectiveness in mind, their adequacy must now be reassessed in the face of extreme events that go far beyond N–1 or N–2 contingencies. Furthermore, deterministic standards have been historically criticized for their lack of economic optimization. Specifically, they do not account for the social costs of post-fault demand curtailment (and corrective control actions) in comparison to the costs of preventive control measures (network congestion and investment). This imbalance fosters a probabilistic approach that explicitly evaluates the trade-offs between preventive actions and the expected costs of energy not supplied (ENS) as well as other costs such as generation tripping actions with the associated revenue losses.

To address these limitations, researchers and academics have proposed probabilistic approaches to network security, which are discussed next.

1.2.2 The More Advanced Probabilistic Network Security

A probabilistic network security approach allows for post-fault demand curtailments by appropriately balancing them against the up-front costs incurred in the pre-fault condition due to network congestion and investments. From the perspective of operation, improving the reliability of a power system through network congestion should continue only until the costs associated with such congestion exceed their benefits in terms of reduction in the unsupplied demand costs. This approach

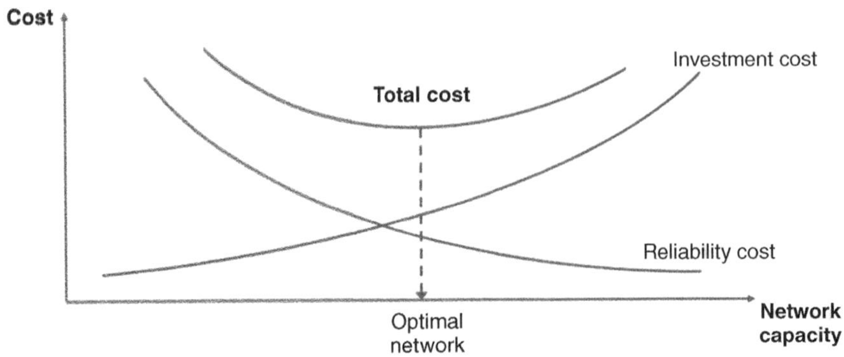

Figure 1.1 Optimal balance between investment and reliability costs.

ensures that operational decisions, such as power transfer levels, are optimized to minimize total costs. These costs include operating expenses, such as congestion and losses, as well as the expected costs of unsupplied demand. The optimization is performed over an array of operating states (i.e. outages), weighted according to their probabilities, resulting in efficient network utilization and a balanced level of risk for specific operating conditions. This means that more congestion can be justified to control reliability levels. Mathematically, the optimization of total cost (operation and unsupplied demand) is performed over an array of operating states (i.e. outages), weighted according to their probabilities, resulting in efficient network utilization and reliability levels.

Similarly, from the perspective of infrastructure planning, network redundancy can be created through new investments rather than simply increasing congestion. In this case, investments are justified only when the marginal savings in demand curtailment costs (and pre-contingency congestion costs) outweigh the marginal costs of the new investments.

In probabilistic reliability assessment, as pioneered by Billinton and Allan [4], the estimated costs from expected ENS are averaged (weighted by probability) across all modeled scenarios and optimized against additional investment and operational costs. Network investments are therefore well justified as a trade-off between economic efficiency and security of supply. If operational costs (e.g. congestion and losses) are simplified or neglected, investments are made until the marginal cost of additional investment equals the marginal benefit of enhanced reliability. This relationship is graphically represented in Figure 1.1, where reliability costs are measured as the product of expected energy not supplied (EENS) and the Value of Lost Load (VoLL). However, while effective for standard reliability challenges, this probabilistic approach has fundamental limitations in addressing HILP events. These limitations and their implications for operational and investment decisions are explored in Section 1.3.

1.3 Need for a (Tail)Risk-Aware Approach

HILP events are, by definition, rare, and their impact on average indicators like EENS is consequently minimal. For example, our analysis indicates that, on average, it would be economically optimal to endure the consequences of very large earthquakes in Chile every 15 years rather than invest in additional infrastructure to reinforce and harden the power system [5]. This raises an important question: Why should we be concerned about events that, on average, have a relatively small effect?

The answer lies in the risk attitudes of electricity consumers and policymakers (and by extension, network planners who serve them). Empirical evidence suggests that both consumers and policymakers strongly dislike the risks associated with the severe consequences of HILP events and therefore seek to minimize them as much as possible. This preference raises critical questions: How can we model this risk attitude? And what assumptions about risk are embedded in the probabilistic assessments currently used?

In risk analysis, attitudes toward risk are generally classified into three categories, which are illustrated as follows:

- *Risk-neutral:* The consumer is indifferent between two options with the same total expected cost.
- *Risk-averse:* The consumer prefers an option with higher predictability and fewer adverse outcomes, even if it is slightly more expensive (on average).
- *Risk-seeking:* The consumer prefers the option with higher variability and larger adverse outcomes, even if it is less predictable.

Consider an electricity consumer choosing between two network services:

1) The first option costs $90, rarely fails, and has minor ENS-related losses totaling $10.
2) The second option costs $50 and fails more frequently, with larger ENS-related losses totaling $50.

In both cases, the total cost is the same, and thus a risk-neutral consumer would view these options as equally attractive. In contrast, most consumers are risk-averse and would prefer the stability and predictability of the first option, even at a slightly higher cost (which is not the case in this example, as both cost the same). This mirrors consumer behavior in other industries, where individuals are willing to pay for insurance to mitigate the impacts of rare but catastrophic events.

This preference for risk aversion is evident even in less extreme scenarios. For instance, during the January 2019 heatwaves in Australia, sporadic rolling load-shedding events occurred in several areas, including central Melbourne. Although these outages had only a minor overall impact, many consumers deemed them unacceptable, illustrating a widespread aversion to uncertainty and disruptions.

Governments, too, often adopt risk-averse stances, as their responsibility extends to ensuring public welfare. For political and societal reasons, they may prioritize actions to hedge against HILP events, regardless of the purely economic risk-neutral rationale. For example, regulatory and market responses were implemented following the South Australia black system event in September 2016, and additional measures were set in response to the widespread bushfires in January 2020.

The key takeaway is that risk-averse approaches and metrics, essential for addressing HILP events, are often absent from current power system planning practices. Incorporating these approaches will require a paradigm shift, moving beyond traditional reliability frameworks to account for the complex interplay of economic, social, and political factors associated with rare but devastating disruptions.

1.4 Inspiration from Other Economic and Engineering Areas

1.4.1 Risk Hedging and Portfolio Optimization

The concept of risk hedging, originating in financial portfolio theory, offers valuable insights for determining an optimal mix of investments to enhance power system resilience. In finance,

Harry Markowitz's modern portfolio theory revolutionized decision-making by proposing diversification as a strategy to maximize returns for a given level of risk [6]. This approach introduced the efficient frontier or Capital Asset Pricing Model (CAPM) balancing risk (represented by the standard deviation) and return (represented by the expected return) across a portfolio of assets.

Building on Markowitz's work, subsequent advancements in risk assessment, such as conditional value at risk (CVaR), have provided robust tools for quantifying and managing risk. CVaR, pioneered by Rockafellar and Uryasev, focuses on minimizing potential losses in the tail-end of a probability distribution [7], making it particularly suited for addressing HILP events. Notably, CVaR integrates seamlessly with linear programming, a critical asset in modern decision-making processes.

Applying these principles to power systems, resilience strategies can be viewed as portfolios of measures – including system expansion, hardening, and smart technologies – that balance cost and risk. Just as financial portfolios are designed to hedge against market volatility, resilience investments aim to mitigate the impacts of extreme events on energy infrastructure. Decisions informed by risk-hedging methodologies ensure that resources are allocated efficiently, targeting the most vulnerable points in the system while optimizing long-term benefits.

For example, investing in microgrids (MGs), distributed energy resources (DERs), and adaptive control systems diversifies the resilience portfolio, reducing reliance on any single strategy and enhancing overall system robustness. By adopting a portfolio optimization approach, decision-makers can better navigate the trade-offs between upfront costs and the potential benefits of reduced downtime, improved recovery, and sustained service delivery during crises.

1.4.2 Risk-Aware Design in Structural and Civil Engineering

The field of structural and civil engineering offers a wealth of practices that can inspire resilience strategies in power systems. One cornerstone concept is the "return period" [8], a metric used to design structures to withstand extreme conditions that may occur once every 50, 100, or even 1000 years. This approach ensures that critical infrastructure remains robust against rare but catastrophic events. Central to this approach is the key question: How rare is the event we need to protect against?

In structural engineering, risk-aware design involves tailoring buildings, bridges, and other infrastructure to endure specific environmental threats such as earthquakes, hurricanes, and floods. These designs integrate both probabilistic and deterministic models, assessing historical data to predict the likelihood and intensity of extreme events. For example, earthquake-resistant buildings are designed to absorb seismic forces based on data about the most severe quakes recorded in a region over centuries.

This philosophy aligns closely with resilience planning in power systems, where strategies must also account for HILP events. By incorporating ideas from structural engineering, power system planners can adopt a similar mindset, designing grids and components to handle worst-case scenarios over long time horizons.

Furthermore, structural engineering emphasizes redundancy and flexibility. Just as bridges are often designed with multiple load paths to prevent collapse in the event of a localized failure, power systems can enhance resilience by creating alternative energy pathways and resources, such as MGs and additional transmission lines. Similarly, flexibility in materials and construction techniques enables buildings to adapt to dynamic forces – a principle mirrored in power systems through adaptive protection schemes and demand response mechanisms.

Another critical parallel is the integration of risk-based decision-making. Structural engineers use tools like fragility curves to assess how different materials and designs perform under stress, enabling cost–benefit analyses for resilience investments. In power systems, similar probabilistic concepts can support the optimization of investments in infrastructure hardening, operational measures, and smart technologies.

By learning from the structured, risk-aware practices of structural and civil engineering, power system resilience can advance toward a more robust and proactive approach. This cross-disciplinary inspiration underscores the importance of designing for long-term challenges, ensuring that both systems and structures are prepared for the uncertainties of an evolving world.

1.5 From Reliability to Resilience Paradigm

The transition from a reliability-centered approach to a resilience-oriented paradigm marks a significant shift in how energy systems are designed, managed, and evaluated. Traditionally, reliability has been the cornerstone of power system planning and operation, ensuring a consistent and uninterrupted supply of electricity to meet consumer demand. While reliability standards have been effective in minimizing routine disruptions, they are insufficient to address the growing complexities and uncertainties of the modern energy landscape. Emerging challenges such as climate change, increased frequency of extreme weather events, cyber-physical threats, and the integration of renewable energy necessitate a broader framework – one that prioritizes not only prevention but also adaptation and recovery.

Reliability in power systems focuses on ensuring that generation, transmission, and distribution networks perform their intended functions under normal operating conditions. Reliability metrics, such as the System Average Interruption Duration Index (SAIDI) and the System Average Interruption Frequency Index (SAIFI) in distribution networks, have traditionally been used to evaluate reliability. These metrics emphasize steady-state performance and the likelihood of equipment or system failures within a given timeframe.

Resilience, by contrast, encompasses the ability of a system to anticipate, absorb, adapt to, and recover from disruptive events [9]. It extends beyond the steady-state analysis of reliability to address the dynamic and often unpredictable nature of threats. Resilience emphasizes three key capabilities [10]:

1) Robustness – the ability to withstand shocks without significant degradation.
2) Adaptability – the capacity to adjust operations in response to changing conditions.
3) Rapid recovery – the ability to restore normalcy quickly and efficiently after an adverse event.

1.5.1 Key Differentiators Between Reliability and Resilience

While reliability and resilience share overlapping goals, the resilience paradigm is broader in scope. Reliability focuses on minimizing the probability of failure, whereas resilience acknowledges that failures are inevitable and seeks to mitigate their consequences. The complementary nature of these concepts is evident in the following areas:

1) *Proactive risk management:* Reliability standards traditionally rely on deterministic approaches, such as the N–1 contingency criterion, which ensures that the system can withstand the failure of a single component without widespread disruptions. However, resilience incorporates

probabilistic and scenario-based analyses to address low-probability, high-impact events. This shift enables planners to account for compounding factors and cascading failures that are not captured by conventional reliability metrics.

2) *System flexibility and redundancy:* Resilience emphasizes building flexibility and redundancy into the system. For example, DERs such as rooftop solar panels and battery storage can enhance resilience by providing localized energy supply during grid outages. Similarly, MGs equipped with advanced control systems can operate autonomously in islanded mode, ensuring continuity of service for critical loads.
3) *Real-time monitoring and adaptive control:* Advances in digital technologies, such as the Internet of Things (IoT), artificial intelligence (AI), and machine learning, enable real-time monitoring and adaptive control of power systems. These technologies enhance both reliability and resilience by providing early warning of potential failures and enabling dynamic responses to evolving conditions.
4) *Interdisciplinary collaboration:* The resilience paradigm requires a multidisciplinary approach that integrates engineering, economics, environmental science, and social dimensions. Unlike traditional reliability standards, which are often narrowly defined, resilience planning engages a broader set of stakeholders, including policymakers, community leaders, and emergency management agencies.

1.5.2 Expanding Resilience Practices

1.5.2.1 Broadening the Scope of Contingencies

Reliability-driven standards focus on managing credible contingencies, such as single- or double-circuit outages (commonly referred to as N–1 or N–2). Resilience, on the other hand, considers a broader range of simultaneous outages that could be triggered by natural events. By encompassing these extreme scenarios, resilience provides a more comprehensive framework for mitigating cascading failures and widespread disruptions.

1.5.2.2 Focusing on High-Risk Scenarios

Traditional reliability metrics often rely on averages, across all scenarios, to quantify system performance. While these averages provide a baseline understanding, they fail to capture the full impact of rare but catastrophic events. Resilience, in contrast, emphasizes a risk-focused approach that considers the tail-end of probability distributions, where the most severe consequences reside. By prioritizing the mitigation of high-risk scenarios, this approach ensures that power systems are better prepared for extreme events.

1.5.2.3 Incorporating Environmental Interactions

Traditional reliability approaches often treat power systems as isolated entities, ignoring the broader environmental context. In contrast, resilience explicitly accounts for the interactions between power systems and their surrounding environment, recognizing that external factors such as storms, floods, or earthquakes directly influence system performance. This holistic perspective enables more effective planning and response strategies tailored to specific environmental threats.

1.5.2.4 Tailoring Strategies to Natural Threats

Resilience is inherently dependent on the nature of the threats faced by a specific region. Areas prone to storms, for instance, require different resilience measures than regions vulnerable to earthquakes. This adaptability ensures that resilience strategies are context-sensitive, addressing the unique risks and vulnerabilities of each location.

1.5.2.5 Integrating Diverse Measures and Technologies

While reliability-driven standards primarily rely on N–1 or N–2 redundancy to ensure security, resilience incorporates a diverse array of measures. These include system expansion, infrastructure hardening, and enhancing responsiveness through advanced technologies and operational practices. This multi-faceted approach provides greater flexibility and adaptability in dealing with extreme events.

1.5.2.6 Emphasizing Dynamics and Recovery

Reliability standards traditionally focus on robustness, aiming to withstand outages through snapshot-type analyses. Resilience, however, emphasizes a dynamic approach that accounts for the entire lifecycle of an event. This includes the ability to "fail gracefully" immediately after a disruption and recover rapidly, as illustrated by the resilience trapezoid curve. By prioritizing recovery alongside robustness, resilience ensures that power systems can adapt and respond effectively to evolving threats.

1.6 Fundamentals of Power System Resilience

The aim of this book is to provide power system engineers and scholars with the fundamentals of power systems resilience. The overview of the book is captured in Figure 1.2.

Before moving to the technical discussions, Chapter 2 first introduces in detail the concept of power systems resilience, outlines the features and characteristics of the concept, and compares it with traditionally used concepts, including reliability and robustness. Further, popular resilience frameworks are presented, including the resilience triangle and trapezoid, where the relation of the time-series performance of a system exposed to an extreme event to the key resilience characteristics is outlined. Hence, Chapter 2 lays the basis for the subsequent book chapters.

Chapter 3, on system resilience assessment and quantification, presents different resilience metrics and provides illustrative examples on real-world systems of how these metrics can be applied to measure the spatial and temporal resilience of a power system. Further, recognizing the need to capture the cascading impacts of weather-induced asset outages, a novel AC cascading fault model is

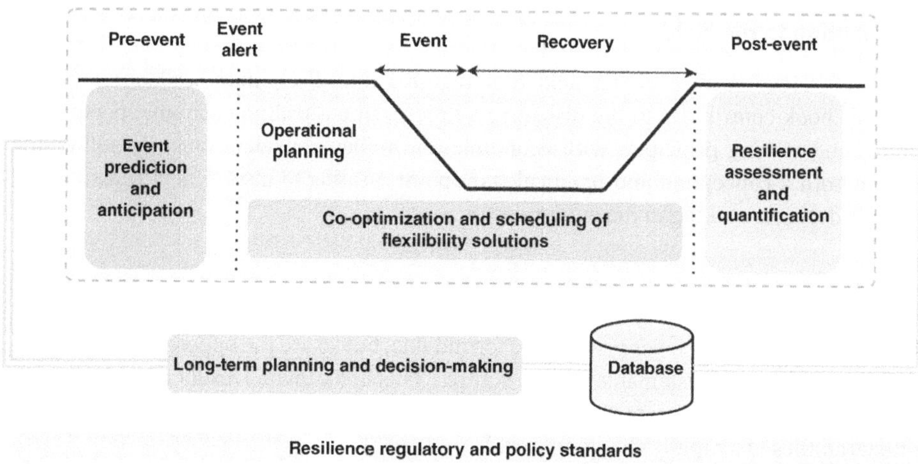

Figure 1.2 Overview of the book context.

presented, highlighting its capability of seamlessly integrating to various resilience metric systems. The resilience assessment models introduced in Chapter 3 are subject to the high uncertainty of the data inputs to these models.

In this context, Chapter 4 next addresses the data granularity and ambiguity, shedding light on methodologies for linking failure rate data, network-branch data, and weather data together for the purpose of visualizing overhead line failure probability for the purpose of providing quantitative assessments of system risk.

Chapter 5 then builds on Chapters 3 and 4 to explore strategies for enhancing power systems' resilience against natural hazards. It incorporates HILP events into investment planning and decision-making, introducing mathematical models for resilient network design. It discusses the need for a balanced investment portfolio between network expansion, infrastructure hardening, and smart grid technologies, emphasizing the significance of probabilistic analysis and risk metrics like CVaR in decision-making. Additionally, it also reviews alternative modeling approaches that use stochastic and robust optimization. Further to investment planning.

Chapter 6 delves into introducing advanced concepts on operational resilience planning and flexibility solutions. This chapter discusses the operational measures applied to enhance power system resilience in all the phases of a progressing extreme event and how the impact on the power system can be modeled to be taken into account during its operation. The use of machine learning techniques and the resilient operation of low-inertia power systems with high penetration of renewable resources during extreme events is also analyzed.

Chapter 7 explores the role of DERs as one of the most effective operational measures for enhancing power system resilience and flexibility. MGs in particular, applied during pre-event, during-event and post-event phases are able to enhance resilience of both transmission and distribution systems, due to their ability to operate connected to the main power grid or islanded. This chapter highlights practical cases in which MGs have been used for enhancing resilience during major natural disasters in the past.

Finally, Chapter 8 aims to discuss, with the aid of several real-world case studies, general economic, market, regulatory, and policy aspects of power system resilience. It also concludes with a discussion on the links between decarbonization policies and resilience, and the potential role and opportunities for DERs and grid digitalization to enhance power system resilience.

1.6.1 Contribution to the Field

Power system resilience is an evolving field that requires interdisciplinary collaboration and innovation. This book contributes to the discourse by presenting a techno-economic perspective that integrates engineering principles with economic and policy considerations. By addressing resilience from both a conceptual and practical standpoint, it aims to inspire new research and inform strategic decision-making in the energy sector.

1.6.2 Closing Remarks

The challenges facing modern power systems are formidable, but so are the opportunities to transform them into resilient and sustainable infrastructures. This book invites readers to engage with these challenges critically and creatively, contributing to a resilient energy future that can withstand the uncertainties of a rapidly changing world.

References

1. G. Strbac, D. Kirschen, and R. Moreno, "Reliability standards for the operation and planning of future electricity networks," *Found. Trends® Electr. Energy Syst.*, vol. 1, no. 3, pp. 143–219, 2016.
2. World Meteorological Organization, Atlas of Mortality and Economic Losses from Weather, Climate and Water-related Hazards (1970–2021), 22 May 2023. [Online]. Available: https://storymaps.arcgis.com/stories/8df884dbd4e849c89d4b1128fa5dc1d6.
3. B. Bednar-Friedl *et al.* "Europe," in *Climate Change 2022: Impacts, Adaptation and Vulnerability*, vol. 2022, G. Kaser and J.M. Moreno, Eds. Cambridge, UK and New York, NY, USA: Cambridge University Press, pp. 1817–1927.
4. R. Billinton and R. N. Allan, *Reliability Evaluation of Power Systems.* New York, NY, USA: Plenum Press, 1996.
5. R. Moreno *et al.*, "From reliability to resilience: planning the grid against the extremes," *IEEE Power Energy Mag.*, vol. 18, no. 4, pp. 41–53, 2020.
6. H. Markowitz, "Portfolio selection," *J. Finance*, vol. 7, no. 1, pp. 77–91, 1952.
7. R. T. Rockafellar and S. Uryasev, "Optimization of conditional value-at-risk," *J. Risk*, vol. 2, no. 3, pp. 21–41, 2000.
8. R. E. Melchers and A. T. Beck "Measures of structural reliability," in *Structural Reliability Analysis and Prediction*, R.E. Melchers and A.T. Beck, Eds. New Jersey, NJ, USA: John Wiley & Sons, 2018, pp. 1–29.
9. IEEE PES IEEE Task Force on Definition and Quantification of Resilience, The definition and quantification of resilience, April 2018. [Online]. Available: https://grouper.ieee.org/groups/transformers/subcommittees/distr/C57.167/F18-Definition&QuantificationOfResilience.pdf.
10. E. Ciapessoni, D. Cirio, A. Pitto, M. Panteli, M. Van Harte and C. Mak, "Defining power system resilience," CIGRE WG C4.47 Power System Resilience, 2019.

2

Conceptualizing and Contextualizing Power Grid Resilience

2.1 Shocks and Stresses on Critical Power Infrastructure

When analyzing critical infrastructure resilience, the focus should go beyond the events that have short-term impact to also consider the effects of longer-term events. In this context, the concept of "shocks and stresses" has been recently emerging in infrastructure system planning, operation and design. A shock is broadly defined as "sudden events that impact on the vulnerability of the system and its components", while stress refers to "long-term trends that undermine the potential of a given system or process". In power systems, for example, a shock can be a flood or a windstorm impacting the system integrity, while a stress can be the longer-term impacts of climate change, or the longer-term impacts of growing penetration of renewable energy sources on system security and resilience. The relation between shocks and stresses can be graphically seen in Figure 2.1. It can be seen that the stronger the shock (e.g. high intensity of windstorm) and the more impactful the stress is (e.g. close to 100% renewable energy integration), the bigger the impact on the infrastructure resilience.

Within this concept of shocks and stresses, it is critically important to differentiate between "typical" outages and extreme events (Table 2.1) [1]. The typical outages are usually more predictable and controllable, and a well-designed, reliable system should be capable of withstanding these events without loss of service. On the other hand, extreme events are very difficult to predict, and they are very often unforeseen events that are not usually considered in any contingency analysis tools. As such, they can have disastrous impacts on the infrastructure, leading to widespread loss of service and longer restoration times. It is very important to recognize this differentiation, which will drive the identification of events that could have high impacts on the system operation, and should be considered in the resilience assessment, quantification and enhancement procedures.

However, it is a very difficult task to clearly classify an event as a "typical" outage or an "extreme" event, as it depends on many factors, e.g., experiences or available data. This, in turn, drives the focus of infrastructure planners and operators, whether it should be on mitigation or response, depending on the available information and experience with an event. In this context, there are mainly three types of extreme or rare events to consider [1]:

1) ***Known-knowns***: This category refers to events that can be reasonably anticipated and how they may manifest themselves. Based on historical data and experiences, their potential future impact can be assessed and hence appropriate mitigation and response measures to be implemented. For example, it is possible from experiences to know where flooding is likely to happen based on environmental and weather data, and therefore flood defences to be built.

Fundamentals of Power System Resilience: Disruptions by Natural Causes, First Edition.
Mathaios Panteli, Rodrigo Moreno, Dimitris Trakas, Magnus Jamieson, Pierluigi Mancarella, Goran Strbac, and Nikos Hatziargyriou.

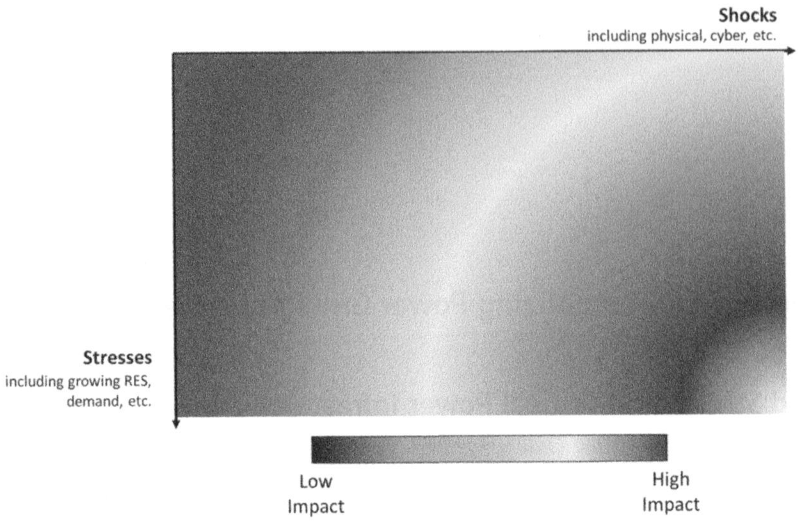

Figure 2.1 Conceptual relation between shocks and stresses.

Table 2.1 Typical outage versus extreme event.

Typical Outage	Extreme Event
Low impact, high probability	High impact, low probability
More predictable/controllable	Less predictable/controllable
Random location and time of occurrence	Spatiotemporal correlation between faults and event
Supported by contingency analysis tools	Unforeseen event
Limited number of faults due to component failures	Multiple faults
Network remains intact	Large portion of the network is damaged/collapsed, with possibly separated areas
Quick restoration	Requiring more time and resources/longer restoration

Source: [1] / IEEE.

2) ***Known-unknowns***: This category refers to events that can be reasonably anticipated, but assessing how they may manifest is reasonably difficult to assess. For this type of event, it is possible to put mitigation plans in place, but primarily, the focus should be on response. For example, in power systems, it is known that a cyber attack might take place, but where it will manifest and which assets might be affected is not known. Hence, it is of course desirable to have cyber defence schemes in place, but response planning and preparation to enable the fast system restoration is very critical.
3) **Unknown-unknowns:** This category refers to events that cannot be reasonably anticipated, nor how they may manifest themselves. In this category, it should be acknowledged that it is difficult to predict all scenarios that may result in service disruption. It could be possible to

assess the impact without knowing the exact cause or initiating event; however, it is more likely for investment and actions to be focused on flexible and quick response and recovery from such events, rather than mitigation. It should be noted though, that mitigation actions to known events may enhance resilience to unknown-unknown events, e.g., building embedded system flexibility to react and adapt to changing, potentially unforeseen events.

2.2 Defining Power System Resilience

Reliability in power systems has been defined differently by various organizations such as CIGRE, IEEE, NERC, and IEC. These definitions are summarized in Table 2.2 [2].

Compared to reliability, the concept of resilience in power systems is a relatively recent development. The term "resilience" (or "resiliency") originates from the Latin word resiliō, meaning the ability to spring back or return to an original state after being subjected to stress – such as bending, compression, or stretching. The first formal definition was introduced by C.S. Holling in 1973, describing resilience as "the persistence of systems and of their ability to absorb change and disturbance and still maintain the same relationships between populations or state variables" [8]. Since then, the concept has evolved significantly across various domains, including safety management, organizational theory, social-ecological systems, and economics. Following Holling's seminal work, numerous interpretations have emerged, leading to a diversity of definitions and an ongoing lack of consensus on a universal understanding of resilience [1, 2].

In the context of power systems, as critical infrastructures, the picture is even more blurred, as the concept of resilience has only emerged as a subject of interest in the last decade or so.

Table 2.2 Definitions of power system reliability from the literature.

Ref	Organization	Definition
[3]	CIGRE	A measure of the ability of a power system to deliver electricity to all points of consumption and receive electricity from all points of supply within acceptable standards and in the amount desired.
[4]	IEEE	Reliability of a power system refers to the probability of its satisfactory operation over the long run. It denotes the ability to supply adequate electric service on a nearly continuous basis, with few interruptions over an extended time period.
[5]	NERC	Reliability of the interconnected bulk-power system is defined in terms of two basic and functional aspects. These aspects are: *Adequacy:* The ability of the electricity system to supply the demand and energy requirements of the end-use customers at all times, taking into account scheduled and reasonably expected unscheduled outages of system elements. *Operational Reliability:* The ability of the bulk-power system to withstand sudden disturbances, such as short-term or unanticipated loss of system elements from credible contingencies, while avoiding uncontrolled cascading blackouts or damage to equipment.
[6]	IEC	The ability of a power system to provide an adequate supply of electrical energy to consumers at all times, taking into account both the continuity and quality of the supply.
[7]	ORNL	The degree to which the performance of the elements in a bulk system results in electricity being delivered to customers within acceptable standards and in the amount desired.

Source: Adapted from [2].

Table 2.3 Widely used resilience definitions.

ID	Authors	Definition
1	UKERC, UK Energy Research Center (UKERC), "Building a Resilient UK Energy System", 2009 [9]	The ability of a power system to withstand extraordinary and high impact-low probability events such as due to extreme weather, rapidly recover from such disruptive events and absorb lessons for adapting its operation and structure to prevent or mitigate the impact of similar events in the future
2	Haimes [10]	The ability of the system to withstand a major disruption within acceptable degradation parameters and to recover within an acceptable time, and to consider composite costs and risks
3	NIAC [11]	(Infrastructure resilience) the ability to reduce the magnitude and/or duration of disruptive events. The effectiveness of a resilient infrastructure or enterprise depends upon its ability to anticipate, absorb, adapt to, and/or rapidly recover from a potentially disruptive event
4	UK Cabinet Office [12]	The ability of assets, networks and systems to anticipate, absorb, adapt to and/or rapidly recover from a disruptive event
5	PSERC [13]	The ability of a system to gradually degrade under increasing system stress, and then to return to its pre-disturbance condition when the disturbance is removed
6	NAURC [14]	Robustness and recovery characteristics of utility infrastructure and operations, which avoid or minimize interruptions of service during an extraordinary and hazardous event
7	Presidential Policy Directive 21: Critical Infrastructure Security and Resilience, 2013 [15]	The ability to prepare for and adapt to changing conditions and withstand and recover rapidly from disruptions. Resilience includes the ability to withstand and recover from deliberate attacks, accidents, or naturally occurring threats or incidents
8	Sandia Lab 2011 [16]	Given a **disruptive event** (or set of events), the resilience of a system to that event (or events) is the ability to reduce **"efficiently"** both the magnitude and duration of the deviation from targeted **system performance** levels The **bold words** of this definition are key components of resilience; further discussion follows below: • **Disruptive event:** Different disruptions may affect a system in different ways and thus necessitate different recovery processes. Hence, a system may have different levels of resilience to different disruptions. This definition considers resilience of a system to a specific disruption • **Efficiently:** Efficiently means using the lowest possible amount of resources during recovery processes; depending on the domain, these resources could be dollars, repair man-hours, infrastructure replacement assets, or time • **System performance:** Given the flexibility of many systems to adjust and reconfigure to a disruptive event, maintaining system structure is not as important as maintaining system performance. Hence, measurement of resilience should evaluate how a disruption affects system performance and causes productivity to decrease relative to targeted system performance levels: that is, how the system should behave during and after disruptive events

Table 2.3 (Continued)

ID	Authors	Definition
9	Consultation document of Ministry of Economic Development, June 12, 2017 [17]	The ability of a system not only to resist stresses that have overcome the withstanding limits of the system itself, but also to come back fast to a normal state of operation. The effectiveness of the resilient system depends on its capability to anticipate, to absorb, to adapt to and/or recover itself from an extreme event
10	IEEE Task Force on Definition and Quantification of Resilience, April 2018 [18]	The ability to withstand and reduce the magnitude and/or duration of disruptive events, which includes the capability to anticipate, absorb, adapt to, and/or rapidly recover from such an event
11	National Security Policy for Critical Infrastructures, Brazilian government, November 2018 [19]	The capacity of the critical infrastructures to be recovered after the occurrence of an adverse situation
12	NATF (North American Transmission Forum) [20]	The ability of the system and its components (i.e. both the equipment and human components) to minimize damage and improve recovery from non-routine disruptions, including high-impact, low-frequency (HILF) events, in a reasonable amount of time. Resiliency includes a diverse range of topics, such as flexibility, hardening, security, and recovery
13	US National Academies of Science [21]	The ability to prepare and plan for, absorb, recover from, or more successfully adapt to actual or potential adverse events

Source: Adapted from [2].

Despite numerous efforts by organizations across the global power and energy engineering communities, a universally accepted definition of resilience has yet to emerge, primarily due to its multi-dimensional and evolving nature. Resilience goes beyond the simple notion of "bouncing back" after a failure. A truly resilient organization must proactively address the potential for multiple failures across all levels, continuously enhance its emergency preparedness and operational procedures before, during, and after major disturbances and service interruptions, and integrate lessons learned from past events to drive ongoing improvement. There have been several attempts by organizations worldwide in the power and energy engineering communities to define resilience, some of which are shown in Table 2.3 [2].

In the light of a systematic review with stakeholders around the globe, CIGRE established the working group (WG) C4.47 "Power System Resiliency" in 2019 with the aim to formulate a standard definition and approach to resiliency [2]. This WG performed an international survey with over 100 respondents from around the globe including industrial, consultancy, governmental, and academic organizations. This survey exposed the participants to the wider interpretation of power system resiliency and invited the participants to comment and provide their views on these definitions. It is therefore believed that the definition provided by the WG includes the key features of existing definitions and provides a well-suited definition to be adopted by the electricity sector. CIGRE WG C4.47 defined resilience as the:

> "**ability to limit the extent, severity, and duration of system degradation following an extreme event**" [22].

It also extends this definition by a set of key actionable measures to be taken before, during, and after extreme events [22]. These include (1) anticipation, (2) preparation, (3) absorption, (4) sustainment of critical system operations, (5) rapid recovery, (6) adaptation, and (7) application

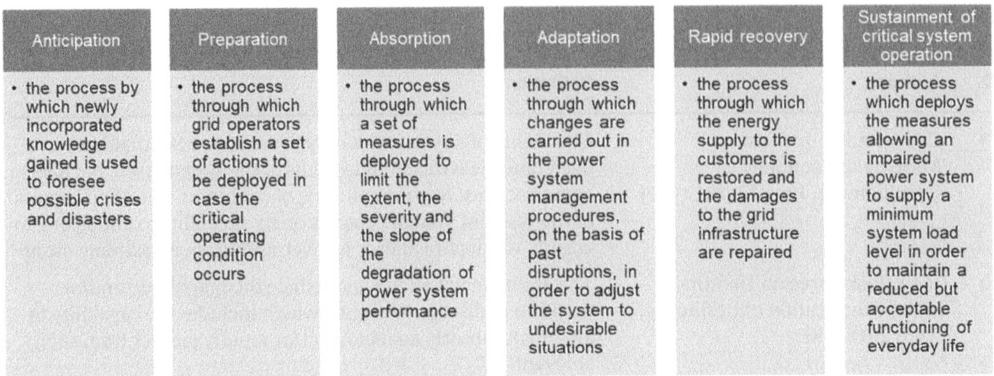

Figure 2.2 CIGRE WG C4.47 features. *Source:* [22] / CIGRE.

of lessons learnt (see Figure 2.2). The terms "extent and severity" in the CIGRE WG definition respectively refer to the geographical extent and the intensity of the effects of the event on the interconnected power system. This assures a more focused characterization of the dimensions of system degradation, while keeping the definition concise and informative. The term "severity" of system degradation must be kept separate from the "severity of the event," which, in general, does not imply any system degradation. "Severity" also depends on the (inter)dependence between essential or mission-critical loads and the disrupted and/or impaired system. The term "degradation" is intended as a deviation from specified target performances. This term refers to the criteria used to apply the resilience concept in system planning and operation. The rationale is to provide a resilience-centric criterion of not exceeding maximum specified deviations of system performances (degradation) in case of extreme events. In simpler terms, "resilience" focuses on the degradation and restoration of a certain "functionality" (that can be defined in different ways) due to an extreme event, and "means to resilience" are case-dependent strategies to minimize the loss and accelerate the recovery of this functionality in such extreme events [22].

2.3 Resilience Capacities and Features

Climate-change-driven extreme weather and natural disasters may have disastrous effects on different parts of the critical power infrastructure, including transmission, distribution, and generation, e.g., [1, 23]:

- High temperatures and heat waves limit the transfer capability of transmission lines, and increase the energy losses and the line sagging.
- High winds during storms and hurricanes can lead to faults and damage to overhead transmission and distribution lines, either by debris being blown against the lines or even due to tower collapse in extremely high winds.
- Cold waves, heavy snow, and the accumulation of ice can also cause failures of overhead lines and towers. Also, under freezing conditions, ice and snow may gather on insulators, which bridge the insulators and provide a conducting path, resulting in a flashover fault.
- Floods do not usually pose a danger to overhead transmission lines, but to substation equipment, such as switchgear and control cubicles. The combination of rain with strong winds or lightning can, however, be a significant threat to overhead lines.

- Earthquakes can have significant impacts on the entire power infrastructure, including overhead and underground transmission and distribution systems and generation stations. It can result in substantial damages to substations, disabling big parts of the infrastructure for long time periods.

In this context, when developing strategies to implement resilient decarbonization pathways, key considerations need to be considered from a reliability and resilience perspective across all parts of the power system infrastructure in a holistic way, which can be grouped as follows:

1) Resource Adequacy: Ensuring sufficient availability and deliverability of energy resources to meet demand across all hours, even under extreme conditions and equipment failures.
2) Transmission Infrastructure: Ensuring adequate transmission capacity to support interregional energy transfers and efficient delivery to load centers within regions.
3) Distribution Infrastructure: Ensuring a robust distribution network capable of accommodating distributed generation, increased demand from electrification, and flexible loads that support grid services – while maintaining established reliability standards.
4) Operational Reliability – Balancing and Flexibility: Maintaining real-time balance between supply and demand under all operating conditions through sufficient system flexibility.
5) Operational Reliability – Grid Stability: Maintaining stable grid performance during normal operations and credible disturbances, while preventing cascading failures and enabling timely restoration following extreme events that exceed planning assumptions.

It is therefore apparent that a power system needs to possess key operational capacities in order to be highly resilient. Within the multi-dimensional concept of power system resilience, it is therefore critical to identify the so-called resilience capacities to enable the quantification, assessment, and enhancement of resilience. Figure 2.3 shows the key resilience capacities, which are defined as:

- *Absorptive capacity* refers to the ability to minimize exposure to shocks and stresses through preventive measures and strategies to mitigate or avoid negative impacts.
- *Restorative capacity* refers to the ability to rapidly recover from a degraded state to the pre-event state, accounting for both infrastructure and operational recovery.
- *Adaptive capacity* refers to the ability to make proactive and informed decisions on strategic, alternative planning based on processing and understanding of new information and changing conditions.

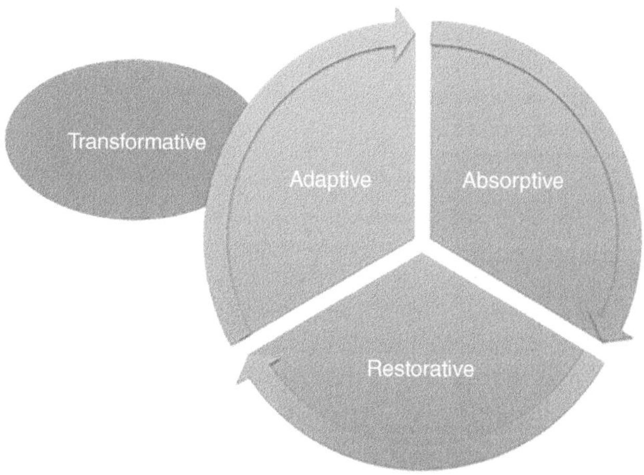

Figure 2.3 The key resilience capacities.

- *Transformative capacity* acts as an enabler to the adaptive capacity as it involves those mechanisms (e.g., governance, policy and regulation) to create the environment to adapt to systemic change.

2.4 Comparing and Clarifying Reliability and Resilience

Over the years of power system evolution, reliability is termed to be the fundamental function for the design and operation of power systems. Based on widely used definitions of reliability, it broadly combines grid adequacy (i.e. having enough generation capacity to satisfy demand) and grid security (i.e. the ability to withstand disturbances). In all the definitions, the grid security is mainly referred from utility viewpoint to ensure a normal state of the system, even during disturbances by taking corrective measures such as load shedding (may be under special contract). This may lead to the possibility of not serving the complete customer demand. For example, the definition of reliability from IEEE explicitly mentions that the power system might have a few interruptions over an extended period of time. Similarly, NERC mentions to have reasonably expected unscheduled outages of system elements (see Table 2.2). In spite of the fact that bulk-power systems recognize reliability constraints such as N–1 contingency security reserves, transmission load reliefs, and security-constrained economic dispatch, they are insufficient for growing smart grids. This is because the framework does not provide a way to measure the real-time risk of power delivery, if the reliability constraints are violated. In general, reliability is well recognized from a planning perspective because of the probabilistic measures of generation adequacy such as loss of load expectation (LOLE); however, these do not address grid security, especially when the system is under threat due to natural hazards or cyber-attack. In other words, the reliability of a power system ensures minimal energy not served (ENS) over the long term; however, this does not guarantee the short-term control of high-impact natural hazards. Alternatively, a reliable power system ensures minimal ENS for low-impact high-probability (LIHP) events, whereas this is not guaranteed for high-impact low-probability (HILP) events. Meanwhile, the aspect of resilience opens the scope for analyzing the system performance that includes the interaction of the power system with threats like extreme events or natural hazards.

Figure 2.4 outlines a conceptual differentiation between reliability and resilience based on system degradation against event probability. There are four segments represented in this figure, which

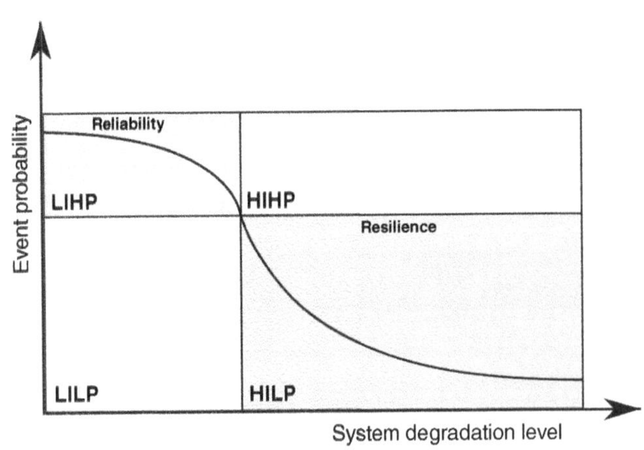

Figure 2.4 Mapping reliability and resilience.

Table 2.4 Comparison of power system reliability and resilience based on key features.

Feature	Reliability	Resilience
Selection of Contingencies	Grid codes established provide a predetermined list of contingencies such as single outages and few multiple outages that could result from the loss of generators and transmission links, i.e., N–1 to some N–k analysis. These events are generally classified under LIHP	Performing an N–k contingency analysis with a high value of k may result in unacceptable impacts. Therefore, selection of multiple contingencies based on experience or from risk-based analysis (where risk can be referred to as a threat from natural hazard or contingency) can be better suitable for resilience analysis. These events are generally classified under HILP
Temporal Dynamics	The adequacy and security requirements are checked at each system state that occurs over the study horizon, considering factors such as load and generation, as well as the failures and return to service of components. This ensures the system remains reliable under different operating conditions	Temporal dynamics is even more important as it includes not only the factors mentioned for reliability, but also the time evolution of the weather threat, operators' behavior during emergency management and restoration phases, and the interactions between power systems. Moreover, special attention is given to restoration and emergency management due to the widespread impact of extreme events.
System Impact	The focus is on identifying the potential risks to the power system's ability to provide sufficient power to meet demand. This includes analyzing the capacity of the generation and transmission systems and evaluating the availability of reserves for use in the event of the most limiting contingencies. Security analysis considers the potential for instability or violations of frequency, current, or voltage regulations in the system's response to contingencies. The primary objective of reliability analysis is to evaluate the quality of the power supply delivered to customers in terms of customer interruption times and ENS	The aim is to evaluate the power system's ability to provide a degraded but still acceptable power supply service to customers even during extreme events. This includes considering the potential effects that severe infrastructure damage can have on the power supply service, as well as the interactions with the human and natural environment during extreme events. Resilience assessment considers not only the technical aspects of power generation and transmission, but also the human and environmental factors that can impact the system's ability to respond to and recover from extreme events
Acceptable Standard	The criteria for adequacy and security must be strictly followed regardless of the event being considered. This means that the system is designed and operated to ensure that there is always enough capacity, and the system remains stable even in the face of the most limiting contingencies	Some level of service degradation is acceptable depending on the severity and scope of the event. This means that the system may not be able to provide the same level of service during extreme events, but it is designed and operated to ensure that it can continue to provide an acceptable level of service even in the face of significant disruptions

Table 2.5 Alternative comparison of resilience and reliability.

Feature	Reliability	Resilience
Event characteristics	Low-impact, high-probability	High-impact, low probability
Contingency	Limited number of faults due to component failures	Multiple and (near-) simultaneous faults
Condition	Binary, i.e., a system is either reliable or not	Continuous, i.e., different levels of resilience are possible and may even be desired
Restoration	Quick restoration	Longer restoration, more time and resources consuming
Impact evaluation	Focuses on measuring the impact to the system	Focus on measuring impact to humans
Improvement measures	Deterministic reliability criteria (N–1, N–2)	Stochastic N–k criteria, where k may be in the order of tens or hundreds

Source: [22] / CIGRE.

include low-impact high-probability (LIHP), high-impact low-probability (HILP), low-impact low-probability (LILP), and high-impact high-probability (HIHP). A LILP could possibly cause a minor disturbance to the system (e.g. frequency perturbation). Therefore, these are given low priority for mitigation and planning studies. However, they are still monitored and addressed to ensure overall system stability. As the power system does not encounter any HIHP events and the LILP events have a lower priority, the degradation curve shown in Figure 2.4 represents HILP and LILP events. LIHP refers to events that result in minimal system degradation but have a high event probability. For example, N–1 contingency, where N refers to the number of elements (like buses, lines, generators, and transformers) that can be affected by a fault. HILP events lead to higher system degradation but have a minimal event probability (e.g. natural hazards).

The major differences between reliability and resilience lie in key features such as selection of contingencies, temporal dynamics, system impact, and acceptable standards. A comparison of reliability and resilience based on these features is tabulated in Table 2.4. Table 2.5 provides an alternative comparison of the two concepts based on similar, but complementary, features.

2.5 Conceptualizing Power System Resilience to External Shocks

A power system might reside in different states when exposed to an external shock, such as natural disaster or extreme weather [26]. It is, therefore, critically important to define these states, in order to enable the systematic resilience assessment and enhancement of power systems to prepare for such events. Under these premises, Figure 2.6 shows a conceptual resilience trapezoid [26], which clearly demonstrates the states (phases) of a power system associated to an external disturbance, as well as the time sequence of these states and related events and the type of available actions. Breaking the event into different phases (namely, Phases I, II, and III) enables the dynamic, multiphase resilience assessment. This is a significant advantage of the resilience trapezoid when compared with the so-called resilience triangle (Figure 2.5) traditionally used and modeled in past resilience studies [25]. The resilience triangle supports only a single-phase assessment, focusing primarily on evaluating an infrastructure's recovery performance after a disruptive event, as illustrated in

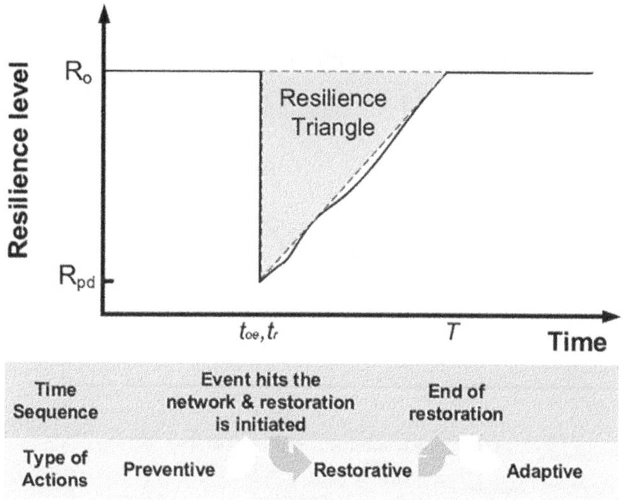

Figure 2.5 The resilience triangle. *Source:* Adapted from [25].

Table 2.6 Resilience triangle versus resilience trapezoid.

Resilience Triangle	Resilience Trapezoid
Single-phase resilience assessment	Dynamic, multi-phase resilience assessment
Lack of corrective actions during the progress of an event	Considers corrective actions
Lack of post-disturbance degraded stated	Considers post-disturbance degraded state and its duration
Threat-specific	Applicable to any threat

Source: Adapted from [26].

Figure 2.5 [25]. In contrast, the resilience trapezoid offers broader applicability, accommodating any type of threat regardless of its nature or duration. The triangle's underlying assumption of an abrupt and immediate drop in resilience makes it suitable mainly for short-duration, high-impact events – such as earthquakes – where performance declines sharply within seconds or minutes. The resilience triangle is not able to capture the evolution of an event, such as a typhoon, which could last from hours to days. Further differences between the two approaches are shown in Table 2.6 [26].

The various phases of the power system state can be accurately represented by the resilience trapezoid model, as follows [1, 26]: if the time and location of the external shock can be sufficiently predicted, the system operator should proceed to the application of preventive actions (e.g. preventive generation rescheduling) before the event hits the network at t_{oe}. The aim of these actions is the boosting of the pre-disturbance resilience of the infrastructure. During the disturbance progress (Phase I, $t \in [t_{oe}, t_{ee}]$), emergency or corrective actions (e.g., generation redispatch) can mitigate the impact of external shock on the power system. Following the event, the system will reside in the post disturbance degraded state (Phase II, $t \in [t_{ee}, t_r]$), where proper and effective emergency coordination and preparedness would enable the quick commencing of the restoration phase (Phase III, $t \in [t_r, T]$, where T is the time that the network fully recovers from its post disturbance resilience level R_{pd} to its predisturbance resilience level R_0, as shown in Figure 2.6, or to a desired resilience

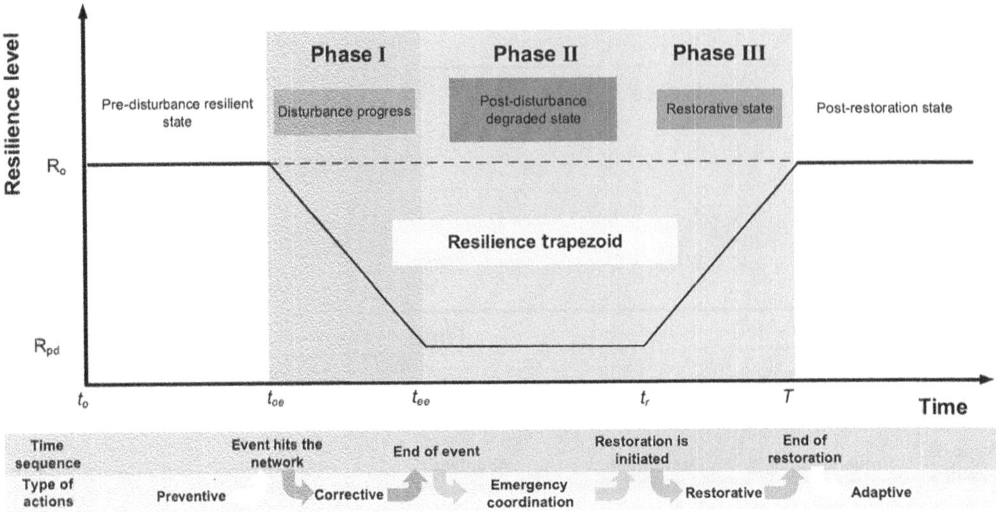

Figure 2.6 The resilience trapezoid. *Source:* [26] / IEEE.

level which can be different than R_0) and the application of the restorative actions. When the system recovers from the post-event degraded state, the impacts of the external shock on system resilience and its performance during the event need to be assessed and analyzed, so that adaptive actions can be taken to increase the resilience of the infrastructure during future similar or unforeseen events.

The resilience trapezoid of Figure 2.6 is based on the assumption that the entire infrastructure is exposed to the event during Phase I and that no restoration of the damaged components is performed during the event [1, 26]. The latter is the usual practice followed in case of disastrous external shocks, like severe windstorms or wildfires, for safety reasons. If it is considered that the restoration of one affected part of the infrastructure is initiated while another part of the infrastructure is still exposed to the event, then a non-flat bottom of the trapezoid might be observed and multiple times would have to be defined, e.g., multiple t_{ee} and t_r corresponding to the times that each part of the infrastructure is no longer subject to the event and to the time the restoration is initiated, respectively. This would be needed to distinguish the phases of the resilience trapezoid.

Within the broad definition of power systems resilience given in Section 2.2 and the resilience trapezoid curve of Figure 2.6, a resilient power system should possess the following key resilience features [1, 26].

- **Pre-disturbance resilient state:** A sufficient estimation of the event's location and severity would enable the application of preventive actions and the network configuration in a state that would help the system operator to deal effectively with the upcoming event. It would also enable the prepositioning of the resources possibly required following the event, e.g., repair and recovery crews, mobile generators, etc. Therefore, preventive operational flexibility is critically important.
- **Phase I (disturbance progress):** High robustness/resistance and redundancy would help boost the resilience to the initial impacts of the external shock and reduce the level of resilience degradation (i.e. R_0–R_{pd}; see Figure 2.6). Further, resourcefulness (supported by smart grid technologies, e.g., advanced monitoring and distributed energy systems) is particularly important as it provides the corrective operational flexibility required for dealing with the prevailing conditions and reducing the slope/speed of the resilience degradation. Also, advanced information

systems would help develop high situation awareness, allowing the system operators to remain adequately informed on the evolving conditions.

- **Phase II (post-disturbance degraded state):** Disaster assessment and priority setting, and proper emergency preparedness and coordination would help the system operator to assess the damage by the event, identify the critical components for the recovery of the system to a resilient state, and initiate as fast as possible the procedures for restoring the damaged infrastructure. This reduces the duration of Phase II, i.e., $t_{ee}-t_r$ (see Figure 2.6).
- **Phase III (restorative state):** Following the actions in Phase II, a resilient system should demonstrate high restorative capabilities in order to first restore the disconnected customers (i.e. operational resilience) and then fully restore the collapsed infrastructure (i.e. infrastructure resilience). Several actions should take place in this phase, such as reenergizing transmission and distribution lines, restoration of damaged components, unit restarting, resynchronization of areas, load restoration, etc. The aim of these actions should be to reduce the duration of Phase III, i.e., $T-t_r$ (see Figure 2.6).
- **Post restoration state:** Following the event and the restoration of the infrastructure to a resilient state, the impact of the event and the performance of the network should be thoroughly analyzed to identify weaknesses or limitations of the network, which could be improved to be better prepared for future (similar or unforeseen) events. Therefore, being adaptive and reflective to the experiences gained through the different events and threats is a key feature of a resilient infrastructure.

An additional strong advantage of the resilience trapezoid is that it can seamlessly be integrated with other performance quantification concepts and definitions, as for example the bowtie and the key features of the CIGRE resilience definition, as shown in Figure 2.7 [24].

Figure 2.8 illustrates the resilience trapezoid for a typical system (solid line) and for a system with enhanced resilience across all phases (dashed line). In reality, however, not all resilience improvement measures impact every phase of the trapezoid. For instance, an optimized restoration strategy would specifically influence the recovery phase. Additionally, the shape of the resilience trapezoid depends on the selected quality indicator. For example, when using the number of

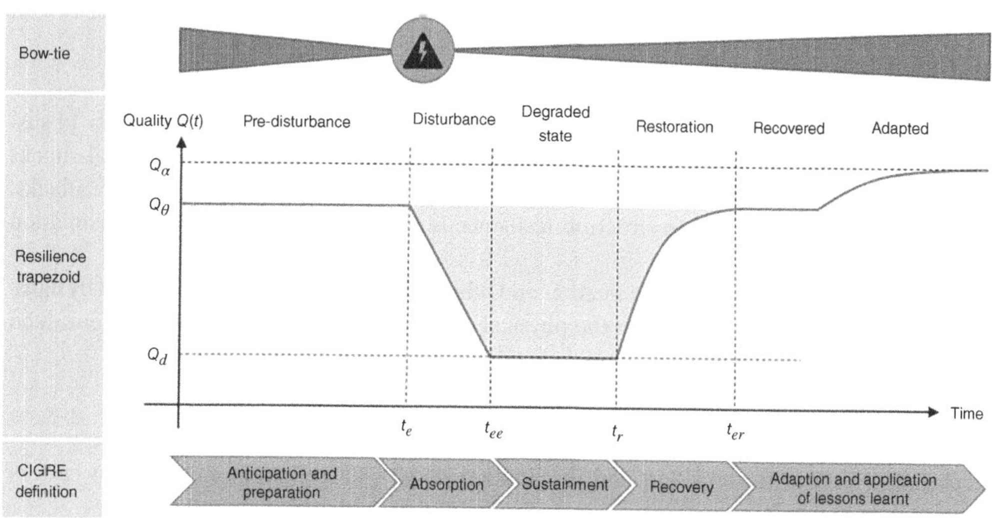

Figure 2.7 Combining the resilience trapezoid, bowtie, and CIGRE definition. *Source:* [24] / with permission of M. Noebels, 2022.

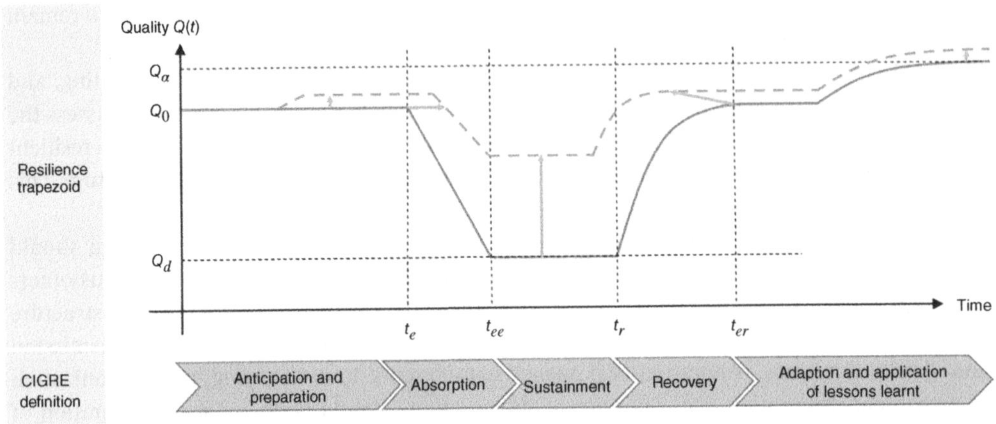

Figure 2.8 Change of the shape of resilience trapezoid for normal (solid) and improved (dashed) resilience. *Source:* [24] / with permission of M. Noebels, 2022.

connected customers as the indicator, the post-event level typically cannot surpass the pre-event state through adaptation. In contrast, if the indicator is the number of available transmission lines, it may increase after the event due to added transmission capacity [24].

Arrows mark the opportunities for enhancing resilience in each phase.

2.6 Domains of Resilience

It is apparent that achieving high resilience is not a simple task, and it goes beyond simply engineering practices to withstand and recover from an event. Hence, it is critical to account for different dimensions of resilience within an organization or a system exposed to external shocks and stresses [24].

2.6.1 Infrastructure Resilience

Infrastructure resilience refers to the physical durability of network assets and the ability to sustain, withstand, and restore them when disrupted. It is primarily influenced by external shocks and stresses, including extreme weather events, natural disasters, accidents, or deliberate attacks. A representative indicator of infrastructure resilience is the number of operational transmission lines [24].

Infrastructure resilience to extreme weather and other natural hazards is most improved by hardening strategies, which aim to prevent the physical damage of the assets of a power system, such as [23, 24]:

- upgrading the poles of transmission and distribution lines to withstand higher forces,
- improving the vegetation management to reduce risk of tree-related outages,
- replacing overhead transmission and distribution lines with underground cables in case of extremely high winds,
- elevating or relocating substations in areas prone to flooding, and
- creating redundancy and backup capacity in the system.

2.6.2 Operational Resilience

Operational resilience refers to the ability to maintain stable and reliable system operations, such as ensuring continuous electricity supply, keeping customers connected, maintaining adequate generation capacity, and enabling rapid and flexible recovery in the event of a blackout [25]. A typical indicator of operational resilience is the amount of load supplied [24].

Digitalization and the transition of the conventional electricity grid to a smart grid bear multiple opportunities for increasing operational resilience. These include, e.g., [24, 27]:

- implementation of distributed generation,
- demand-side management,
- improved location and isolation of faults as well as service restoration,
- reliability-centered maintenance based on condition monitoring, and
- splitting the network into islands or microgrids.

2.6.3 Organizational Resilience

Organizational resilience refers to the capacity of an organization to maintain business continuity during extreme events or periods of uncertainty [28]. These situations can directly impact service or product providers. While traditional risk management and business continuity practices are designed to address predictable risks, they may prove insufficient in rapidly evolving scenarios – such as natural disasters, critical infrastructure failures, or supply chain disruptions [24].

In the electricity sector, organizations particularly involved are utility companies, such as generation plant operators, transmission system operators, and distribution network operators, but also supply chains, such as parts or fuels, and contractors. Organizations involved with transport infrastructures are also relevant, especially during the restorative state, for the transfer of crews and equipment.

The organizational resilience of the electricity sector proved to be of critical importance during the Coronavirus pandemic [29, 30]. The complexity of decision-making processes and human resource management posed significant challenges, often hindering timely responses to routine operations. For example, workforce shortages and the need to reprioritize tasks during the crisis led to delays in both unexpected and planned maintenance activities, such as repairs to cables or transformers. Moreover, the pandemic fundamentally altered electricity consumption patterns. Public health measures – such as widespread remote working and the full or partial shutdown of economic sectors – had a marked impact on electricity networks. National Grid, for instance, reported an increase of up to 30% in residential demand, while overall electricity consumption fell by nearly 20% [31]. These shifts also influenced the markets in which utilities operate. According to an Eurelectric report on the pandemic's impact on customers and society, electricity prices fell to around 20 €/MWh in most EU countries [31]. Despite these challenges, organizational resilience played a key role in safeguarding the continuity and security of electricity supply in most regions worldwide [24].

2.6.4 Community Resilience

Community resilience refers to the capacity of a social system to effectively respond to and recover from extreme events [32]. This is particularly important, as power outages often trigger secondary risks that impact both physiological and psychological well-being. For instance, blackouts have been linked to increased mortality rates, as well as mental health issues such as stress, anxiety, depression, heightened public dissatisfaction, and even crime [32]. Existing literature on community resilience highlights that, compared to infrastructure or organizational resilience, social resilience is often overlooked, with the social capacities of communities receiving insufficient attention [24].

The value of community resilience during blackouts has been demonstrated and analyzed by several researchers over the years [24]. Notable examples where community resilience had a positive impact include the 2007 power outage in the Netherlands [33], a series of storm-induced blackouts in Sweden in 2005 and 2007 [34], and the 2010 earthquake and tsunami in Chile [35]. Community-led responses in these cases involved the provision of food, warmth, information, shelter, mutual support, alternative energy sources, and coordination of local efforts. While the benefits of community resilience are widely recognized, fostering it in a proactive and sustained manner remains a significant challenge. Increasingly, governments and other stakeholders are working to strengthen this aspect of resilience. A recent example – although not directly related to power outages – is the NHS Volunteer Responders program, launched in 2020 to address the societal impacts of the Coronavirus pandemic [36].

References

1 M. Panteli, D. N. Trakas, P. Mancarella, and N. D. Hatziargyriou, "Power systems resilience assessment: hardening and smart operational enhancement strategies," *Proceedings of the IEEE*, vol. 105, no. 7, pp. 1202–1213, 2017, doi: 10.1109/JPROC.2017.2691357.

2 E. Ciapessoni, D. Cirio, A. Pitto, M. Van Harte, and M. Panteli, "Power system resilience: definition, features, and properties," *CIGRE Science and Engineering*, vol. 30, 2023.

3 CIGRE WG C1.27, "The future of reliability - Definition of reliability in light of new developments in various devices and services which offer customers and system operators new levels of flexibility," 2018.

4 P. Kundur, "Definition and classification of power system stability IEEE/CIGRE joint task force on stability terms and definitions," *IEEE Transactions on Power Systems*, vol. 19, no. 3, pp. 1387–1401, 2004, doi: 10.1109/TPWRS.2004.825981.

5 North American Electric Reliability Corporation (NERC), "Reliability Terminology," 2013. Accessed: Feb. 16, 2023. [Online]. Available: https://www.nerc.com/AboutNERC/Documents/Terms%20AUG13.pdf.

6 IEC TR 62511:2014, "Guidelines for the design of interconnected power systems," 2014. Accessed: Feb. 16, 2023. [Online]. Available: https://webstore.iec.ch/publication/7144.

7 J. D. Kueck, *Measurement Practices for Reliability and Power Quality A Toolkit of Reliability Measurement Practices.* Oak Ridge, Tennessee, USA: Oak Ridge National Laboratory, 2004.

8 C. S. Holling, "Resilience and stability of ecological systems," *Annual Review of Ecology and Systematics*, vol. 4, no. 1, pp. 1–23, 1973.

9 UKERC, *Building a Resilient UK Energy System.* United Kingdom: UK Energy Research Center, 2009.

10 Y. Y. Haimes, "On the definition of resilience in systems," *Risk Analysis*, vol. 29, no. 4, pp. 498–501, 2009.

11 Alfred R. Berkeley III, Mike Wallace, "A Framework for Establishing Critical Infrastructure Resilience Goals Final Report and Recommendations by the Council", National Infrastructure Advisory Council (NIAC) report", October 19, 2010.

12 UK cabinet office, "Keeping the Country Running: Natural Hazards and Infrastructure - A Guide to improving the resilience of critical infrastructure and essential services", October 2011.

13 T. J. Overbye, V. Vittal, I. Dobson, "Engineering Resilient Cyber-Physical Systems", PSERC (Power Systems Engineering Research Center) report, May 2012.

14 M. Keogh and C. Cody, *Resilience in Regulated Utilities*. Washington, DC, USA: The National Association of Regulatory Utility Commissioners (NAURC), 2013.

15 Presidential Policy Directive 21: Critical Infrastructure Security and Resilience, Washington, DC: Feb. 12, 2013.

16 E. D. Vugrin, D. E. Warren, and M. A. Ehlen, "A resilience assessment framework for infrastructure and economic systems: quantitative and qualitative resilience analysis of petrochemical supply chains to a hurricane," *Process Safety Progress*, vol. 30, no. 3, 2011.

17 Italian Ministries of Economic Development and of Environment and Land and Sea Protection, Strategia Energetica Nazionale (SEN 2017), 10 November, 2017 (in Italian).

18 IEEE PES IEEE Task Force on Definition and Quantification of Resilience, "The Definition and Quantification of Resilience", Technical report PES-TR65, April 2018.

19 Brazilian Government, "Política Nacional de Segurança de Infraestruturas Críticas," *Decree*, vol. 9753, 2018.

20 NATF, "Transmission system resiliency: an overview", report, 2017.

21 N. Academies, *Disaster Resilience: A National Imperative*. Washington, DC: The National Academies Press, 2012 http://www.nap.edu/catalog/13457/disaster-resiliencea-national-imperative.

22 E. Ciapessoni, D. Cirio, A. Pitto, M. Pantell, M. Van Harte and C. Mak, "Defining Power System Resilience", Reference paper, CIGRE, October 2019.

23 M. Panteli and P. Mancarella, "Influence of extreme weather and climate change on the resilience of power systems: impacts and possible mitigation strategies," *Electric Power Systems Research*, vol. 127, pp. 259–270, 2015.

24 M. Noebels, "Assessment and Enhancement of Power Grid Resilience to Shocks and Stresses", PhD Thesis, The University of Manchester, May 2022, Available: https://research.manchester.ac.uk/en/studentTheses/assessment-and-enhancement-of-power-grid-resilience-to-shocks-and.

25 K. Tierney and M. Bruneau, "Conceptualizing and measuring resilience: a key to disaster loss reduction," *TR News*, vol. 250, pp. 14–18, 2007.

26 M. Panteli, P. Mancarella, D. N. Trakas, E. Kyriakides, and N. D. Hatziargyriou, "Metrics and quantification of operational and infrastructure resilience in power systems," *IEEE Transactions on Power Systems*, vol. 32, no. 6, 2017.

27 J. Jasiūnas, P. D. Lund, and J. Mikkola, "Energy system resilience – a review," *Renewable and Sustainable Energy Reviews*, vol. 150, 2021.

28 S. Duchek, "Organizational resilience: a capability-based conceptualization," *Bus Res*, vol. 13, pp. 215–246, 2020.

29 E. Hollnagel *et al.*, *Resilience Engineering: Concepts and Precepts*. United Kingdom: Ashgate Publishing, Ltd, 2007.

30 Deloitte, "Increasing Organizational Resilience in the Face of COVID-19", 2020.

31 Eurelectric, "Impact of Covid-19 on Customers and Society", 2020.

32 J. Moreno and D. Shaw, "Community resilience to power outages after disaster: a case study of the 2010 Chile earthquake and tsunami," *International Journal of Disaster Risk Reduction*, vol. 34 2018, pp. 448–458, 2019.

33 I. Helsloot and R. Beerens, "Citizens' response to a large electrical power outage in the Netherlands in 2007," *Journal of Contingencies and Crisis Management*, vol. 17, no. 1, pp. 64–68, 2009.

34 J. Palm, "Emergency management in the Swedish electricity grid from a household perspective," *Journal of Contingencies and Crisis Management*, vol. 17, no. 1, pp. 55–63, 2009.

35 D. Platts-Fowler and D. Robinson, "Community resilience: a policy tool for local government?" *Local Government Studies*, vol. 42, no. 5, pp. 762–784, 2016.

36 GoodSAM, NHS Volunteer Responders, 2020. [Online]. Available: https://www.goodsamapp.org/NHS (visited on May 17, 2021).

3

System Resilience Assessment and Quantification

3.1 Needed Resilience Metrics for Power Systems

Regulators set the expected target performance of network operators using well-established reliability criteria, such as energy not supplied (ENS), system average interruption frequency index (SAIFI), and system average interruption duration index (SAIDI). For example, Ofgem (the UK electricity and gas regulator) sets a performance target of 147MWh/year for National Grid (the UK transmission system owner and operator). This benchmarking is usually related to the reliability performance of a power system, that is, the ability of the system to keep the lights on at all times, as part of everyday operations. However, benchmarking the network's expected performance using average indicators across all events (both "normal" and more extreme) fails to ensure that transmission networks are resilient to non-ordinary, low-frequency/high-consequence events such as extreme weather and natural hazards. These events may lead to load curtailments on a much larger scale (>GWh) than the current reliability-driven performance targets. New sets of metrics and performance targets need to be holistically designed and adopted by the power system industry to measure and ensure reliability to typical outages as well as resiliency to severe, rare events [1].

Risk-based metrics can provide a more accurate estimation of the so-called "tail" risks, that is, the risks that lie in the tail of a probability distribution, i.e. high-impact, low probability events. The conditional values of a performance metric, such as ENS, can provide its average value across a specific percentage (e.g. 5%) of the worst events that can be anticipated. Hence, it provides a power system performance assessment that is complementary to the traditional reliability one, in the sense that the average value across all events quantifies the reliability performance, and the tail average across the worst events quantifies the resilience performance. There is an increasing trend in academia [2] and industry [3] of adopting risk-based metrics (including value at risk [VaR] and conditional value at risk [CVaR]) for resilience quantification and differentiation from reliability quantification (see Section 3.3).

Risk-driven metrics provide insights by processing large amounts of data and generating and analyzing probability distributions of different metrics, but they do not provide any information on the actual performance of the system during and/or after the event [4]. Complementing risk metrics, resilience metrics can provide such information, and they are usually either attribute-based (qualitative) or performance-based (quantitative) [5, 6]. Attribute-based resilience metrics aim to identify the properties of a system that make the system more or less resilient, such as robustness, resourcefulness, adaptivity, and recoverability. Resilience is then assessed as the degree to which these properties are present within a system. Methods for determining attribute-based resilience metrics include, e.g., site visits and questionnaires [7]. This may also

Fundamentals of Power System Resilience: Disruptions by Natural Causes, First Edition.
Mathaios Panteli, Rodrigo Moreno, Dimitris Trakas, Magnus Jamieson, Pierluigi Mancarella, Goran Strbac, and Nikos Hatziargyriou.

include scoring based on qualitative assessment, where direct quantitative assessment is not possible. The Argonne National Laboratory's Resilience Measurement Index, which assesses a system's resilience based on scoring of four major components (preparedness, mitigation measures, response capabilities, and recovery mechanisms), is an example of an attribute-based resilience metric [8]. Performance-based resilience metrics (i.e. metrics that measure the behavior, response and performance of a business, organization, or infrastructure) are based on quantitative data probabilistically describing the system's behavior in the event of a disruption. The data may come from historical observations, estimates, or modeling, including data on both the system performance and the asset performance (e.g. through fragility curves) under stressed conditions. Performance-based metrics are particularly useful for cost–benefit and planning analyses, as they provide a direct evaluation of potential benefits and costs of enhancement measures or investments [9]. Performance metrics are directly related to the performance of the power system, including the connected load or generation, duration, frequency and probability of disruptions, and time-development of these characteristics. The latter is often done in form of a resilience triangle or resilience trapezoid (i.e. resilience curves [10]) (see Chapter 2, Figures 2.5 and 2.6 respectively). The resilience of an infrastructure can be measured by quantifying the area of the resilience triangle and trapezoid, respectively, enabling the targeted resilience enhancement for minimizing this area. For example, robustness measures for hardening the transmission system can be applied to reduce the performance degradation in Phase 1 of the trapezoid (hence reducing the overall area of the entire resilience trapezoid), or more efficient restoration strategies can be adopted to accelerate the restorative phase of the trapezoid (hence reducing the area under Phase 3). The US Department of Energy has recently funded the "North American Energy Resilience Model" project, which aims to improve the understanding of modeling, quantifying and evaluating the resilience performance of the US critical power infrastructure [11].

Attribute-based and performance-based resilience metrics are subject to different trade-offs [4]. Both types can be retrospective or forward-looking, but significant differences may lie in their level of complexity and consistency. Attribute-based metrics may require less complex mathematics, but often depend on expert knowledge for subjective scoring of qualitative data. This may subsequently also affect consistency and validity in assessing how a system will operate under different conditions. Performance-based metrics tend to be more complex as they often rely on system modeling. However, this can also make performance-based metrics more informative. They may also provide meaningful and consistent results under different conditions, and make them valuable for decision-making on planning and investment activities [2]. Nevertheless, regardless of the selected metrics, it should be stressed that assessment models of high quality are essential in order to be meaningful for practical applications and decision-making [4].

3.2 Quantifying the Resilience Trapezoid

3.2.1 The FLEP and Area Metric Systems

A power system might reside in different states when exposed to an external shock, such as natural disaster or extreme weather. It is, therefore, critically important to define these states, to enable the systematic resilience assessment and enhancement of power systems to such events. Figure 3.1 expands the resilience trapezoid of Figure 2.6 by clearly differentiating and depicting the infrastructure and operational resilience (introduced in Section 2.5). In Figure 3.1 [12], it is considered that the resilience indicators used for expressing and quantifying the operational and infrastructure resilience levels in the pre-disturbance resilient state (i.e. R_{0o} and R_{0i}, respectively) are 100%

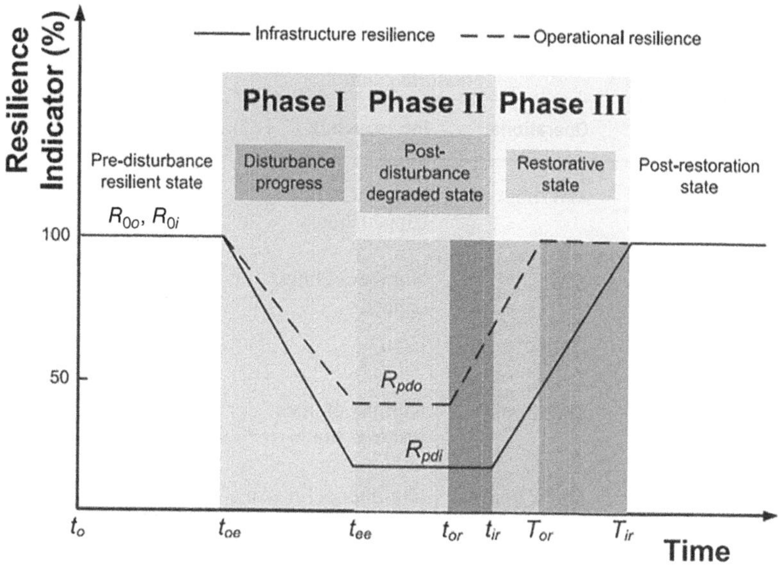

Figure 3.1 The multi-phase resilience trapezoid: infrastructure and operational resilience [12].

before the event occurs at t_{oe}, e.g., 100% of the demand and of the transmission lines are online, respectively. However, it must be clarified that these resilience levels might differ.

In order to quantify the resilience of a power system, it is critical to define a set of metrics capable of capturing its performance during the different phases of the resilience trapezoid. Table I presents the key resilience metrics proposed here to characterize the resilience trapezoid, specifically for how fast (Φ) and how low (Λ) resilience drops in Phase I, how extensive (E) is the post-event degraded state (Phase II) and how promptly (Π) the network recovers to its pre-event resilient state (Phase III), considering both operational and infrastructure resilience in each phase. This set of four metrics is defined in this work as the ΦΛΕΠ resilience metric system ("ΦΛΕΠ" is pronounced like "FLEP") [12].

As seen in Table 3.1, the Φ-metric in Phase I is evaluated by estimating the slope of the resilience degradation during the event (where $t_{ee} - t_{oe}$ is the duration of the windstorm, i.e., $t_{\text{windstorm}}$), while the Λ-metric is defined by the resilience degradation level, i.e., $R_{0i} - R_{pdi}$ and $R_{0o} - R_{pdo}$ for the infrastructure and operational resilience, respectively. The E-metric showing the time that the network remains in the post-disturbance degraded state (Phase II) is given by $t_{or} - t_{ee}$ and $t_{ir} - t_{ee}$ for

Table 3.1 The ΦΛΕΠ resilience metric system [12].

Phase	State	Resilience metric	Symbol
I	Disturbance progress	How *fast* resilience drops?	Φ
		How *low* resilience drops?	Λ
II	Post-disturbance degraded	How *extensive* is the post-disturbance degraded state?	Ε
III	Restorative	How *promptly* does the network recover?	Π

Table 3.2 Mathematical formulation of the ΦΛΕΠ metric system [12].

	Mathematical expression		Measuring unit	
Metric	**Operational**	**Infrastructure**	**Operational**	**Infrastructure**
Φ	$\frac{R_{pdo} - R_{0o}}{t_{ee} - t_{oe}}$	$\frac{R_{pdi} - R_{0i}}{t_{ee} - t_{oe}}$	MW/hours	Number of lines tripped/hours
Λ	$R_{0o} - R_{pdo}$	$R_{0i} - R_{pdi}$	MW	Number of lines tripped
E	$t_{or} - t_{ee}$	$t_{ir} - t_{ee}$	Hours	Hours
Π	$\frac{R_{0o} - R_{pdo}}{T_{or} - t_{or}}$	$\frac{R_{0i} - R_{pdi}}{T_{ir} - t_{ir}}$	MW/hours	Number of lines restored/hours
Area	$\int_{t_{oe}}^{T_{or}} R_{op}(t)dt$	$\int_{t_{oe}}^{T_{ir}} R_i(t)dt$	MW × hours	(Number of lines in service) × hours

Table 3.3 Mathematical formulation of the linearized trapezoid areas [12].

	Mathematical Expression	
Trapezoid Area	**Operational**	**Infrastructure**
$Area_I$	$\frac{\Lambda_{\text{operational}} \times t_{\text{windstorm}}}{2}$	$\frac{\Lambda_{\text{infrastructure}} \times t_{\text{windstorm}}}{2}$
$Area_{II}$	$\Lambda_{\text{operational}} \times E_{\text{operational}}$	$\Lambda_{\text{infrastructure}} \times E_{\text{infrastructure}}$
$Area_{III}$	$\frac{\Lambda_{\text{operational}} \times (T_{or} - t_{or})}{2}$	$\frac{\Lambda_{\text{infrastructure}} \times (T_{ir} - t_{ir})}{2}$

the operational and infrastructure resilience, respectively. The Π-metric in Phase III is defined by the slopes of the operational and infrastructure recovery curves, which consider both the original pre-event resilience level and the time required for reaching this resilience level. Further to the ΦΛΕΠ resilience metrics, an additional metric is used, i.e., the area of the trapezoid. As shown in Table 3.1, the area metric would be expressed as the integral of the trapezoid for the duration of the event, again taking into account the operational and infrastructure resilience. Considering piece-wise linearity for the different phases of the resilience trapezoid of Figure 3.1 (which results in two right triangles for Phases I and III and one rectangular for Phase II), the areas corresponding to the three phases of the resilience trapezoid (I, II and III) can be approximated as shown in Table 3.2. The total area of the operational and infrastructure resilience trapezoid is given by the sum of Areas I, II, and III. It can be seen that the area metric is dependent on the ΦΛΕΠ metrics (mainly on the Λ and E metrics) defined and mathematically expressed in Tables 3.2 and 3.3 [12].

Different resilience indicators can be used to express and quantify these resilience metrics. It is critical to use indicators that capture both the prevailing operational and infrastructure resilience during the event. In this study, which focuses on assessing the resilience of transmission networks to extreme weather events, the following indicators are used: (1) the amount of generation capacity

(MW) and load demand (MW) that are connected during the event are used as indicators for the operational resilience; and (2) the number of online transmission lines is used as an indicator for the infrastructure resilience. Based on these indicators, the measuring units of the resilience metrics are also defined in Table 3.2.

Hence, a set of five resilience metrics is proposed in this work for effectively and systematically quantifying the resilience of a power system subject to an external shock. This quantitative resilience metric framework includes the four independent ΦΛΕΠ metrics, and the area metric, which is dependent on the ΦΛΕΠ metrics. The resilience assessment of a power system using this framework would enable an in-depth understanding of the resilience level of the power system and drive the utilization of targeted resilience enhancement strategies for improving the resilience of the system if and during the specific phase of the resilience trapezoid it is deemed inadequate [12].

3.2.2 Fragility-Driven Resilience Assessment Against Windstorms

In order to model the fragility of a transmission network to a weather event, the concept of fragility curves is used here, which expresses the failure probability of a component as a function of the weather parameter, e.g., wind speed. By mapping the weather profile at each simulation step to these curves, the weather- and time-dependent failure probability of a transmission corridor can be obtained. A generic fragility curve is shown in Figure 3.2, whose shape varies and depends on the application, but it can be expressed as follows [12]:

$$P(w_i) = \begin{cases} 0, & \text{if } w < w_{\text{critical}} \\ P(w), & \text{if } w_{\text{critical}} \leq w < w_{\text{collapse}} \\ 1, & \text{if } w \geq w_{\text{collapse}} \end{cases}$$

where $P(w_i)$ is the failure probability of the component as a function of the weather parameter (w) at simulation step i, w_{critical} is the value of the weather parameter at which the failure probability picks up and w_{collapse} represents the value of the weather parameter at which the failure of the component is certain, i.e., $P_i(w) = 1$. The transmission corridor will trip at the next simulation step if $P_i(w) > r$, where r is a uniformly distributed random number ($r \sim U[0,1]$). If $P_i(w) < r$, the corridor will not trip. The restoration time is randomly generated if a transmission corridor outage occurs, which is also weather-dependent, as will be discussed later.

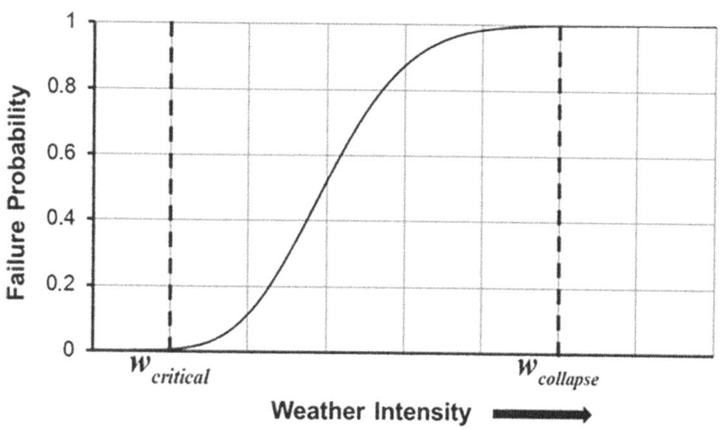

Figure 3.2 A generic fragility curve showing the failure probability of a component as a function of the weather intensity [12].

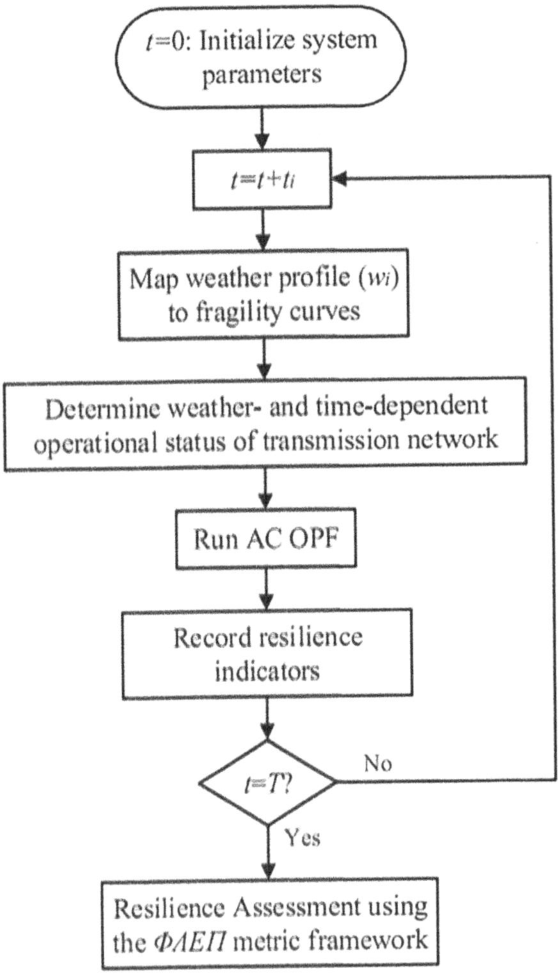

Figure 3.3 System approach for assessing the resilience of transmission networks to weather events [12].

The weather- and time-dependent operational modes of the corridors are then fed to a Sequential Monte Carlo simulation-based engine to capture the multi-temporal and multi-spatial impact of the weather front as it moves across the transmission network. Figure 3.3 shows the generic approach followed for obtaining the weather-dependent resilience indicators and applying the ΦΛΕΠ resilience metric framework. It is worth mentioning here that even though AC optimal power flow (OPF) is considered an appropriate dispatch tool for this specific application, where the focus is on developing and illustrating a complete resilience quantification framework, it does not consider other important relevant issues, such as, e.g., the ramping rates of the generators. However, the proposed tool offers the capability and flexibility to include such constraints for modeling additional dimensions of the problem [12].

In particular, regarding the calculation of the operational resilience indicators, the amount of the generation capacity connected at each simulation step is taken as the capacity that can be provided, given the prevailing transmission network topology, connectivity, and constraints during the event. It has to be clarified that this indicator is different from the generation output at each simulation step. Also, for the purposes of the AC OPF implementation, the load shedding that might occur at each simulation step (which enables the calculation of the connected load) is calculated using

Figure 3.4 The test version of the Great Britain transmission network [13].

an objective function for the minimization of the load shedding, which considers the power flows (PFs) and network constraints, such as generation and transmission capacities and voltage operational limits. These resulting non-linearities are considered in the calculation of the operational indicators. Finally, the infrastructure resilience indicator is calculated by recording the status of the transmission lines at each simulation step [12].

The test version of the GB transmission network of Figure 3.4 is used here, for which further details can be found in [13]. The node demands are fluctuating and are time-varying in the simulations, but it is assumed that there is no impact of the weather event on the demand profiles. The generation is then dispatched at each simulation step through the AC OPF to meet this demand, taking into account the network constraints. As the focus is on assessing the impact of windstorms on the transmission network resilience, and in particular on transmission lines and towers, the wind fragility curves of these components are required. Figure 3.5 shows the transmission tower and lines' wind fragility curves used in this application (for the base and robust case scenarios) [14, 15]. Three windstorm profiles are generated, with maximum wind speeds (w_{max}) equal to

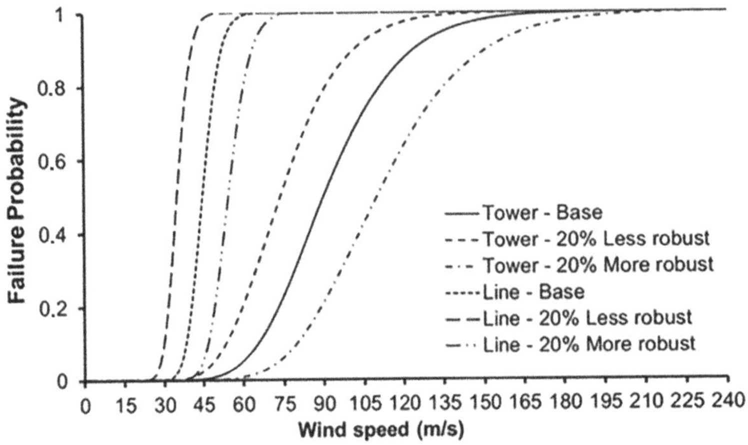

Figure 3.5 Wind fragility curves in the base and robust case scenario [14, 15].

40 m/s, 50 m/s, and 60 m/s, respectively (referred as $WS_{40\,m/s}$, $WS_{50\,m/s}$, and $WS_{60\,m/s}$ in the rest of the chapter), which according to the Beaufort wind force scale by the Met Office, UK [16] represent hurricanes. This approach is followed in order to model the worst conditions that the test network might experience.

Different operational and infrastructure aspects can affect the resilience performance of a power system subject to extreme weather. Within this context, several case studies have been systematically developed to evaluate the impact of these aspects on the resilience of the test network, as this is quantified using and to test the suitability of the proposed ΦΛΕΠ-based resilience metric framework. In particular, the following illustrative scenarios are considered [12]:

1) Base: The resilience trapezoid is modeled with the base data and for $WS_{40\,m/s}$, $WS_{50\,m/s}$ and $WS_{60\,m/s}$.
2) Robust: In this case scenario, the transmission lines and towers are made 20% more and less robust to the windstorm, by adjusting the fragility curves (see Figure 3.5).
3) Response: As in the robust case scenario, the responsiveness to the weather event is made 20% better and worse, by decreasing and increasing the $MTTR_{normal}$ assumed for the transmission lines and towers by 20%.
4) Resources: A critical aspect is the amount of resources, i.e., repair crews (RCs), for responding quickly and effectively to the collapsed components during weather events. Hence, in this case scenario, the effect of an unlimited number of RCs, 5, 10, and 15, is evaluated. It is considered that one repair crew is needed for each damaged transmission corridor, which introduces a limitation on the number of corridors that can be repaired at the same time.

It is also assumed that no restoration takes place during the event (i.e., $t \in [t_{oe}, t_{ee}]$). These case scenarios have been based on engineering judgments and discussions with National Grid, the GB transmission system operator. It must be noted here that the case scenarios robust response and resources are applied only for $WS_{50\,m/s}$; similar conclusions can be obtained by applying these scenarios for the other windstorms [12].

1) Obtaining the Time-Dependent Resilience Indicators [12]: Figure 3.6 shows the time-dependent resilience indicators for the base scenario and the three windstorms modeled, i.e., $WS_{40\,m/s}$, $WS_{50\,m/s}$ and $WS_{60\,m/s}$. The shape of these curves clearly recalls the resilience trapezoid of

Figure 3.1, enabling its proper modeling and quantification; that is, the three phases can be clearly distinguished: disturbance progress, post-event degraded state, and restoration phase. The duration of Phase I is known and equal to the duration of the extreme event (i.e. 24 h in this work), while the duration of Phases II and III would be the output of the simulation engine and, as demonstrated later, would depend on the emergency and preparedness plans in place by the electric utility (i.e. the response and RCs case studies considered in this work). In general, Phase II starts at the end of the event (t_{ee}) and ends when the operational and infrastructure resilience start recovering (t_{or} and t_{ir}, respectively), e.g., when some demand or a damage component is restored. Similarly, Phase III starts at t_{or} and t_{ir} and ends at T_{or} and T_{ir}, respectively, i.e., when the operational and infrastructure resilience fully recover.

Further, as illustrated in Figure 3.6, it can be observed that the operational resilience degradation (R_{pdo}) as obtained by the operational resilience indicators in Figure 3.6 (b) and (c) is significantly lower than the infrastructure resilience degradation (R_{pdi}) in Figure 3.6 (a). It can also be seen that the operational indicators of generation and load connected recover much faster than the infrastructure indicator of the transmission lines online. This proves the importance and the need for distinguishing the concepts of infrastructure and operational resilience, as well as the use of different resilience indicators for their quantification.

Application of the ΦΛΕΠ Resilience Metric Framework [12]: Based on the above resilience indicators, it is now possible to quantify the system resilience via the proposed metrics. The first metric to evaluate is the Φ-metric, i.e., how fast the operational and infrastructure resilience levels degrade. Table 3.4 shows the Φ-metric of the resilience degradation curves of Phase I of the resilience trapezoid for the infrastructure and operational resilience indicators. These slopes are negative, as resilience degrades. It can be clearly seen that the degradation slopes increase

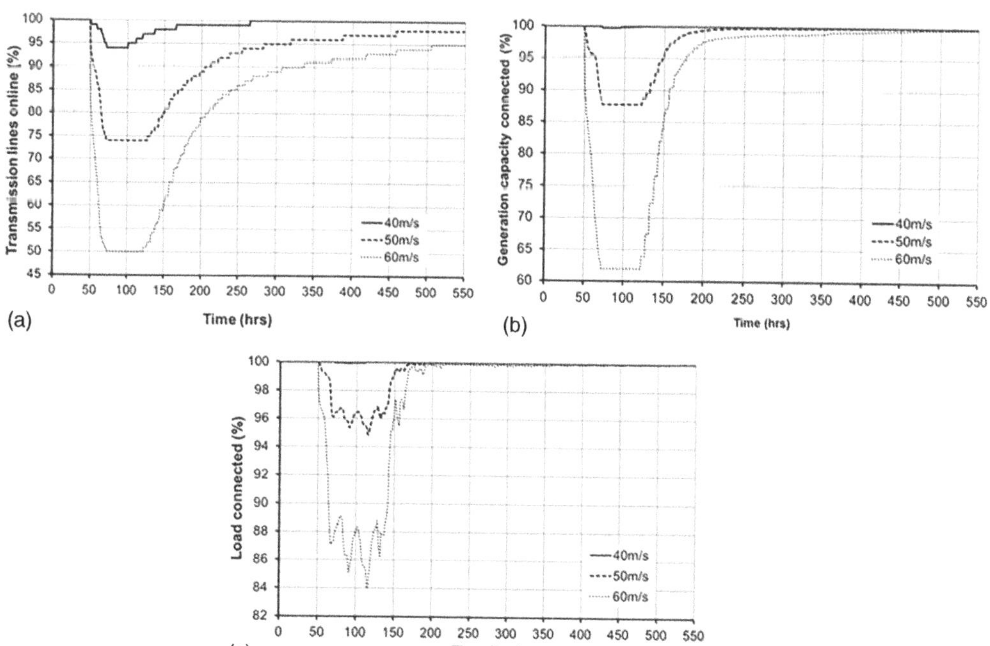

Figure 3.6 Resilience indicators as a function of time for the base case scenario: (a) Transmission lines online (%), (b) Generation capacity connected (%), (c) Load connected (%).

Table 3.4 The Φ-metric for the base case scenario.

Event	Resilience indicator		
	Trans. lines	Gen. connected	Load connected
$WS_{40\,m/s}$	-0.2500	-0.0125	-0.0024
$WS_{50\,m/s}$	-1.0833	-0.521	-0.249
$WS_{60\,m/s}$	-2.0833	-1.5876	-0.6668

Table 3.5 The Φ-metric for the robust case scenario and $WS_{50\,m/s}$.

Event	Resilience Indicator		
	Trans. lines	Gen. Connected	Load Connected
20% Less Robust	-2.2083	-1.8083	-0.7132
Base	-1.0833	-0.521	-0.249
20% More Robust	-0.2500	-0.0121	-0.0117

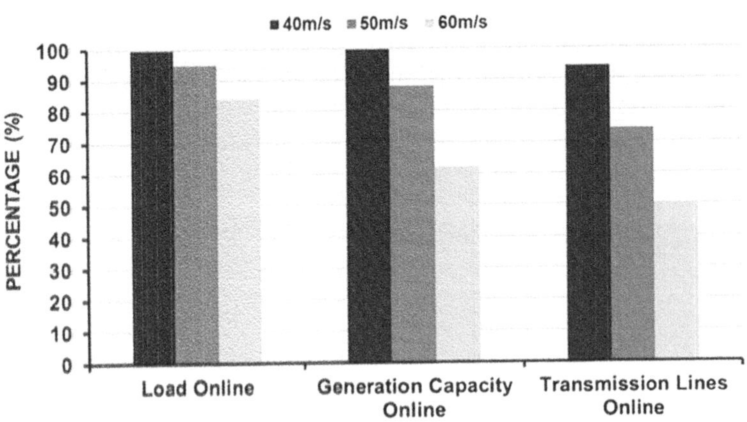

Figure 3.7 The Λ-Metric for the base case scenario.

for higher wind speeds for all the resilience indicators. Further, the highest slopes occur for the infrastructure indicator of transmission lines online, followed by the operational resilience indicators.

The second metric to evaluate within the ΦΛΕΠ framework is the Λ-metric, i.e., how low resilience drops in Phase I. Figure 3.7 shows this metric for the base case scenario, as expressed using the infrastructure and operational resilience indicators. As expected, the resilience degradation increases for higher wind speeds. Also, as shown in Figure 3.6 as well, the highest impact occurs on the infrastructure resilience for all the windstorm intensities evaluated here.

Table 3.5 shows the Φ-metric and Figure 3.8 the Λ-metric for the robust case scenario, which is the scenario with the largest effect on these metrics, for $WS_{50\,m/s}$. It can be seen that boosting (reducing) the robustness of the transmission corridors improves (worsens) the slope and level of resilience degradation in Phase I.

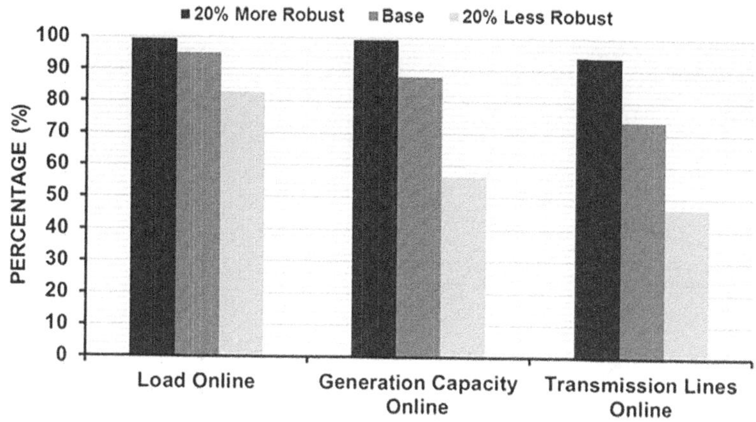

Figure 3.8 The Λ-metric for the robust case scenario and $WS_{50\,m/s}$.

Table 3.6 The E-metric for the response case scenario and $WS_{50\,m/s}$.

Case study	Duration of post-event degraded state (hours)		
	Transmission lines	Generation capacity	Load
20% More Response	44	47	48
Base	53	54	57
20% Less Response	76	80	83

Table 3.6 demonstrates the E metric, i.e., how long is the duration of the post-disturbance degraded state (Phase II), for the base and response case scenarios for $WS_{50\,m/s}$. It is shown that improving (resp. worsening) the responsiveness to the event results in lower (resp. higher) durations of the post-event degraded state for all the resilience indicators. This scenario does not have an impact on the Φ and Λ metrics of Phase I. Also, the robust and resource scenarios do not have a great effect on the E-metric.

The last metric to evaluate within the ΦΛΕΠ framework is the Π-metric, i.e., how fast the operational and infrastructure resilience levels recover in Phase III. For this purpose, Table 3.7 shows

Table 3.7 The Π-metric in Phase III for the response and resources case scenarios and $WS_{50\,m/s}$.

Case study	Resilience indicator		
	Trans. lines	Gen. connected	Load connected
5 RCs	0.0128	0.0060	0.0026
10 RCs	0.0137	0.0069	0.0039
15 RCs	0.019	0.0148	0.0076
20% Less response	0.0455	0.0266	0.0111
Base	0.0578	0.0330	0.0724
20% More response	0.0925	0.0354	0.0925

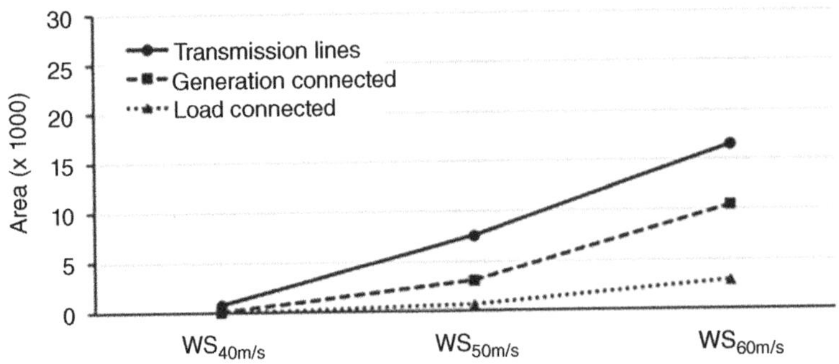

Figure 3.9 The area metric for $WS_{40\,m/s}$, $WS_{50\,m/s}$, and $WS_{60\,m/s}$.

the recovery slopes of the response and resources scenarios for $WS_{50\,m/s}$ (the robust scenario does not affect this metric). As can be seen, different recovery slopes correspond to the infrastructure and operational indicators. Further, it is apparent that the limitation of the RCs has the greatest effect on the restoration capability of the restoration slopes of both infrastructure and operational resilience. Also, the higher the number of RCs, the higher the number of transmission corridors that can be fixed at the same time and hence the higher the restoration slopes (i.e., Π -metric) of all the resilience indicators.

2) Calculation of the Area Metric [12]: Based on the resilience metrics estimated within the ΦΛΕΠ resilience framework in the previous subsection, the area metric is next estimated. Figure 3.9 shows the total area metric for $WS_{40\,m/s}$, $WS_{50\,m/s}$, and $WS_{60\,m/s}$ for the different resilience indicators. The larger the area, the larger the overall impact of the windstorm. Therefore, as aforementioned, the highest impact is observed on infrastructure resilience. Further, the area metric increases significantly for higher wind speeds.

Figures 3.10 and 3.11 show the area metrics for the different phases of the resilience trapezoid using the indicator of transmission lines online for $WS_{50\,m/s}$ and for the robust, response and resources (RCs) case scenarios, respectively. Similar trends are observed for the operational indicators. It can be seen that the robust case scenarios (Figure 3.10) result in a large difference in Area II due to the large difference in the resilience degradation. However, this variance is lower in the case scenarios of response and RCs (Figure 3.11) due to the smaller difference in the duration of Phase II in these scenarios. Further, Figure 3.11 shows that Area I is equal for the response and

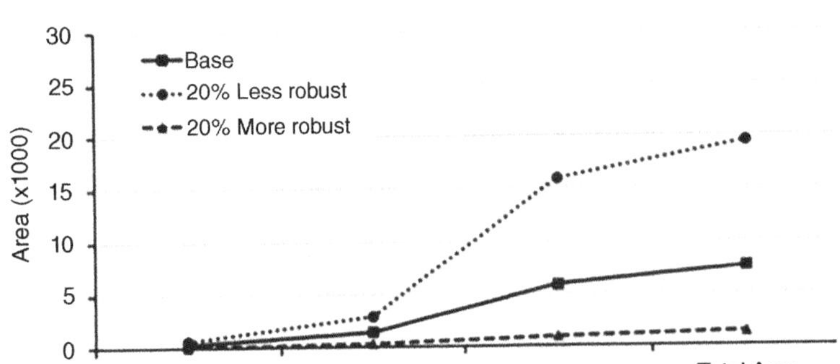

Figure 3.10 The area metric of the indicator transmission lines online for $WS_{50\,m/s}$ for the robust case scenario and the phases of the trapezoid.

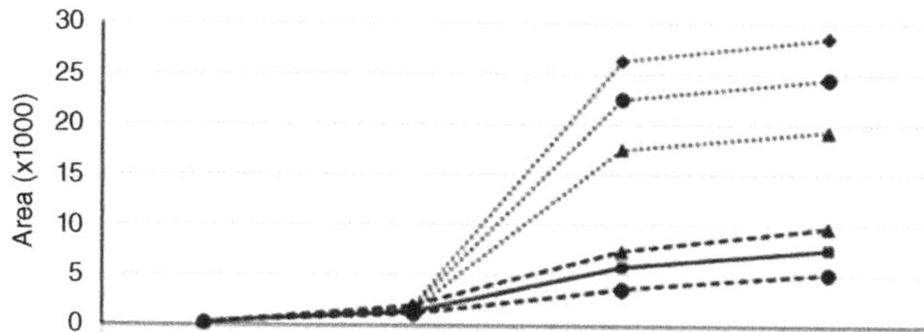

	Area I	Area II	Area III	Total area
Base	0.31	1.43	5.85	7.59
5 RCs	0.31	1.95	26.29	28.55
10 RCs	0.31	1.69	22.51	24.51
15 RCs	0.31	1.51	17.51	19.33
20% Less response	0.31	2.03	7.42	9.76
20% More response	0.31	1.20	3.65	5.16

Figure 3.11 Area metric of the indicator transmission lines online for $WS_{50\,m/s}$. Results shown for the case scenarios, base, response and resources (RCs).

RCs scenarios, as the resilience degradation is the same for these scenarios; only the post-event restoration time is influenced. It can also be seen that Area III is significantly larger than Areas I and II. This is due to the much longer duration of Phase III compared to Phases I and II. Further, this is because the value of the Total Area is close to the value of Area III. Also, in Figure 3.11 it is shown again that the limitation of RCs has a much higher effect on Area III than the response case scenarios.

3.3 CVaR-Driven Resilience Assessment

As aforementioned, quantifying and benchmarking the resilience of the power system using average value indicators does not yield a realistic estimation. Therefore, as an alternative, conditional or risk-based indicators, such as CVaR, are being introduced to quantify the effect of HILP events. In this method, the conditional indicators are complemented with resilience indicators from the ΦΛΕΠ metric system to provide a full picture of network resilience and reliability, respectively. CVaR is a risk-based indicator that provides a more realistic estimation of the potential impact of HILP events on the power system that represent the so-called tailed risks. CVaR (or CVaRα) is calculated as the expected value of losses beyond a certain threshold, known as the VaR threshold (associated with the 1-α worst cases). In the case of the power system, the losses could be in terms of the amount of ENS to customers, the cost of restoring power, or any other relevant metric.

In this context, CVaR indicates the average value across the worst-case scenarios as illustrated in Figure 3.12, which takes the example of ENS as the resilience metric. It can be seen that the value of CVaR provides an explicit risk indication depending on α. The expected value of ENS is evaluated by considering all the credible sets of scenarios that could overshadow, by averaging, the non-credible worst-case outages. Therefore, in the proposed framework, the mean value is used for reliability assessment and its CVaR for resilience assessment. The mean and conditional values of the selected metrics are given below (also shown in Figure 3.12 for better illustration):

$$\text{Mean} = \int_0^D x \cdot f(x)\, dx$$

$$\text{CVaR} = \frac{1}{1-\alpha} \int_v^D x \cdot f(x)\, dx$$

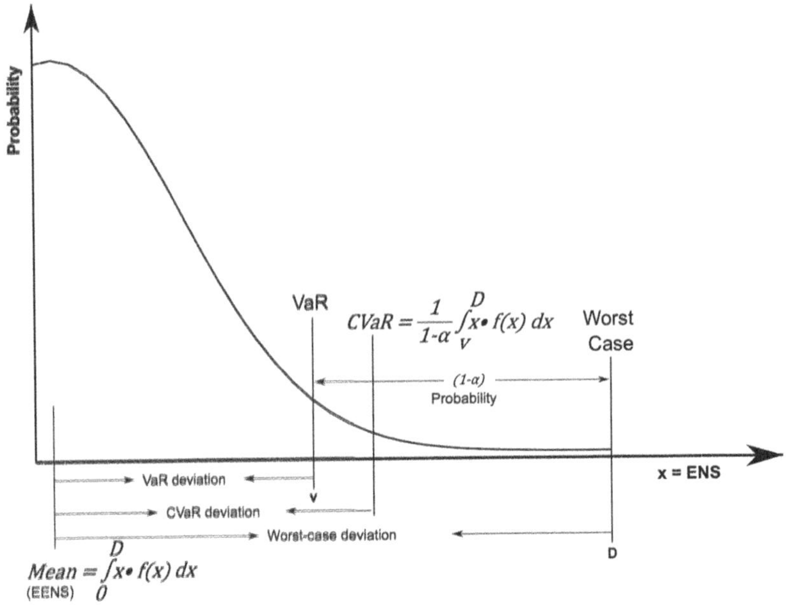

Figure 3.12 VaR and CVaR resilience assessment.

3.3.1 Illustrative Example of CVaR Resilience Assessment Against Wind Events

This section presents a detailed explanation of the performance of the proposed optimal resilient planning framework using a 4-bus system shown in Figure 3.13. This radial system is at 12.5 kV and consists of four nodes with a total peak load of 1.2 MW. Each node has an energy requirement of 8.8 MWh/day. This network is exposed to 100 windstorm scenarios and its corresponding ENS is evaluated. For a better understanding of the intersection points of windstorms and the distribution network line, a sample windstorm from the 100 windstorm scenarios is considered for illustration. This windstorm intersects the line connecting node-1 and node-2 during the simulation step $\boldsymbol{i = 7}$ at a wind speed of 45 m/s shown in Figure 3.14 (b). Further, this windspeed is mapped on the fragility curve to obtain the failure probability. In this example, the failure probability obtained by mapping the windspeed of 45 m/s is $\boldsymbol{P_L(45_7) = 0.52}$ as shown in Figure 3.14 (c). Later, the procedure of impact analysis is carried out by generating a random number $\boldsymbol{r}$ between 0 and 1 to decide whether the line is tripped or not. In this case, the random number generated is, say, 0.3. Since the wind-dependent failure probability $\boldsymbol{P_L(45_7) = 0.52}$ is greater than the random number $\boldsymbol{r = 0.3}$, the

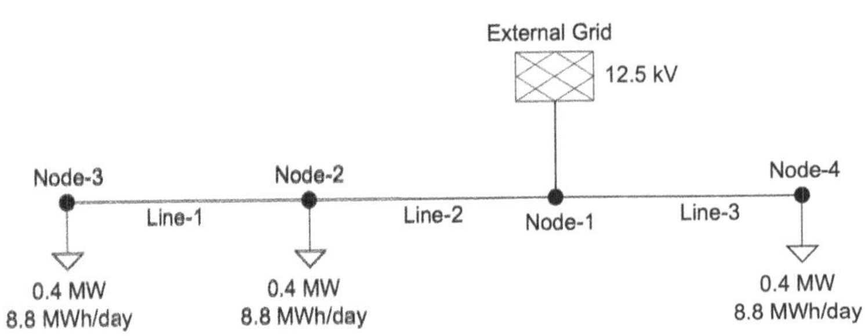

Figure 3.13 The 4-bus illustrative test system.

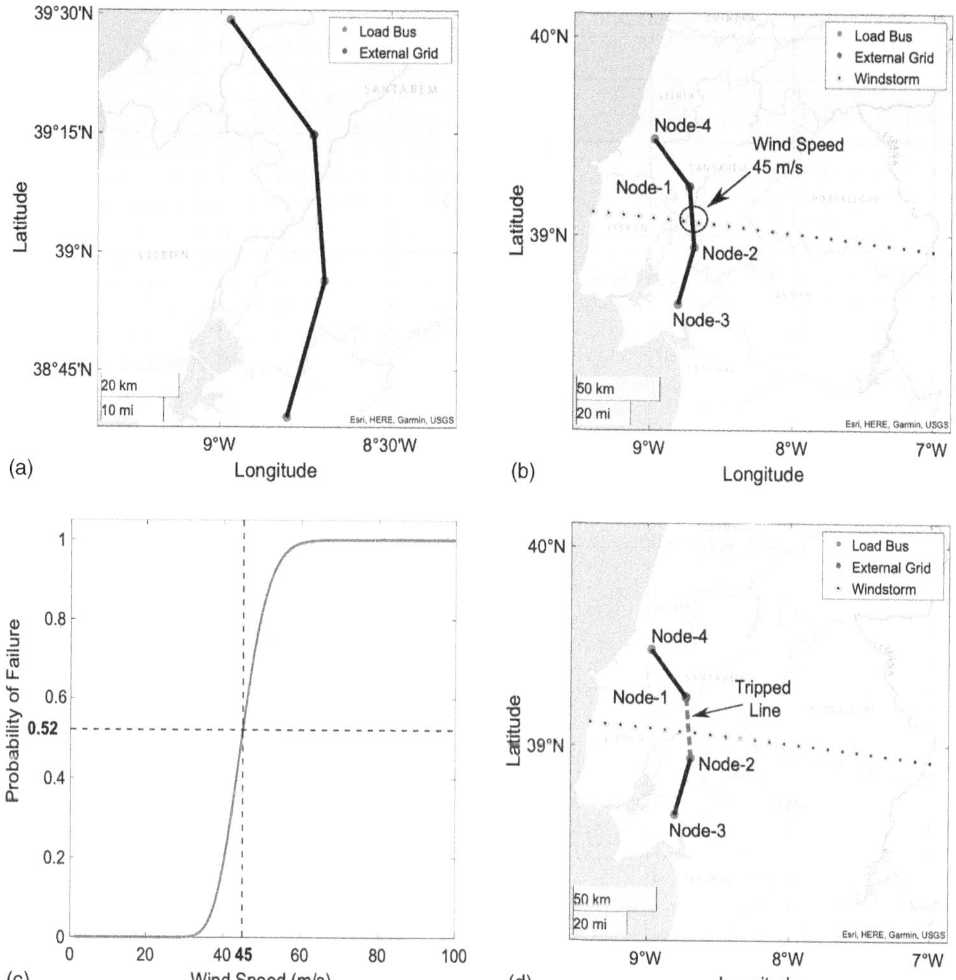

Figure 3.14 (a) The test system randomly projects to a geographic area, (b) Identification of wind storm speed that the line is exposed to, (c) Mapping to overhead-line fragility curve, (d) Wind-dependent status of affected line.

line will be tripped and a random restoration time will be generated. Hence, the line connecting node-1 and node-2 is tripped by the windstorm as shown in Figure 3.14 (d). The temporal line status of the windstorm is shown in Figure 3.14 for the complete time horizon considered in this study is shown in Table 3.8. Figure 3.15 shows the probability distribution of ENS for the 100 windstorm scenarios generated, where the resulting expected energy not supplied (EENS = 3.77 MWh) and CvaR ($\text{CVaR}_{0.9}$ = 7.85 MWh) values can be calculated using the formulation presented in Section 3.3.

3.4 AC Cascading Modeling for Resilience Applications

The majority of existing resilience tools, including the tools presented in the previous sections in this chapter, do not consider the cascading propagation following the initial set of outages (N–k, where k can be a big number beyond N–1 or N–2). This section will introduce a new AC cascading

Table 3.8 Temporal line status of windstorm is illustrated in Figure 3.14.

Hour	Line-1	Line-2	Line-3
1	1	1	1
2	1	1	1
3	1	1	1
4	1	1	1
5	1	1	1
6	1	1	1
7	1	0	1
8	1	0	1
9	1	0	1
10	1	0	1
11	1	0	1
12	1	1	1
13	1	1	1
14	1	1	1
15	1	1	1
16	1	1	1
17	1	1	1
18	1	1	1
19	1	1	1
20	1	1	1
21	1	1	1
22	1	1	1
23	1	1	1
24	1	1	1

model capable of simulating the impacts of large disturbances and of seamlessly integrating to resilience metric systems.

Following the initial shock of the external event, it is possible for a cascading event to be initiated and propagate across and between infrastructures. Cascading is a series of dependent outages or failures of individual components that successively weaken the system. It occurs during the Phase I of the resilience trapezoid (see Figure 3.1) and, in power systems, cascading is the main way that blackouts become widespread and lead to significant socio-economic impacts. Cascading outages can also propagate between infrastructures, and this interdependence is critical in assessing the cascading impact, infrastructure interactions, and vulnerabilities and hence valuing resilience.

In resilience studies, of particular interest are the common-cause failures caused by extreme shocks. These failures can propagate in the system and create further outages via interactions with infrastructure assets. Cascading is usually observed from simulations or real blackout data systematically collected over time. Its impact can be quantified with mainly two types of cascading

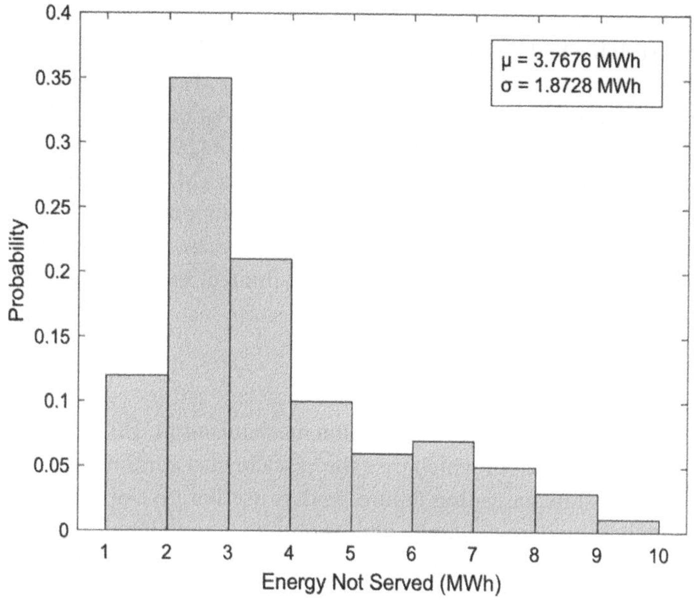

Figure 3.15 Probability distribution of ENS for the 100 windstorm-generated scenarios.

metrics: metrics of blackout impact outside the power system and metrics of the cascading process inside the power system. The former category primarily refers to measuring the blackout size and impact on the connected customers, and it includes the number of customers disconnected, the blackout duration, the power shed, and energy unserved. These can also be translated to monetary cost to society, using, e.g., load cost indices such as Value of Lost Load. The latter metric category refers to internal cascading metrics and mainly includes the number of line branches or other components outaged. These metrics are particularly important as they provide a clear picture of the evolving topology of the system as the cascade propagates, which can be represented by the so-called influence graph or interaction graph [17].

Resilience analysis to extreme events requires cascading failure models that reliably converge and thus provide meaningful results even for large contingencies [18]. Additionally, models often have to be applied to large datasets and networks and, therefore, need to be computationally fast. Dynamic models provide comprehensive details about cascades but require extensive and often unavailable input data describing the dynamic characteristics of a power system. Additionally, dynamic models are often computationally expensive, which makes them impractical in large networks. DC-based models are, hence, frequently used in resilience studies [19–24]. However, past outages have shown the significant role of voltage deviations and reactive PFs, such as during the 2003 blackout in the United States and Canada [25] or the 2009 blackout in Brazil [26]. While DC PF models are fast and numerically stable, they fail to incorporate these aspects. AC PF models usually suffer from non-converging PFs, which regularly occur when considering stressed networks and large contingencies. Resilience analysis, however, depends on the analysis of such extreme conditions and thus requires dedicated cascading failure models. Some AC PF models, such as [27], do not address the matter of non-converging PFs at all. Other models, such as [23], particularly address non-converging PFs, but do not consider reactive power and voltage limits and lack subsequent reactions by protection mechanisms, such as excitation limiters and undervoltage load shedding (UVLS). These mechanisms play a crucial part in large cascading failures [27]. Additionally, resilience analysis requires a whole-systems approach; thus, models need to be able to link seamlessly to established resilience evaluation frameworks.

This section describes the implementation of AC cascading fault model (CFM) and presents an internal validation of the modeling assumptions by performing dynamic simulations [4]. AC-CFM is implemented in MATLAB and uses the PF solvers of the MATPOWER toolkit [28]. The reference also contains the mathematical descriptions of the solvers. The model requires a MATPOWER network case structure and the initial contingency set as inputs. Any network component modeled in MATPOWER, i.e., buses, generators, loads, lines, transformers, and shunt devices, can be a part of the initial contingency set. The size of the initial contingency set is not limited, as it is with other models [29, 30].

3.4.1 Recursive Application of Protection Mechanisms

A cascading failure is governed by successive activation of protection mechanisms [4, 18]. This can cause disintegration of the network into islands, in which case the cascade may continue within each island independently. Most steady-state cascading failure models use iterative approaches to detect the islands at any discrete cascade generation and apply the necessary protection mechanisms. However, iterative approaches fail to identify causalities between generations and cannot address different time scales of protection mechanisms, because cascade generations are determined for the entire network, even if it has disintegrated into independent islands. Instead, AC-CFM uses a recursive approach (see Figure 3.16) to handle cascades within each island individually until the cascade comes to a halt. Beginning with the initial network, the PF within every island is calculated and protection mechanisms are applied (see Figure 3.17) [18]. The protection mechanisms are explained in detail in the subsequent sections. If the protection mechanisms have changed the conditions within an island, such as loads, generators, or operating lines, the recursion is applied to the island again (induction case). If the conditions have not changed or are within the specified tolerances, the cascade within the island comes to an end, and the model proceeds with the next island (base case). This implementation is a tree traversal, depth-first search, in which the cascade in one island is handled until its termination before continuing with the next island. Figure 3.17 illustrates the succession of protection mechanisms in AC-CFM. Protection mechanisms require knowledge of the generators and loads as well as bus voltages and line currents. The initial step in the routine, therefore, is to obtain a solvable AC PF. Afterward, protection mechanisms are applied.

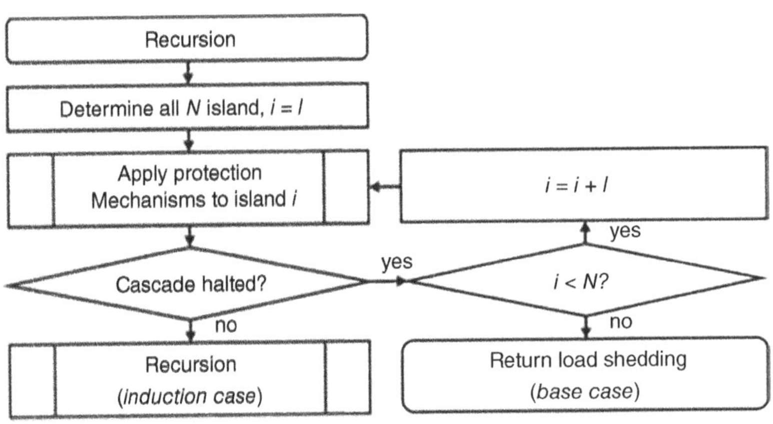

Figure 3.16 Recursive approach of AC-CFM [18].

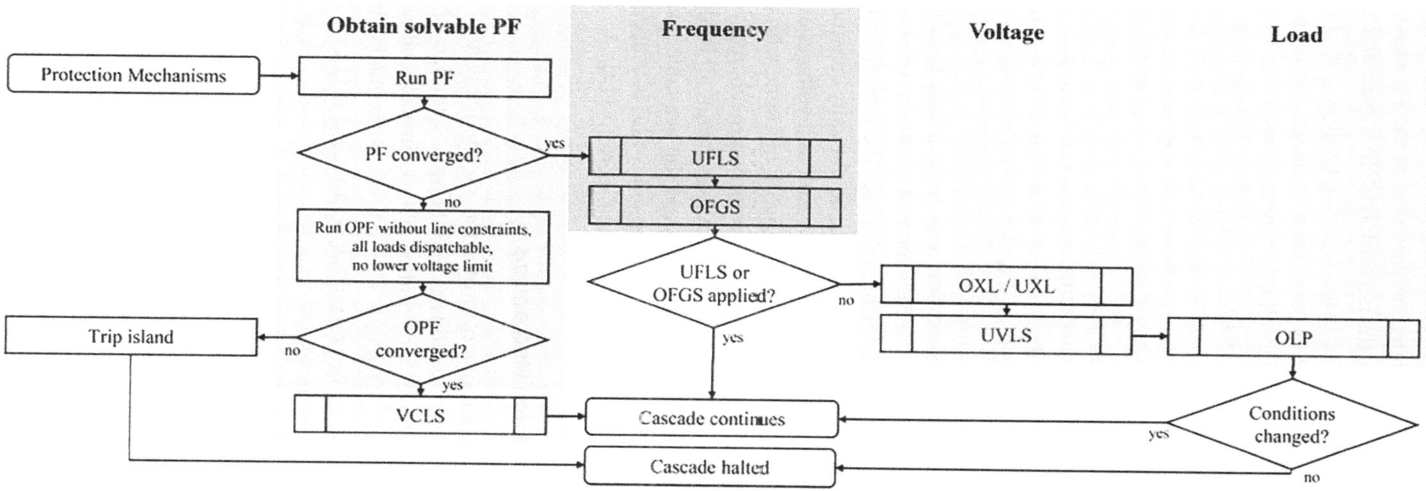

Figure 3.17 Implementation and succession of protection mechanisms in AC-CFM [18].

3.4.2 Obtaining a Solvable PF

A common problem when performing AC contingency analysis is that the AC PF solver may not converge. Non-convergence is mainly a mathematical issue, as the PF equations do not provide a mathematical solution for the physical state the network is in. In some cases, this indicates that the network is beyond its physical capability. For instance, if the reactive power demand in a network is significantly larger than the reactive power generation capacity, a voltage collapse occurs. In a real power network, this would lead to undervoltage at some buses. If a UVLS scheme exists, loads would automatically be shed to support voltages. A cascading failure model needs to overcome unsolvable PFs by moving the system back into a solvable region, usually by shedding some loads. Previous models have addressed this either by assuming that non-convergence leads to a blackout of the entire system, load shedding at all buses [28], reverting to a DC PF [29], or sensitivity factors identifying the loads that move the system most effectively back into the solvable region. The challenge is to minimize the required load shedding, because this load shedding is a purely mathematical way of obtaining a solvable PF and does not correspond to any response of an actual power network. Furthermore, excessive load shedding may prevent the revelation of underlying voltage issues. AC-CFM converts, in case of a non-convergence, all loads to being dispatchable and uses an OPF solver to identify the minimum load shedding required to make the PF solvable again (voltage collapse load shedding, VCLS). Different established OPF solvers, such as MIPS [29] or IPOPT [30], can be integrated into AC-CFM, which further increases the range of solvable OPFs. Line and lower voltage limits are, at this point, disabled to not mathematically constrain the solution space. Loads in the island are then shed according to the OPF result and voltages at generating buses are set as calculated by the OPF. Often, VCLS leaves the network in a state with undervoltage at some buses or load violations. The modeling of subsequent protection mechanisms, hence, continues as usual (see Figure 3.17). In rare cases, OPF does not converge even if all loads are reduced to zero, usually for small islands where reactive line losses exceed the available reactive power capacity. The island is then assumed to be beyond its physical capability and is tripped. Further elaboration on cases with non-converging OPF is given in subsequent sections [18].

3.4.3 Implementation of Protection Mechanisms

The different stages and implementations of protection mechanisms are explained in detail in the following sections. First, frequency stability is maintained. Second, over- and underexcitation are resolved and resulting undervoltages are handled. Third, lines exceeding their load rating are tripped by overload protection (OLP). Dynamic simulations, performed using DIgSILENT POWERFACTORY 2019 SP3 on the IEEE 39-bus test network, are used to internally validate the implementation of the individual mechanisms [4, 18].

1) **Under- and Overfrequency:** When considering synchronous generators, sudden mismatches between electrical and mechanical power, for instance, caused by unintentional islanding, result in a change in rotor frequency. Protection schemes such as underfrequency load shedding (UFLS) or overfrequency generator shedding (OFGS) can reestablish a balance of mechanical and electrical power and ensure that frequency stays within limits. For instance, standards by the North American Electric Reliability Corporation (NERC) require UVLS schemes to be capable of handling mismatches of up to 25% of the connected load [35]. Generator protection must be coordinated to prevent excessive shedding. Algorithms 1 and 2 describe how UFLS and OFGS are implemented in AC-CFM. The maximum generation imbalance δG_{max} reflects the limited ability of generators to quickly adjust to load changes before the frequency exceeds

the set limits. The dispatch overhead $\delta G_{overhead}$ caters for network losses. Assuming a single network frequency, UFLS sheds the same percentage of load at every bus, but prioritization of load shedding can be implemented in the proposed algorithm. Coordination of OFGS is done so that shedding starts with the smallest connected generator, for instance, because those are most likely to lose synchronism. UFLS and OFGS are repeatedly applied in the model until δG_{max} is no longer exceeded and a balance is established. In the model, it is assumed that frequency returns to nominal between cascade generations, i.e., that there is sufficient time for automatic generation control to act between successive trips of lines. To demonstrate the behavior of the UFLS protection scheme, Figure 3.18 shows the frequency response and connected active generation and demand of an island with insufficient generation that has been disconnected from the network at $t = 1$ s. After islanding, the electric active power generation steps up significantly in order to meet the demand, which leads to a decreasing system frequency. If the imbalance between generation and demand is only small, the frequency stabilizes within the allowed frequency range (solid black). If the imbalance between generation and demand is large, the frequency decreases gradually and will eventually exceed the allowed frequency range eventually (dashed black). In this example, reducing the load by 17% lets the frequency return to the allowed range (dotted grey) [4, 18].

Algorithm 1: UFLS.

1: **if** generation increased by less than δG_{max} but does not exceed the available capacity, **then**
2: Distribute slack generation so that it is shared by each generator proportional to its capacity, while no generator exceeds its capacity.
3: **else**
4: Reduce loads to $(1 - \delta G_{overhead})$ of the previously dispatched generation increased by δG_{max}, but not exceeding the available capacity.
5: **end if**

Algorithm 2: OFGS.

1: **if** generation decreased by less than δG_{max}, **then**
2: Distribute slack generation so that it is shared by each generator proportional to its capacity, while no generator exceeds its capacity.
3: **else**
4: Trip generators, beginning with the smallest, so that the dispatched generation does not exceed the total load by more than δG_{max}.
5: **end if**

2) **Over- and Underexcitation:** If the reactive power demand from a generator changes, its excitation system reacts by adjusting the excitation voltage in order to keep the terminal voltage stable. While the excitation system can adjust the field within a certain range to cater for different reactive power outputs, over- and under-excitation leads to overheating and eventually damage to the generator, and needs to be prevented. Over- (OXL) and underexcitation limiters (UXL) constrain the field current, and as a result, the terminal voltage is adjusted. The implementation of OXLs and UXLs in AC-CFM is shown in Algorithm 3. In AC PF, buses are either specified as PV (active power and voltage) or PQ (active and reactive power) buses, identifying

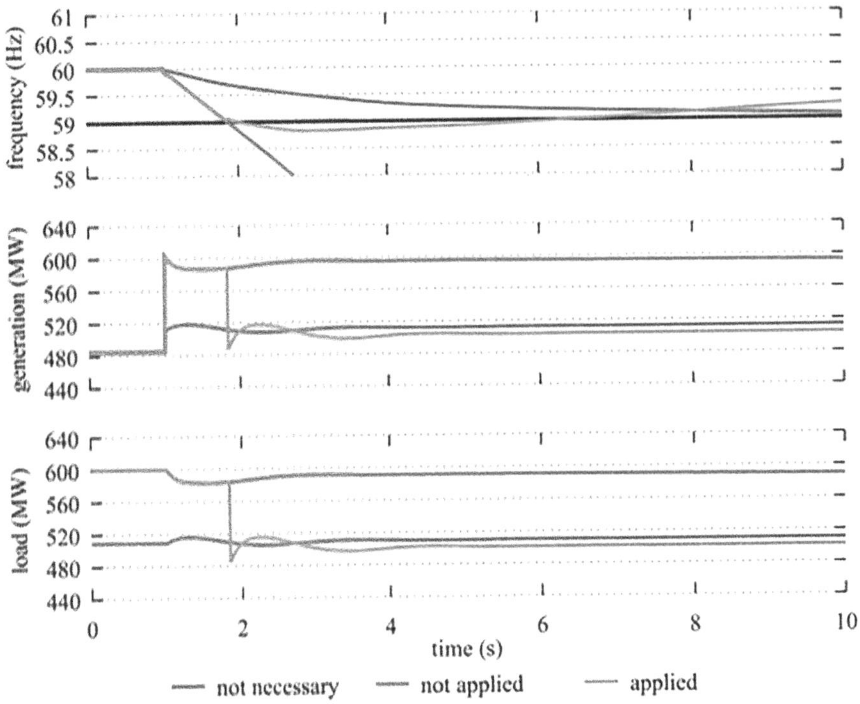

Figure 3.18 Dynamic modeling of UFLS.

which are the fixed parameters at the bus. Demand buses are generally PQ buses, while generating buses are usually PV buses. In Algorithm 3, buses with generators that exceed their reactive power limits are converted to PQ buses, meaning that their reactive power output is now fixed to its closest, upper or lower reactive power limit, and the bus voltage now becomes variable. This effectively mimics the impact of OXLs and UXLs. The behavior of an OXL is demonstrated in Figure 3.19. After a line fault at $t = 1$ s and subsequent islanding, the reactive power output of the shown generator increases. Depending on the demand, the reactive power output can be within the reactive power limits of the generator (solid) or outside (dashed). In order to prevent the generator from damage, the OXL effectively reduces the terminal voltage of the generator after a time delay ($t = 10$ s, dotted). While actual implementations of OXLs and UXLs differ and act on different components of the generator, Figure 3.19 illustrates in simple terms how overexcitation can successfully be prevented. If the terminal voltage exceeds the voltage limits after triggering the OXL, UVLS can be applied, which is part of the following sections [18].

Algorithm 3: OXL / UXL.

1: Determine all generators exceeding their reactive power limits as defined in the MATPOWER case structure.
2: Convert the buses of these generators to PQ buses.
3: Set their reactive power output equal to their closest reactive power limit.

3) **Undervoltage:** Reactive PF leads to a voltage drop at the receiving end of a line. In order to prevent voltages below the lower voltage limit, UVLS can be applied, which gradually decreases the

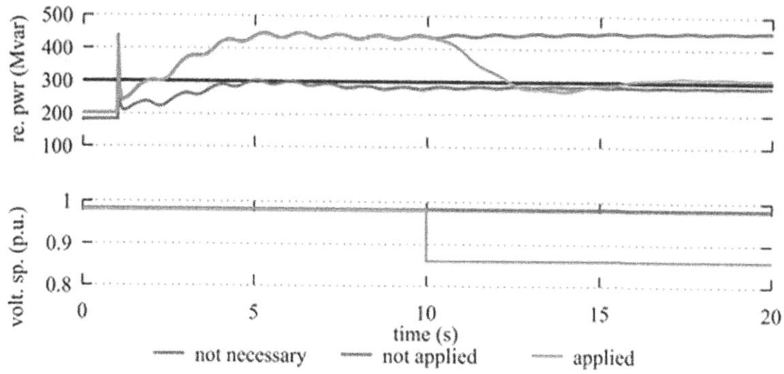

Figure 3.19 Dynamic modeling of OXL.

demand at the bus until the voltage is again within limits. UVLS is usually applied in blocks of predefined sizes. Algorithm 4 shows the implementation of UVLS in AC-CFM. The load shedding block size is defined as δP_{ls}, and the maximum block-wise load shedding before shedding all loads at the bus is $\delta P_{ls,max}$. Figure 3.20 demonstrates the behavior of UVLS, in which a line fault at $t = 1$ s and subsequent islanding causes a voltage drop at one bus. Depending on the reactive power demand at the bus, the voltage at the bus can stay within limits (solid) or outside (dashed). With UVLS, 5% of load is shed att = 5, 6, 7, 8, 9 s (dotted). As the voltage is still below the lower voltage limit, at $t = 10$ s all loads at the bus are shed as a final measure to reestablish acceptable voltages [18].

4) **Overload:** The OLP in AC CFM trips all lines that exceed their rating as defined in the MATPOWER case structure. This is to prevent lines from damage due to overheating and ensure safe operation. In real OLP, there is usually a time delay to allow short overcurrents. However, as AC-CFM does not model with respect to time, this is not necessary to model [18].

Algorithm 4: UVLS.

1: Shed δP_{ls} of load at buses falling below the UVLS trigger voltage U_{UVLS}.
2: **if** applied load shedding at a bus exceeds $\delta P_{ls,max}$ **then**
3: Trip all loads at this bus
4: **end if**

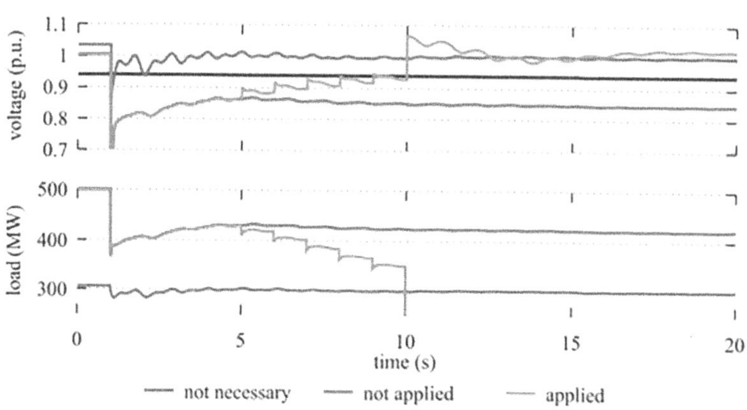

Figure 3.20 Dynamic modeling of UVLS.

3.4.3.1 Cascade Visualization

AC-CFM provides a novel way of visualizing cascading failures as a tree-like graph (see Figure 3.21). The graph expresses causalities in cascading failures and helps mitigate the impact of large and wide-spread blackouts. It holds all model output parameters, including the succession of protection mechanisms, the available loads, generators and lines, the disintegration of the network into islands, and the amount of load shedding at each generation. It can easily be accessed and searched by subsequent analysis procedures depending on the requirements of the researcher, even for very large networks or contingencies with thousands of buses and lines. The example in Figure 3.21 shows, beginning with a fully connected network and a single line fault, how a network first disintegrates into four islands due to OLP. After that, the protection mechanisms in each island try to stabilize frequency and voltage. At the end, each island either remains alive or blacks out. The full model output is shown in the Appendix. The graph reveals how the overloading of a single line triggers the entire cascade after the initial contingency. Reducing the loading of such critical lines or increasing their capacity can, for instance, prevent major blackouts. The graph can, thus, be used to identify key mechanisms that are crucial in the propagation of a failure cascade [12, 18].

3.4.4 Integration into Resilience Metric Frameworks

The straightforward and seamless integration with established resilience metrics enhances comparability and applications of AC-CFM to existing works and is a main advantage over other models, which have not addressed this need. The ΦΛΕΠ framework introduced in Section 3.2 will serve as an example for such an integration in this study. The following indicators are used in this study: lost load, lost lines, lost buses, and lost generators. Figure 3.22 shows the lost load in a network for an exemplary cascade triggered at $t = 0$ [4, 18].

Sensitivity analysis is completed for the following input parameters and ranges: initial contingency size (lines) $ci \in [1, 200]$, maximum generation imbalance $\delta G_{\max} \in [0.0, 0.5]$, UVLS trigger

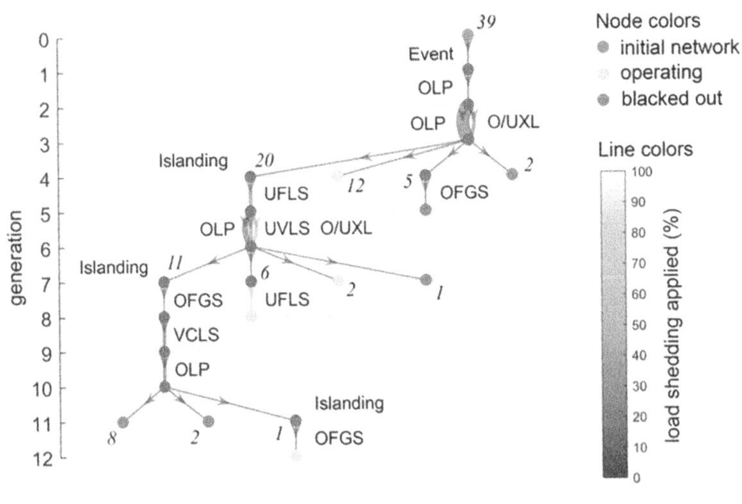

Figure 3.21 Visualization of a cascade in the IEEE 39-bus network. Each circle represents the network or a part of it at a certain generation, starting with the initial network, and leading to four operating islands and five blacked-out islands. The number at each island gives the number of buses within the island. The connecting lines represent the protection mechanisms that have been applied between generations. The thickness of the lines indicates how many components have been affected, while the color indicates the percentage of load shedding [18].

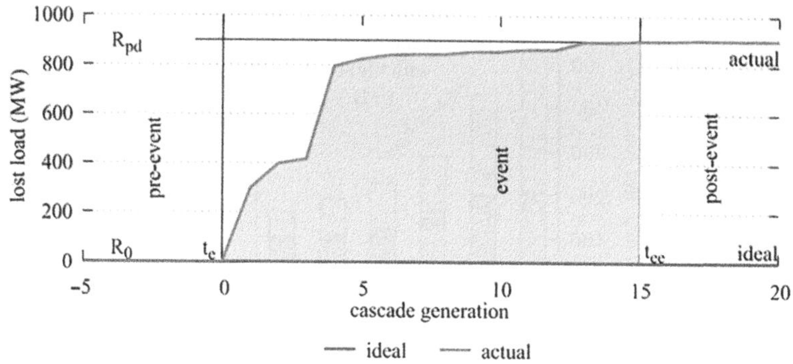

Figure 3.22 Lost load over cascade generation with actual and ideal performance [18].

voltage (p.u.) $UUVLS \in [0.80, 0.95]$, maximum block-wise load shedding $\delta Pls, \max \in [0.05, 1.00]$, and maximum number of load shedding blocks$(\delta Pls,\max)/(\delta Pls) \in [1, 10]$. The output parameters are the indicators and Φ and Λ metrics introduced in Section 3.2. Simulation results are obtained from the summer 2004 peak Polish power system containing 2736 buses and 3504 lines [36]. The network has a total generation capacity of 28.9 GW and a total demand of 18.1 GW. Reactive power capacity of the generators in the test case has been relaxed to +80% and − 40% of the maximum active power output. This relaxation is purely to prevent the otherwise almost always instantaneous occurrence of voltage collapses due to excessively stringent reactive power limits. FAST was performed using the GSAT for MATLAB [36]. A sample size of 10,000 was sufficient for sensitivities to converge. The resulting sensitivities are shown in Figure 3.23. Initial contingency size has the largest impact on lost load, lost lines, and lost buses. The impact on lost generators is negligible. The Spearman coefficient reveals a very strong positive correlation (coefficient is larger than +0.8) between initial contingency and lost load, lost lines, and lost buses. Increasing initial contingency size, thus, leads to larger and faster loss of load, lines, and buses. Maximum generation imbalance only has a significant impact on lost generators. The Spearman coefficient reveals very strong negative correlation (coefficient is less than −0.8). Increasing maximum generation imbalance, thus, leads to less and slower loss of generators. UVLS trigger voltage has no significant first-order impact. Maximum block-wise load shedding and number of load shedding blocks have

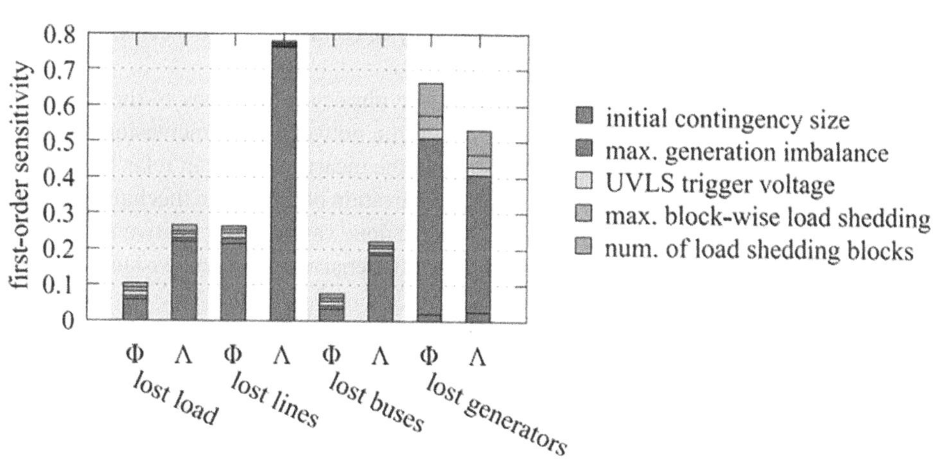

Figure 3.23 First-order sensitivity coefficients calculated by FAST [18].

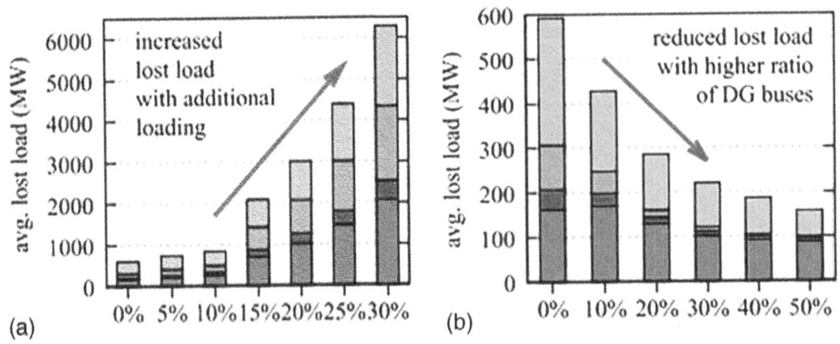

Figure 3.24 Causes of lost load: UFLS (drak grey), UVLS (black), VCLS (grey), and tripped buses (light grey). (a) Additional load, (b) Ration of DG buses [18].

a slight impact on lost generators. The Spearman coefficient reveals moderate negative correlation (coefficient between −0.6 and − 0.4) of the number of load shedding blocks with Φ and Λ. A small number of load-shedding blocks increases the load-shedding per block, which leads to more rapid load shedding. This rapid load shedding causes generation imbalances and increases loss of generators [18].

A key question in resilience studies is how resilience changes with increasing network loading due to additional loads and generation. Beginning with the default loading, all loads are increased in steps of 5%. The generation capacity is adjusted likewise. Until an additional loading of 10%, the average lost load does not increase significantly (see Figure 3.24 [a]). From an additional loading of 15% on, an approximately linear increase of average load shedding can be observed. Lost load is caused by UFLS, UVLS, VCLS, and tripped buses, i.e., buses that become part of an island without generation or insufficient ramping capability. AC-CFM can, thus, be used to determine the critical loading, beyond which mean lost load sharply increases. The UK National Infrastructure Commission emphasizes the need for thorough stress testing and determination of system breaking points, as it can be performed by AC-CFM [4], in its report on resilient infrastructure systems [37]. For further understanding of this behavior, the mean values of the Φ and Λ metrics from the ΦΛΕΠ resilience metric framework are calculated. While metrics do not change for additional loading of up to 10%, both metrics for all indicators increase significantly from an additional loading of 15% on (see Figure 3.25). This means that loads, lines, buses, and generators are lost more and faster with increasing network loading. Additional loads can, thus, only be connected up to a certain loading level without significantly reducing resilience [18].

Following the results, critical loading is governed by the physical limitations of the network and its protection settings. For instance, adding inertia to the network, thus increasing δGmax, or deploying shunt devices for voltage support, are some of the measures that can relax the physical limitations. Easing of protection settings can delay the activation of protection mechanisms, but might trigger subsequent damage to the system and must be done carefully. Preventive measures, building on the capabilities of a modern smart grid, such as defensive or corrective islanding, can also improve critical loading of networks [4, 18].

A further trend is the increasing amount of distributed generation (DG) connected to the network, while large and conventional generators are being disconnected [4, 18]. To model this, the initial 2736-bus Polish network is modified by adding DG with a capacity of 10 MW each, a typical maximum size for generators connected to the distribution system, to a certain ratio of DG buses. In a real network, DG can be accumulations of a number of small-scale generators connected to the same bus. It is assumed that DG is grid-forming. The generation capacity of conventional generators

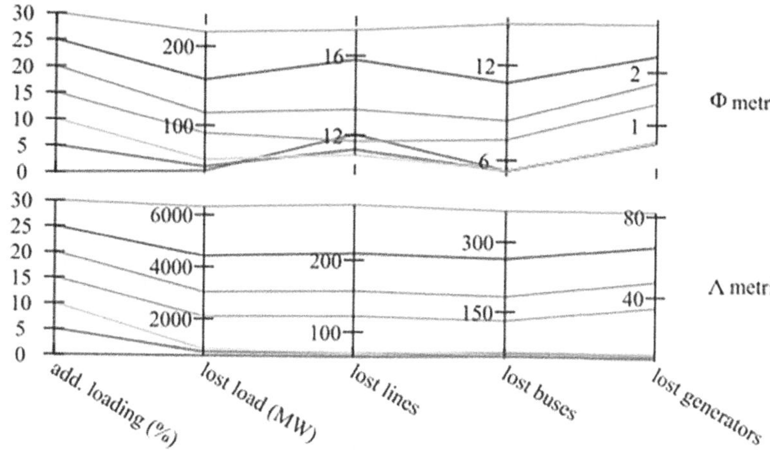

Figure 3.25 Average resilience metrics depending on additional loading [18].

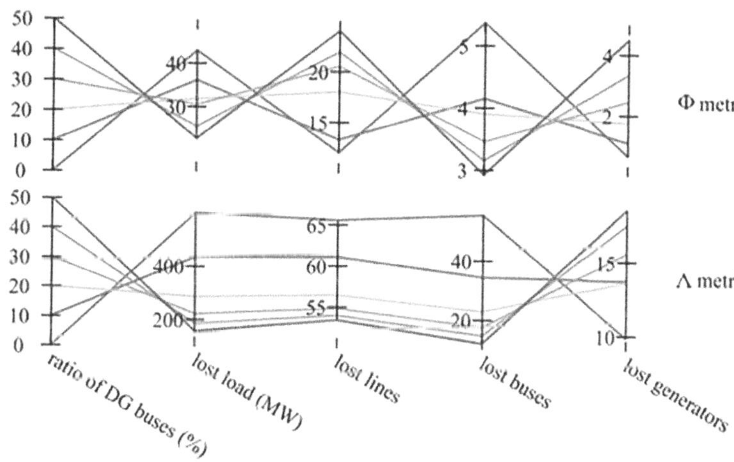

Figure 3.26 Average resilience metrics depending on ratio of DG buses [18].

is downsized so that the overall generation capacity does not change. The model is applied to various ratios of DG buses, and the mean values of the Φ and Λ resilience metrics are calculated. With increasing ratio of DG buses from 0% to 50%, lost load decreases approximately exponentially (see Figure 3.24 [b]). Particularly, the amount of UVLS, VCLS, and lost load due to tripped buses is reduced, while the reduction in UFLS is not as significant. This is further analyzed by calculating the mean values of the Φ and Λ metrics from the ΦΛEΠ resilience metric framework. With increasing ratio of DG buses, loads, and buses are lost less and slower (see Figure 3.26). Larger ratios of DG buses also decrease the number of lines lost, but lines are lost faster. Generators are lost more and faster; however, this is mainly controlled by the coordination of OFGS. Increasing the ratio of DG can, thus, improve power network resilience and limit the lost load caused by cascading failures in several ways, if it is connected and controlled adequately. First, a larger number of DG decreases the number of buses that are part of islands formed during a cascade and lack generation capacity, and are, thus, tripped. Second, DG stabilizes bus voltages and reduces the amount of UVLS. However, recent cases such as the 2019 blackout in the United Kingdom have shown that large amounts of

DG in combination with inadequate protection settings can increase the risk of blackouts and, thus, reduce power network resilience [38]. Deployment of DG, hence, requires additional protection mechanisms to particularly deal with large frequency fluctuations and loss of mains.

References

1 B. Unel and A. Zevin "Toward resilience; Defining, measuring, and monetizing resilience in the electricity system," in *Institute for Policy Integrity*. NY, USA: New York University School of Law, 2018, p. 50p.

2 R. Moreno, M. Panteli, P. Mancarella, H. Rudnick, T. Lagos, and A. Navarro, "From reliability to resilience: Planning the grid against the extremes," *IEEE Power Energy Mag.*, vol. 18, no. 4, pp. 41–53, 2020.

3 Forward Resilience Measures, https://smarter.energynetworks.org/projects/nia_ngt0049

4 M. Noebels, "Assessment and Enhancement of Power Grid Resilience to Shocks and Stresses", PhD Thesis, The University of Manchester, Available: https://research.manchester.ac.uk/en/studentTheses/assessment-and-enhancement-of-power-grid-resilience-to-shocks-and

5 E. Vugrin, A. Castillo, and C. Silva-Monroy, "Resilience metrics for the electric power system: a performance- based approach," sandia national laboratories, Albuquerque, New Mexico and Livermore, California, Tech. Rep. February 2017.

6 H. Raoufi, V. Vahidinasab, and K. Mehran, "Power systems resilience metrics: A comprehensive review of challenges and outlook," *Sustainability*, vol. 12, no. 22, pp. 1–24, 2020.

7 F. Petit *et al.*, ""Resilience measurement index: an indicator of critial infrastrucutre resilience," Argonne National Laboratory, Argonne, IL," *Technical Report*, 2013.

8 F. Petit *et al.*, ""Resilience measurement index: An indicator of critial infrastrucutre resilience," Argonne National Laboratory, Argonne, IL," *Technical Report*, 2013.

9 E. Vugrin, A. Castillo, and C. Silva-Monroy, ""Resilience metrics for the electric power system: a performance- based approach," sandia national laboratories, Albuquerque, New Mexico and Livermore, California," *Technical Report*, 2017.

10 M. Panteli, D. N. Trakas, P. Mancarella, and N. D. Hatziargyriou, "Power systems resilience assessment: Hardening and smart operational enhancement strategies," *Proceedings of the IEEE*, vol. 105, no. 7, pp. 1202–1213, 2017.

11 US Department of Energy, "North American Energy Resilience Model", 2019

12 M. Panteli, P. Mancarella, D. N. Trakas, E. Kyriakides, and N. D. Hatziargyriou, "Metrics and Quantification of Operational and Infrastructure Resilience in Power Systems," *IEEE Transactions on Power Systems*, vol. 32, no. 6, pp. 4732–4742, 2017, doi: 10.1109/TPWRS.2017.2664141.

13 M. Belivanis and K. Bell, *Representative GB Network Model: Notes*. Glasgow, U.K.: Univ. Strathclyde, 2011.

14 R. Dawson and J. Hall, "Adaptive importance sampling for risk analysis of complex infrastructure systems," *Proc. Roy. Soc. Amer.*, vol. 462, pp. 3343–3362, 2006.

15 F. Casciati and L. Faravelli, *Fragility Analysis of Complex Structural Systems*. Taunton, U.K.: Research Studies Press, 1991.

16 UK Met Office, UK climate-extremes. 2016. [Online]. Available: www.metoffice.gov.uk/public/weather/climate-extremes

17 K. Zhou *et al.*, "A Markovian influence graph formed from utility line outage data to mitigate large cascades," *IEEE Trans. Power Systems*, vol. 35, no. 4, 2020.

18 M. Noebels, R. Preece, and M. Panteli, "AC cascading failure model for resilience analysis in power networks," *IEEE Systems Journal*, vol. 16, no. 1, pp. 374–385, 2022, doi: 10.1109/JSYST.2020.3037400.

19 M. J. Eppstein and P. D. Hines, "A "random chemistry" algorithm for identifying collections of multiple contingencies that initiate cascading failure," *IEEE Transactions on Power Apparatus and Systems*, vol. 27, no. 3, pp. 1698–1705, 2012.

20 H. Sabouhi, A. Doroudi, M. Fotuhi-Firuzabad, and M. Bashiri, "Electrical power system resilience assessment: a comprehensive approach," *IEEE Systems Journal*, vol. 14, no. 2, pp. 2643–2652, 2020.

21 S. Yang, W. Chen, X. Zhang, C. Liang, H. Wang, and W. Cui, "A graph-based model for transmission network vulnerability analysis," *IEEE Systems Journal*, vol. 14, no. 1, pp. 1447–1456, 2020.

22 M. J. Eppstein and P. D. Hines, "A "random chemistry" algorithm for identifying collections of multiple contingencies that initiate cascading failure," *IEEE Transactions on Power Apparatus and Systems*, vol. 27, no. 3, pp. 1698–1705, 2012.

23 H. Sabouhi, A. Doroudi, M. Fotuhi-Firuzabad, and M. Bashiri, "Electrical power system resilience assessment: a comprehensive approach," *IEEE Systems Journal*, vol. 14, no. 2, pp. 2643–2652, 2020.

24 S. Yang, W. Chen, X. Zhang, C. Liang, H. Wang, and W. Cui, "A graph-based model for transmission network vulnerability analysis," *IEEE Systems Journal*, vol. 14, no. 1, pp. 1447–1456, 2020.

25 A. Muir and J. Lopatto, "Final report on the august 14, 2003 blackout in the United States and Canada: causes and recommendations," U.S.-Canada Power System Outage Task Force, Rep. AR-1165, 2004.

26 J. M. O. Filho "Brazilian blackout 2009: Blackout watch," in *Proc. PAC World Mag*, 2010, pp. 36–37.

27 M. Vaiman *et al.*, "Risk assessment of cascading outages: Methodologies and challenges," *IEEE Transactions on Power Apparatus and Systems*, vol. 27, no. 2, pp. 631–641, 2012.

28 R. D. Zimmerman, C. E. Murillo-Sanchez, and R. J. Thomas, "MATPOWER: steady-state operations, planning, and analysis tools for power systems research and education," *IEEE Transactions on Power Apparatus and Systems*, vol. 26, no. 1, pp. 12–19, 2011.

29 M. J. Eppstein and P. D. Hines, "A "random chemistry" algorithm for identifying collections of multiple contingencies that initiate cascading failure," *IEEE Transactions on Power Apparatus and Systems*, vol. 27, no. 3, pp. 1698–1705, 2012.

30 J. Song, E. Cotilla-Sanchez, G. Ghanavati, and P. D. Hines, "Dynamic modeling of cascading failure in power systems," *IEEE Transactions on Power Apparatus and Systems*, vol. 31, no. 3, pp. 2085–2095, 2016.

31 S. Mei, Y. Ni, G. Wang, and S. Wu, "A study of self-organized criticality of power system under cascading failures based on AC-OPF with voltage stability margin," *IEEE Transactions on Power Apparatus and Systems*, vol. 23, no. 4, pp. 1719–1726, 2008.

32 Q. Chen and L. Mili, "Composite power system vulnerability evaluation to cascading failures using importance sampling and antithetic variates," *IEEE Transactions on Power Apparatus and Systems*, vol. 28, no. 3, pp. 2321–2330, 2013.

33 R. D. Zimmerman, C. E. Murillo-Sanchez, and R. J. Thomas, "MATPOWER: steady-state operations, planning, and analysis tools for power systems research and education," *IEEE Transactions on Power Apparatus and Systems*, vol. 26, no. 1, pp. 12–19, 2011.

34 A. Wächter and L. T. Biegler, "On the implementation of an interior-point filter line-search algorithm for large-scale nonlinear programming," *Mathematical Programming*, vol. 106, no. 1, pp. 25–57, 2006.

35 North American Electric Reliability Corporation, "Automatic Underfrequency Load Shedding (PRC-006-3)," Atlanta, GA, USA, 2017.

36 R. D. Zimmerman, C. E. Murillo-Sanchez, and R. J. Thomas, "MATPOWER: steady-state operations, planning, and analysis tools for power systems research and education," *IEEE Transactions on Power Apparatus and Systems*, vol. 26, no. 1, pp. 12–19, 2011.

37 National Infrastructure Commission, "*Anticipate, React, Recover. Resilient Infrastructure Systems*," London, U.K., 2020

38 National Grid ESO, "*Technical Report on the Events of 9 Aug. 2019*," Warwick, U.K., 2019.

4

Addressing Data Granularity and Ambiguity

4.1 Understanding Outage Models on Power Systems

Much of the content discussed within this textbook has concerned how faults and outages on power systems are represented in general terms, and how this is important for operators and planners to understand how to respond to external threats affecting the power system. In large-scale power systems, however, conventional models, which have historically been used to quantify failure probability, and hence risk associated thereof, have not typically been designed in a way that best suits analyses in the context of resilience. Chapters 2 and 3 have discussed how to define, qualify and quantify resilience and discussed some recent advancements in this area, particularly fragility curves and their application. This chapter will discuss how resilience data such as this can be productively visualized and the potential impacts of the granularity of this data.

That is, in considering failure mechanisms of overhead lines (OHL), there may be a large number of different factors that could cause outages. Foundational works such as [1, 2], in quantifying the reliability of power systems, have leaned on using homogeneous failure rates applied to a power system for different asset types. That is, a single defined failure rate is applied across an entire system to model the potential impact of outages caused by a natural hazard, and to determine power system performance using appropriate performance metrics.

To illustrate this, imagine a given line l_n has a failure rate λ_c where n is the identifier of the line in some set. The failure rate λ_c may be dependent on an external factor such as lightning strike rates per km^2, or maximum wind gust speed. This external feature is typically called a *natural hazard*. Most resilience studies study the impact of an individual natural hazard on a power system, and investigate different strategies for mitigating these impacts, for example the effect of extreme wind [3], or the impact of lightning [4].

If we have this single line, subject to a given natural hazard, to calculate the failure rate in a given timestep, conventionally an hour, we need some approximation for:

- The magnitude of the natural hazard affecting the asset on the system
- The exposure of that asset to the natural hazard
- The consequences associated with an outage of that asset

An example is demonstrated in the work presented in [5], and will be described here for illustrative purposes.

Let us consider the wind. There are many features of wind which could be of consequence to a power system: for instance, whether the wind flow is *turbulent* or *laminar*. That is, whether the flow of wind is steady and predictable, or chaotic and variable. We must also consider whether the wind speed being investigated is a time-averaged value or a "maximum gust" value, and over what

Fundamentals of Power System Resilience: Disruptions by Natural Causes, First Edition.
Mathaios Panteli, Rodrigo Moreno, Dimitris Trakas, Magnus Jamieson, Pierluigi Mancarella, Goran Strbac, and Nikos Hatziargyriou.

time-period that time-averaging is taking place, and whether it is appropriate for the application in which we are investigating. Further, we may also have to consider the local geography of the lines under investigation, and, potentially, the direction of that wind. Not only that, but other features such as the height above ground at which the wind speed is being measured may be important, especially in the context of a power system that may have assets of varying types, construction, strength, and age.

Sometimes, assumptions must be made to make models functional. Some example basic assumptions are made herein to illustrate some of the potential weaknesses in analyses that do not consider these features. These shall be discussed in turn here.

Firstly, an assumption may need to be made that weather cannot affect the reliability of electricity generation facilities directly. Or, more specifically, that wind cannot. This is a reasonable assumption, as most types of generators involve the generation of electricity within a confined, sealed building, which will not be exposed to the elements. In reality, features such as ambient temperature or water availability may affect the availability or efficiency of power generation facilities. That is not considered here. The vulnerabilities of these generators do not stem from wind itself acting on the generator, but rather how the wind impacts the wider system and the infrastructure that connects the generator to the main power system.

The outage rate of generators on large power systems is routinely tracked, and standardized values are used by researchers for power systems research. These values might vary based on the age of equipment, types of generation, complexity of technology, and so on. Some examples, typical failure rates are provided in the Table 4.1.

These failure rates are derived from work such as [1] and resources such as the Generation Availability Data Service (GADS), similar to the equivalent Transmission Availability Data Service (TADS) [6] from the National Electricity Reliability Corporation (NERC) in the USA. MTTF refers to "mean time to failure," which is the expectation of the time one would expect to elapse on average until there is a fault or outage on a generator in a given simulation or scenario.

What they tell us has varying usage depending on the nature of the power system simulation we wish to undertake. Resilience and reliability studies that investigate power system adequacy may use this failure rate in stochastic Markov Chain Monte Carlo (MCMC) simulations to generate outage scenarios.

An MCMC simulation involves presuming that there are some stochastic, or random, properties of a power system, and drawing from a presumed distribution or relationship. In more complicated simulations, more specialized sampling may be used if datasets are large or simulations are computationally expensive.

Table 4.1 Example failure rates for different types of generators.

Generator type	Failure rate (/hr)	MTTF (hr)
Wind	2.00×10^{-5}	50,000
Hydro	5.05×10^{-4}	1980
CCGT/OCGT	9.09×10^{-4}	1100
Nuclear	8.70×10^{-4}	1150
Coal	8.70×10^{-4}	1150
CHP/other	2.22×10^{-3}	450
Interconnector	2.51×10^{-5}	39,825

The "Markov Chain" implies that the status of assets on the system, which are generated, can change – e.g., a generator could come in or out of service with the sampling related to whether the asset's state changes or not.

Monte Carlo simulations are used to determine relationships or quantify aspects of models that may be difficult or impossible to compute directly, or analytically. A simple model illustrative of how a Monte Carlo simulation works is to consider the roll of a die.

Example 4.1: Determining the mean value of a die roll

In determining the mean value of a dice roll, there are two possible approaches. The *analytical* approach, whereby the standard form of calculating the mean of a set can be used:

$$\mu = \frac{1}{N}\sum_{i=1}^{N} x_i$$

where μ is the mean, x is the item of i index (typically what is colloquially implied when the term "average" is used), and N is the sample size. Put simply, all values are summed and divided by the sample size.

For a single, six-sided, unbiased die numbered 1 to 6, with an equal 1/6 probability of any side of the die being rolled, this resolves to 3.5.

An alternative approach is, knowing the probability of each individual face of the die occurring, to "roll" the die many times, and calculate the mean value of the results that emerge. You can then derive, first, the mean value of the roll of a die, and the error associated thereof.

The rate of occurrence of any individual side of the die, λ_n is assumed to be 1/6.

The pseudocode for calculating the mean value of a roll of the die can be given as:

For n in sample:
  Sample from (1,2,3,4,5,6)
  Record roll
Calculate average of rolls
Calculate error of rolls by dividing total by the square of sample

Where *sample* represents the size of sample being taken. A value of 10,000 is a typical used, as it represents a good balance between accuracy and computational efficiency. In this method, the mean will tend to 3.5 as more samples are taken, and the larger the sample size, the smaller the error will be.

Exercise 4.1:

1) Determine the average if a 12-sided die is used

2) Determine the mean value of a roll if three 12-sided die are used

In typical power system simulations, the "failure" would amount to taking the generator out of service, and simulating an associated consequence caused by that outage. However, there are multiple modes of "failure" that are possible in a generator, which may take different amounts of time to recover from.

Mechanical failures such as cracked rotors or pump failures may require long shutdowns and replacement parts, taking the generators out of service for days at a time. Alternatively, a short circuit or debris damage to connecting infrastructure off-site could still be counted as a "fault" at the generator site but would take a much shorter period of time to recover from. Therefore, while an average, or expectation, value of repair time may be assumed, in real-world analyses, the types of failure you simulate must be clearly stated and the scope of any simulation problem clearly defined.

Similarly, typically in such stochastic simulations, it is assumed that a "fault" on a generator takes it out of service in a binary manner. That is, the generator is either functional or it is not. In reality, a generator facility may lose a single reactor or burner but may still be able to operate at partial functionality. This was the case in the Three Mile Island incident in 1979, in which one reactor fully melted down, and in pumping clear radioactive gases from the facility, some leaked into the atmosphere causing a localized radiation leak, but in time the facility was able to resume using one of its reactors and continued to supply electricity [7].

Power system infrastructure such as OHL or distribution stations/transformers, may find themselves forced to operate at reduced capacity due to damage caused by a natural hazard, but may still retain some functionality and quantifying this merely as a binary may not adequately capture the failure mechanics of the associated infrastructure.

Fragility curves are used not just to indicate if a piece of transmission infrastructure is online or offline, but to what extent it retains functionality as a result of damage incurred [8]. This is done by reducing the capacity of the substations, dependent on the results of the stochastic sampling. These were introduced and discussed at length within Chapters 2 and 3, but a brief summary is provided here to contextualize the methodology.

To clarify, it is necessary to describe what a *fragility curve* actually is and represents [9]. Definitions and discussions relating to fragility curves can be found in extensive detail within [9] with a comprehensive review of fragility curves and the principles underlying them also in [10], and some of the work necessary to determine their formulation is explored in [11].

Assume a fragility curve represents a relationship between a given natural hazard and the probability of a certain outcome on a power system asset being acted upon by that natural hazard. Significant research has gone into the application of these in the context of quantifying power system resilience in GB when subject to extreme winds [12]. They can also be used in simulation frameworks to optimize responses to system outages and the placement of assets such as repair teams under staffing constraints [13].

A typical fragility curve has the magnitude of the examined natural hazard as the x-axis, and the probability of a given outcome (e.g. damage to asset, asset outage) as the y asset as a cumulative probability function. An example is shown in Figure 4.1, taken from [3].

A similar logic can be extended to investigate the potential impacts on asset restoration of a given natural hazard. A simplified correction on asset restoration rates was used in [5] to correct asset restoration rates for a given wind speed acting upon that asset. Research has also been conducted in which weather conditions were modeled to consider the potential impact on the restoration of OHL

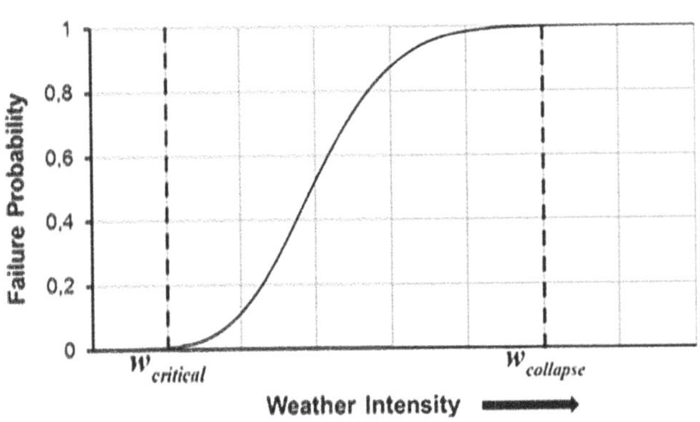

Figure 4.1 Example fragility curve, taken from [3] / IEEE.

of extreme weather in Italy [14], with features such as hill gradients, transportation links considered. These are all factors which, in reality, can have a significant effect on power system restoration but that carry corresponding difficulties when it comes to incorporating them into simulations.

It is reasonable to presume, for example, that in an extreme wind storm, it is not only difficult but unsafe to perform complicated engineering works to restore power system infrastructure, and as such, performing any restoration at all during extreme weather may not be possible.

A proxy, in any performed simulation, for this could be either to reduce the restoration rate for power system assets (and thus increase the expected time to repair) in simulations [5]; to perform power system simulations in timescales where restoration is not a factor and try to plan to restore or sustain as much of the power system as possible before external repairs can be performed to optimize initial responses; to set some limits on when restoration can occur based on some heuristics e.g., "restoration rate set to zero above wind speed w". These are all defensible approaches but depend heavily on the nature and complexity of the simulations being undertaken as to which approach is most appropriate.

Typically, extreme natural hazards present a major challenge for operators not simply because they cause outages on assets, but because they cause damage to multiple assets at the same time, and, frequently, in the same areas. This makes it important to understand where outages are most likely to occur but also the nature of those outages, the potential impact they may have, and what resources are available to intervene – and whether it is practical to intervene at all.

A weakness with fragility curves is that, in the context of resilience, we may be concerned with particularly rare or extreme events, and different natural hazards may not have the required volume of data to form data-driven relationships from which fragility curves can be built. Further, it may simply not be possible to form these parametric relationships at all – e.g., if we were to model the risk of sabotage or risk of cyberattack, what possible x-axis would be appropriate?

For some external threats, therefore, fragility curves can be a useful mechanism in simulations to generate credible scenarios for subsequent simulations and modeling to quantify power system resilience. However, they should preferably be used where the volume of data is appropriate to build reasonable fragility curves – correspondingly, the damage functions used should also be appropriate.

That is, multiple different scales of "damage functions" can be used with different fragility curves to quantify the extent of damage to a given system component (such as a substation or tower) – an example given in [8] as previously mentioned. For example, the capacity of a line may be reduced if it has been damaged by a natural hazard, or a generator may be derated. Fragility curves can be used to represent a simple binary state of an asset, also – that is, "in service" or "out of service".

Something that also needs to be considered is that different natural hazards may impact the power system and assets in different ways – a flooded substation may be rendered completely non-functional, for example, but if the substation can be protected with sandbags etc., and water pumps deployed the substation may be able to be restored to functionality more quickly than if, for instance, the substation is destroyed by a seismic event of landslip.

Similarly, if debris from a windstorm causes a transient outage, a line may only be out of service for a matter of seconds as auto-recloser action isolates the fault, the debris is cleared by wind, and the connection is restored. Models of such outages, therefore, may need to be classified by different fragility curves, or a secondary sample may need to be taken from a distribution of the types of faults that may occur, given a fault *has* occurred.

What a fragility curve also cannot tell the user is the speed of restoration from the initial outage if one has occurred; this also will need separate assumptions about what is represented by the fragility curve. Further, with fragility curves, the actual classification of the data used for different assets also presents a challenge. That is, if we are examining wind speed at a natural hazard, features such as wind direction may be significant; the wind speed may need to be corrected for different asset heights; the grouping together of different asset types and judging whether a given fragility curve is appropriate or not may itself be problematic. Again, in the context of OHL and towers, it may be inappropriate to, for example, gather all 132 kV lines on a system together if there is significant variance in the construction of the towers, condition, or age, and further corrections may be necessary, or different fragility curves considerate of these features.

In summary, fragility curves are a useful tool for stochastically generating outage scenarios provided the problem boundaries are clearly defined and they are applied in appropriate contexts. These requirements may relate to data availability, appropriateness, and the types of scenarios the user wishes to investigate. Further complications will be discussed in the following sections, particularly related to the potential data corrections necessary for fragility curves.

4.2 Homogeneous Versus Distributed Representations of Failure Hazard on Overhead Lines

A power system is fundamentally a means of transporting power from a machine or storage mechanism to an end-user service or device. At every stage of the process, it is necessary to make assumptions about the sub-processes involved to make the models used tractable and appropriate for the purpose of the analysis being performed.

For example, if modeling the spot market in Denmark, it would clearly be absurd to model the thermodynamic properties of a gas-fired turbine in the South of England, even though, because of the nature of the European power system, there is an indirect relationship between the two phenomena.

Similarly, over-simplification can also undermine the usefulness of the results that can be attained from running power system simulations by omitting important information that may be critical for operators to understand and plan the running of power systems they are responsible for. During a large cascading outage, for example, correlated protection settings on distributed generation and vector shift protection can lead to widespread disconnection of supply during system perturbations which can exacerbate frequency excursions, in turn leading to more severe system disruption and cascading outages, ultimately resulting in, as was the case in August 2019 in the United Kingdom, widespread emergency disconnection of loads [15] or the South Australian blackouts caused by cascading outages in 2016 [16].

Simulations which concerned the future resilience of the power system in the face of events similar to these, therefore, would as a matter of necessity have to be improved to consider the potential impacts of loss of supply from distributed generation across the system, if appropriate plans were to be put in place to mitigate or avoid similar outages happening again. The data and model being used must be appropriate to the context in which the analysis is being performed.

Two mantras therefore emerge to this effect: *GIGO*, or *Garbage In Garbage Out* – a term frequently used in computer science and data analysis – and "all models are wrong, but some are useful", coined by George Box. What these should be understood to mean is: a model can only be as useful as the quality of data being used to produce it, but even then, there will always inherently be some manner of error or uncertainty associated thereof.

To return to the subject of power system resilience, therefore, going forward extreme weather and high-impact low probability events are going to be an emergent concern on the power system, not simply because of how they can occur, but because of *where* they may occur, and *what manner of fault* actually occurs.

This section intends to demonstrate the value of distinguishing between these values and to illustrate the benefits of increased data granularity, particularly in the context of understanding the relationship between natural hazards, particularly wind, and risk on the electricity system. First, by describing at a high level the qualitative nature of the distinction, and then by modeling directly a toy model comparing the two representations to illustrate the potential benefits to be gained by more granular representations of failure risk on power systems.

4.2.1 Homogeneous Representations of Outage Risk on Lines

In a *homogenous* representation of an OHL, in effect, it is presumed that, or rather the net effect is, the natural hazard acting upon a given asset does so identically over the entire region over which it is exposed to that hazard identically. For smaller distribution systems, this is likely to be a reasonable assumption in the context of typical weather datasets used for the analysis such as wind-related impacts on the power system. For example, NASA MERRA-2 weather data can be as gross as approximately 50 km by 50 km [17].

For a microgrid covering, for example, a small settlement, it may be entirely reasonable, therefore, to treat lines within such ranges homogeneously. However, at greater scales of system, this rapidly becomes problematic as not only the amount of assets exposed to natural hazards increases, but so too does the potential range of natural hazards a given asset or network of assets may experience will also increase.

However, there are significant benefits to this approach also; otherwise, such methods would not have been used for so long. It is simple to treat and assume a given asset as a point asset because it significantly reduces the computational expense of modeling failures associated with it. It is very convenient to reduce a long piece of equipment spanning many different regions and biomes to a point asset with a given length and an assumed failure rate.

Over a long enough period of time, in a stable, predictable environment, it is not inappropriate to presume that the failure mechanics which may affect assets in that region should be predictable, and that over long enough periods of time such relationships can be established. This is particularly the case for reliability analyses, which generally rely on these kinds of assumptions and models.

To summarize, a homogeneous representation of a line makes the inherent assumptions, either implicitly or explicitly, and includes but is not limited to the features discussed in Table 4.2.

These presumptions will have different impacts dependent on the nature of study being undertaken. In particular, concerns about how long it takes to restore systems will be significantly impacted by the assumptions about restoration times on lines. Over a large enough sampling size, it can be presumed that for predictable, mundane, outages that operators are experienced with handling, then repair times may well "average out".

The challenge with modeling resilience is that frequently these assumptions no longer hold and it is more difficult to build statistical relationships because the data is not available. This is where distributed representation of failure risk on power systems becomes attractive.

4.3 Distributed Failure Risk Representation on Lines

If it is understood that outage risk varies across individual branches and across the system itself, methods of translating the system into a medium that represents this distribution and variety become a significant concern to most accurately and appropriately represent the underlying risks on that system. This inherently carries a series of necessary assumptions and abstractions. Take,

Table 4.2 Summary of properties of homogeneous line modeling features.

Assumption	Consequence
Natural hazards act identically across a given asset	• Simplified simulation approaches, line only needs sampled once with respect to each natural hazard • Not able to identify specific regions online where threat is most acute or regional variations in wind speed, direction • No information regarding the actual exposure of the line to the natural hazard at any given region, if an individual branch includes both over- and below-ground sections, this may be significant • Line represented only by a single length value (typically km), allowing faster simulations • Not able to identify high-risk areas related to temperature; problematic in extreme hot or cold or where primary constraint on line capacity is sagging associated with thermal loading
Length of line has no impact on sampling requirements	• Fast and efficient simulations but less information on location, nature of outage
Failure mode impacts entire line equally	• Depending on the nature of a fault – an outage that occurs near an urban center will likely be much more straightforward to access and repair than one on a line in a more distant location or in a geographically awkward location to reach. Therefore, it may significantly underestimate repair times of assets in extreme scenarios • May also significantly underestimate repair times of line if, for example, the weather is so extreme as to cause multiple failures across the line, or damage multiple sections or assets on the connecting branch
Branch has homogenous electrical properties and construction	• Older lines or lines on hybrid systems may not be accurately represented if, for example, there are different types of tower construction. This would mean certain line sections have greater risk profiles than others and such information may be lost. If different sections of line have different ages, may have different risk profiles also

for example, the following node-branch representation of the Northern Scottish transmission system, known locally as SHETL – or Scottish Hydro Electric Transmission Ltd. This is presented within [18].

Multiple features of this representation should first be observed. This is a standard way of representing grids at a transmission scale, called "node-branch" because it aggregates demands, generators, and other power system infrastructure to a series of, in effect, as they would be referred to in network theory as "nodes" and "edges". Nodes being the points representative of aggregated features of the point architecture – generators, demand, transformers, etc. – with branches representing the connecting lines and cables.

There is no demarcation between what kinds of branches there are visually (e.g. whether branches are single or dual circuit, cable or OHL). Further, there is no indication as to what kinds of assets are connected at each point. Further, lines are straight from point-to-point, whereas, naturally, in reality, they will follow convoluted paths dictated by local geography, use-of-land agreements, and planning permissions across each route. No consideration is given to, for example, switching stations on nodes and how power is actually routed through nodes with connections to multiple branches. This may end up being problematic if attempting to model faults on, in

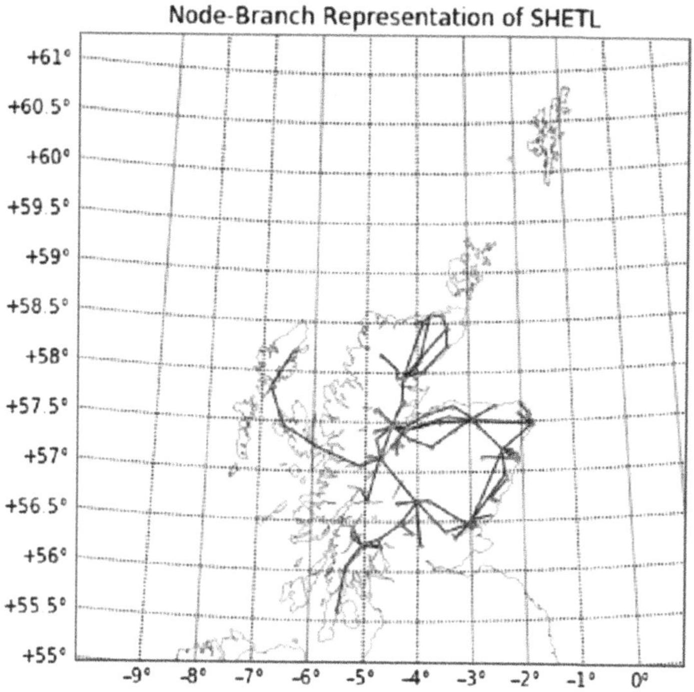

Figure 4.2 Node-branch representation of transmission system of Northern Scotland.

particular, substations that connect multiple branches. Without representation of the switches and breakers within such nodes, this may elide crucial information necessary in quantifying the resilience of systems.

That is, one may model a fault on a substation as disconnecting all branches associated with that node, but that is likely a conservative assumption (unless the substation is affected by a catastrophe such as being inundated with floodwater or seismic damage, which renders it inoperable). Damage to a substation may limit capacity across connections or connections of different pathways passing through that substation, but is less likely to render the entire station inoperable.

Alternative representations of networks are available, which take into consideration these details, known as "node-breaker" representation [19]. This is not used in this implementation for simplicity.

It should be noted that, though there are many weaknesses in the representation, that does not render the representation in and of itself useless. It makes the models comprehendible rapidly, and the data tractable and accessible, and can facilitate rapid decision-making.

Using this representation, it is possible to convert the node-branch representation into a two-dimensional data array to illustrate how much "network" there is distributed across the system. This, therefore, requires a consistent set of coordinates that can be used to determine a standard resolution for the two-dimensional array given. In the work produced in [18], modern era retrospective analysis for research and applications 2 (MERRA-2) data is used, due to its completeness, consistency, and availability [17].

The implementation of this methodology for determining the amount of a system "exposed" to a given natural hazard is demonstrated in [18]. This work is adapted slightly for reproducability.

First, some method needs to be derived to calculate the exposure in a two-dimensional context, though in specific contexts it may be desirable to extend this to three dimensions, particularly if one wanted to understand the dynamics of, for example, wind direction's impact on the risk of outage on particular lines in geographically complex areas where there may be "wind tunnel" effects.

4.4 Method for Representing Spatial Risk of a Natural Hazard

Various software packages have emerged in recent years for the visualization of data, particularly that which is related to geographic datasets. Examples of which are packages such as *QGIS* [20], though many other packages exist to handle GIS data. The method described herein is useful, however, because of its reliance typically on open source packages, widely available datasets, and deployment of skills that can be deployed on many different styles of data, as well as being able to incorporate itself with larger-scale *python*-derived software packages.

Multiple stages shall be described herein, though not necessarily under the given headings:

- Data acquisition
- Data processing into suitable formats
- Combining network data and weather data to derive failure probabilities
- Calculating failure probabilities and how they are geographically distributed
- Applying these to an appropriate basemap for visualization of results
- Interpretation of results

4.5 Data Acquisition

Acquiring data, particularly meteorological or network data, can be challenging. Amongst other reasons is because online sources of data may be subject to government funding of the programs which provide the data, or commercially sensitive data having controlled access. Methods deployed in resilience studies, therefore, should be applicable to different types and formats of data so long as they demonstrate consistency in what the data *represents* – that is, whether the data illustrates wind gusts or mean wind speeds, the elevation at which data is measured, etc.

The data source intended to be used for illustration herein is MERRA-2 [17], and walkthroughs for access and data provision are provided therein. This is preferred due to its availability, its geographic coverage, the variety of data provided, and the possibility of automating the data extraction process. That is, using software such as *wget* [21] to extract large datasets which can be leveraged in software such as QGis [20] and *using command prompt* terminal commands, data for decades worth of weather parameters is accessible while the user leaves the machine to download the data from the main servers.

Other services such as the Copernicus Climate Data Store [22] which also offer large series of data with significant variety and choice available for users; the specific data users choose may come down to as simple a matter as personal taste or choice based on the API or GUI used by the services being chosen. In the case of [18], MERRA-2 datasets were chosen because the data extraction methods were deemed convenient and at the time of writing offered the best options available. In the future, it is likely that more choices will emerge and better datasets as techniques and computing power continue to improve.

For datasets that do not pertain to extreme weather – that is, networks, system data, or data on other natural hazards such as seismic activity, data may be more difficult or awkward to access. Individual applications may be necessary to data providers or holders such as transmission owners –

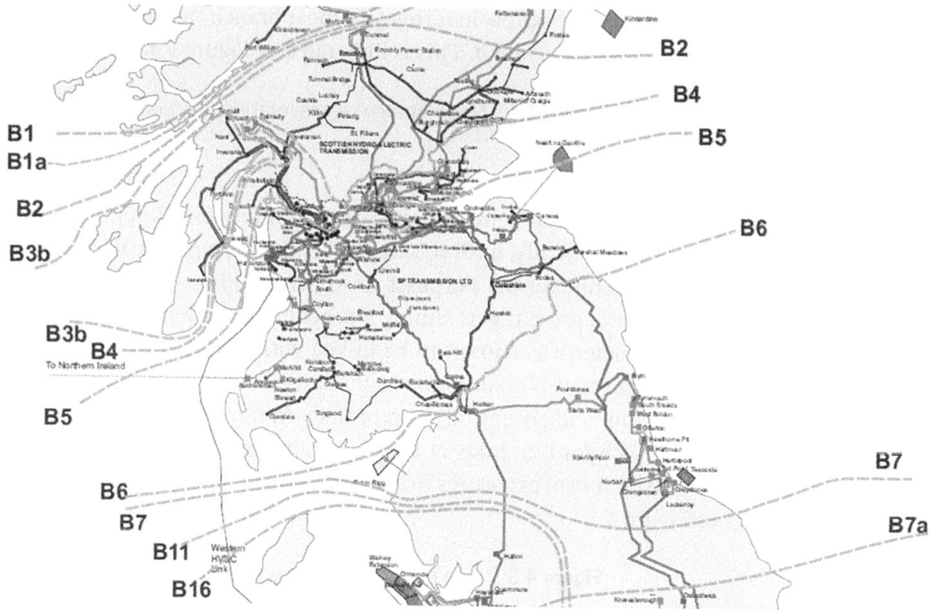

Figure 4.3 Excerpt from NG ETYS 2018 showing Scottish Borders and boundaries. Adapted from [23].

in countries such as the United States of America, power system infrastructure is considered strategically important and release of technical data is typically protected via non-disclosure agreements or simplified models of representative networks.

In the United Kingdom, typically data is more guarded via commercial agreements with companies such as *National Grid TO* (who own the transmission system in much of the mainland United Kingdom) or distribution companies such as *SP Energy Networks*. Data can be acquired from sources such as the National Grid Ten Year Statement [23], which also provides schematics of the system such as Figure 4.3, which shows sections of the transmission system around the Scottish Borders, with datasheets containing the parameters of connections on the system also included. This is extracted directly from the statement itself to be representative.

Other data sources can be from academic literature, which provides data tables of representative test systems, such as the 1996 Reliability Test System, which is very popular and used extensively in research [1].

Assuming a consistent resolution for the data has been ascertained and network data acquired calculating the actual exposure of the power system to external hazards can be undertaken. A method for doing so is proposed in the next section.

4.5.1 Calculating Exposure of System to Natural Hazards

Converting a vector to a visual representation of an overhead line intuitively presents as a straightforward problem – lines can be measured, and distances computed rather easily. However, in when programming, doing so in a manner that can handle different resolutions of data and different natural hazards is non-trivial. A method is presented here that is applicable to many different software packages and is tractable to different network scales and natural hazards that may affect power systems.

First, we must consider a power system line purely in terms of a linear branch "as the crow flies". That is, a straight line connecting two points, a and b. This is illustrated in Figure 4.4.

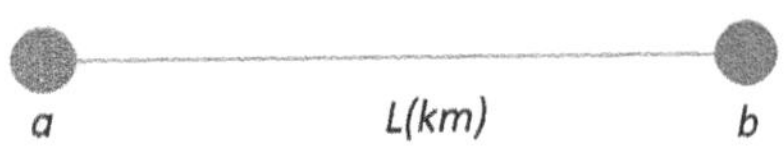

Figure 4.4 Node-branch representation of a line.

This can be understood, in effect, as the line being a point asset with property L representing its length between buses a and b. Conventionally, until recent years, this was typically deemed an adequate representation of an overhead line for the purposes of security studies. The exposure of the asset to whatever natural hazard was present was simply a function of the length of the line and the magnitude of whatever parameter was chosen to be linked with that line. As discussed previously, there are obvious advantages to this assumption as it allows faster simulations and for shorter branches or distribution systems is likely entirely appropriate. At transmission scale, however, where long HVAC lines can stretch for hundreds of kilometers, it is likely far too simplistic. Instead, we should see lines as a distribution of exposures to a natural hazard. In the simplest terms, this can be shown as follows in Figure 4.5.

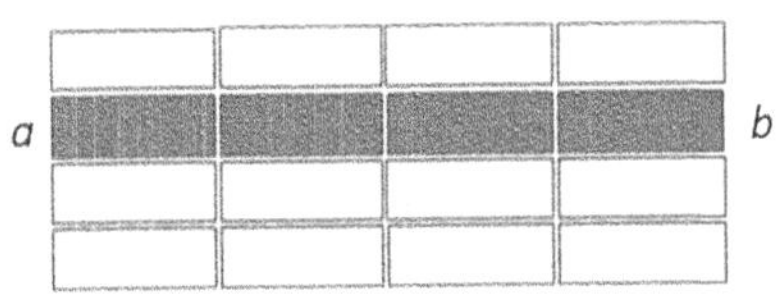

Figure 4.5 Line represented as two-dimensional array.

This shows a representation of a line that extends from the far left of cell a to the far right of cell b in a perfectly straight line, hence the exposure is consistent across all cells. The exposure to any given natural hazard is then a function of how much "line" is present in each cell and the magnitude of the parameter acting upon it.

For non-straight lines, calculating how much "line" is actually exposed in each cell can be challenging. An algorithm is used in this implementation, which is as follows:

- For each line
  - Define a vector representing line x
  - Divide line x into n steps
    - Iterate from point a to point b in steps which are the total length L divided by n
    - Count how many steps occur in each cell
  - Correct for the size of each step to determine exposure in each cell

The number of steps n may be optimized for different scales, but $n = 10{,}000$ should be presumed herein. The benefits of this method are that, for example, the methodology can be linked with an arbitrary coordinate system, and in progressing through each step, it is also potentially possible to link this with other datasets such as topological data to generate more complex models of the system. For now, let us assume the world is smooth, if not flat.

Implementing this in *python* can generate the following heatmap of an arbitrary line connecting two points, as in Figure 4.6.

The co-ordinates used are based on the MERRA-2 data to be used in the case studies, which approximates to 50 × 50 km. The data in each pixel represents the amount of line exposed to external conditions (in km).

As can be seen, clearly, some sections of the line will be more exposed to hazards than others within the scale used. If there are geographical features such as hills or mountains, this too could

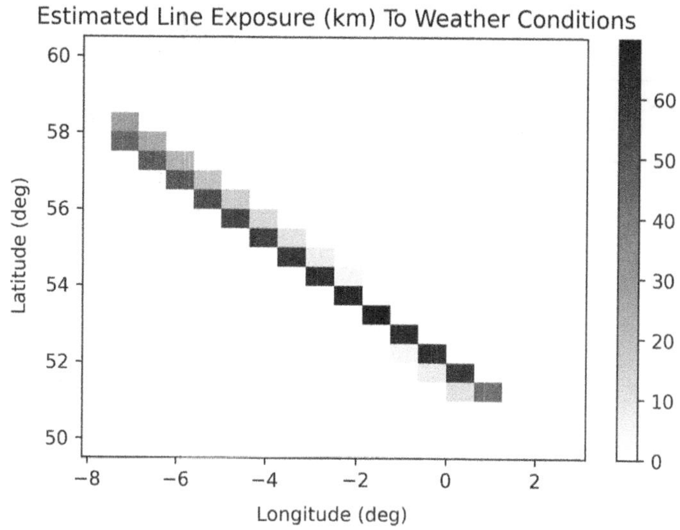

Figure 4.6 Graphical representation of a line progressing through space.

affect not only local weather conditions, but also the "amount" of line within each cell as the line will traverse the axes not only across the surface of the earth but perpendicular to it (that is, the z axis coming out of the page). This is, for the purposes described here, not directly considered, but should be considered for more mountainous regions or if wishing to perform more complicated analysis of wind.

4.5.2 Modeling Failure Probability

Now that a method for establishing how exposed the asset is to the natural hazard in question (wind), a relationship between that exposure, the natural hazard, and failure probability has to be established. The following fragility curves are to be used.

For reproducibility and illustrative purposes, the curves used in the original work from [18] are to be replicated in the form of sigmoidal curves representing similar values. These are more straightforward to replicate and do not contain the "stepped" features of the original data used in [18]. A relationship is postulated such that:

$$y = \frac{l}{1 + e^{-k(x-x_0)}} + b \tag{4.1}$$

where l scales the sigmoid, b can be used as a bias to "shift" the sigmoid, x_0 defines a center of the sigmoid curve, and k is a scaling value.

If we choose the values shown in Table 4.3, we can produce the representative fragility curves to be used in subsequent studies.

These fragility curves look as follows, in Figure 4.7.

Note that the y-axis represents failure rate, not probability. This is so that, in each cell, the exposure of the line within that cell can be corrected using the per-hour per 100 km failure rate for the corresponding wind speed. This can then be converted to a failure probability to be used in sampling using the presumption of an exponential distribution of faults. The following equation is therefore used.

$$p(\text{fault}) = 1 - e^{-\lambda \delta t} \tag{4.2}$$

Table 4.3 Values used for generation of sigmoidal curves used in examples.

Parameter	Case a – 132 kV line	Case b – 275 kV/ 400 kV lines
l	0.1	0.032
x_0	36	36
k	0.55	0.55
b	0	0

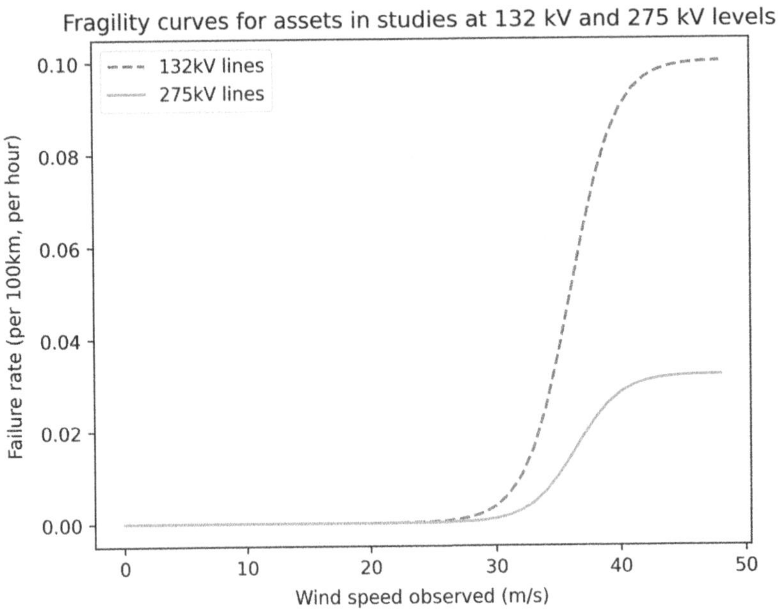

Figure 4.7 Fragility curves to be used in examples, synthesized values comparable to those used in. Adapted from [18].

Where λ is the failure (or, conversely, restoration) rate of an asset being sampled, and δt corresponds to the time-step size being used (this can be assumed to be 1, since the source data and the sampling data are both being considered per hour).

Wind speeds above 50 m/s are incredibly uncommon in the United Kingdom, and therefore, building data-driven relationships between failure probabilities and incident wind speeds becomes difficult because the data simply does not exist in sufficient scale, and the sample sizes are too small to derive meaningful relationships. Simulation-driven relationships based on mechanical models of the lines may therefore become preferable in these circumstances. Studies in the UK have shown however that failure rates can increase by several orders of magnitude during storm conditions [24].

It should also be noted that the data used to derive these curves used in [18] were ERA-Interim weather data from Copernicus, with the maximum wind speed observed over a three-hour window used to infer the causal wind speed for the fault. This in itself may be a conservative assumption – factors such as wind direction, debris, or mechanical wear at sustained high wind speeds may have been just as consequential in causing a fault but once again the interpretation of the data is key to defining what parameters are able to be used for subsequent analysis.

Now that we have methods for determining the line exposure and for using that exposure to determine the failure probability for a given wind speed, we actually need to determine the wind speed that the lines on the system will be subjected to.

"MERRA-2" data, which is the type used throughout this section, represents a broad spectrum of meteorological *hindcasted* weather data covering parameters such as solar irradiance, ambient temperature, and, most relevantly, wind speeds. Hindcast data represents an estimation of the weather conditions at a given region for a point in space and time, which is typically estimated from a mix of satellite data and local historical meteorological measurements, combined to give a complete map of estimated weather conditions. Though other sources are available and are discussed at length in literature such as [25], the subsets of MERRA-2 used were chosen because they were deemed the most appropriate for the analysis being undertaken and was judged able to provide, at worst, an indicative representation of historic risk during different events and a variety of data types for analysis.

Two specific subsets of data were chosen: u and v (East–West and North–South) resolved wind speed data at heights of 2 m, 10 m, and 50 m, and single-level maximum wind speed data [26, 27], henceforth referred to as "three-level" and "single-level" data, respectively. The latter dataset is treated as if it is the wind speed active at 10 m.

The MERRA-2 data effectively treats the globe as a "bumpy-sphere" – that is, a grid of blocks at fixed horizontal displacement with homogenous weather conditions within those 50×50 km blocks. For geographically uniform, flat regions, as stated, this may be an acceptable approximation for shorter line branches. However, the United Kingdom is a geographically diverse country with rugged mountain ranges in Scotland, with lower-lying floodplains in England, creating distinct climatological biomes.

These may be defined as falling into three broad categories – the windy North, the windy and sunny Southeast, and the sunny Southwest Peninsula [28]. Geographic diversity within the scale of 50×50 km ranges may therefore be profound. Homogeneous representations of OHL in system studies cannot directly explain the diversity of weather conditions that may affect network branches; further, the exact mechanisms that drive OHL failures will vary across different geographical and weather conditions. That is, a storm hitting the entire United Kingdom at the same time may manifest as extreme winds in Scotland and Northern England, while simultaneously causing flooding and inundation of substations in Southern England and Wales.

In Scotland, therefore, in such a context, there would be a significantly increased risk of OHL faults at a time when there would be increased risk of flooding on substations in England. This represents one of the major challenges in resilience analysis: during HILP events, a single causal weather pattern could manifest as a variety of natural hazards affecting different areas to different magnitudes simultaneously. These *correlated outages* will be discussed further in the subsequent section.

As has been discussed, data cannot always be directly taken from the source and applied to the fragility curve – or, at least, not without explicit acknowledgement of the compromises and assumptions necessary to apply that data. For example, if a fragility curve has been formulated from maximum gust data, it may not be appropriate to use average speed data to extrapolate failure probabilities from that curve in subsequent analysis. You may choose to do so, but this must be acknowledged. The scenarios that then emerge should you choose to do this may not necessarily be an ideal representation or projection based on the source data, but they could still be used as relative indicators of risk on the system, so long as they are applied consistently across the system and the user is transparent about the compromises that have to be made.

Given that the two fragility curves being used represent wind speeds acting on OHL of different voltage levels, the implicit assumption is that the wind speed must act at different heights on the assets, given 132 kV lines will tend to be smaller than, e.g., 275 kV or 400 kV lines. The next section shall therefore discuss the next step of data processing necessary, correcting the wind speed data for height above ground.

4.5.3 Correcting Wind Speed Data for Different Heights of Asset

For the purposes of subsequent analysis, let us assume 132 kV towers have heights of 30 m, whereas 275 kV or 400 kV towers in GB have heights of 50 m. These are based on typical values used in transmission systems such as the northern Scottish SHETL grid, but naturally, there may be different requirements for tower constructions in different countries. Correcting wind speeds for height assets is important for contexts such as studying potential wind farm outputs, as the wind speed measured at meteorological sites may be significantly different from those which is experienced at hub-height of wind turbines in wind farms, for example.

As an observer moves further away from the ground, measured wind speed will increase significantly. Wind speeds may be corrected using three wind speeds at various heights to extrapolate the wind speed at a given height [29]. A fuller illustration and description of this method can be found in [29]'s supplementary materials but a similar approach is utilized here.

First, the u and v orthogonal components of wind speed at the $(2+d)$m, $(10+d)$m, and 50 m are taken, as is done in the original methodology, with the d a value representing vertical displacement d stored in the raw data. That is, a presumed height from which the data is measured, with the 2 m and 10 m measurements being measured from this point vertically. For the vast majority of observed data in the case studies performed d was at or near zero. This is a replication of the methods used, which are described as a "virtual wind farm model", used for extrapolating wind speeds for different turbine heights in a large-scale study on wind power output in different scenarios in the United Kingdom.

Regressing these against the log of their altitude, then means the wind speed at height h can then be estimated using the following formulation.

$$(h) = A\log(h-d) - A\log(z) \tag{4.3}$$

This presumes a standard linear equation of form $y = mx + c$, with known x values log $(h-d)$ taken from the 2 m, 10 m, and 50 m heights with the d values, and known y values, the resolved wind speeds at these altitudes. Given that y and x are known, the other parameters can be simply calculated.

$w(h)$ represents the incident wind speed $w(\mathrm{ms}^{-1})$ at height h(m), or the anemometer height from which the wind speed can be measured. The gradient A can then be computed via regression, as can $log(z)$, which is the y-intercept. This can then be rearranged to extrapolate the wind speed at unknown height, with z determined using the relationship;

$$z = e^{-c/A} \tag{4.4}$$

where c is simply the y-intercept of the equation determined from the known data and the regression.

This further emphasizes the potential significance of disaggregating fragility curves by voltage levels – 132 kV OHL, due to being typically smaller and shorter than 275 kV/400 kV towers, may be subject to less extreme weather and better protected by wind shadow. Conversely, they are also more vulnerable to inclement weather when it does occur.

The taller 275 kV/400 kV lines, being used for transmission scale and hence being less typically found directly next to heavily populated areas, can reasonably be expected to be subject to more extreme weather conditions with less sheltering from the wind associated with wind shadows and adjacent vegetation. Now that methods for correcting wind speeds for different heights have been established, and a relationship between these parameters and failure probabilities on lines has been established, as well as the actual exposure on the system to extreme wind, the next step is to address the granularity of the data being used. How to do this will be described in the Section 4.5.4.

4.5.4 Interpolation of Datasets for the Purposes of Resilience Studies

MERRA-2 is a relatively coarse dataset in the context of power system studies. On a global or regional scale, or for climatological analyses, this is not inherently problematic. However, in the context of power system analyses, a resolution of 50 × 50 km may not be appropriate for systems as diverse as GB.

For context, it is approximately 100 km between Glasgow and Perth. Using a (50 × 50)km resolution would mean the assumption that an OHL connecting these two cities would experience 2 or 3 different wind speeds across the entire line – while this is an improvement on using a homogeneous representation of that line (which would represent, of course, the entire line experiencing only a single weather condition), it undermines the value of disaggregating the representation of the line in such a manner to use such a coarse dataset.

Another example of an interpolation technique could be LOESS regression [29], a "non-parametric locally weighted scatterplot smoother". Two-dimensional interpolation and regression itself presents a significant computational and academic field in its own right. In this context, an accessible and easily deployable solution was deemed desirable. The *python* function *griddata*, part of the *SciPy* library, was used to perform the interpolation of the incident weather data to improve model granularity [30].

This alone cannot account for local topological factors which may affect wind speed and direction, such as valleys or mountaintops, where phenomena such as the *Venturi effect* and wind shadow may have profound localized impacts on wind if the terrain is particularly variable or there are large man-made structures affecting local wind flows (for example, wind farms). Nonetheless, it allows a first-order improvement on the raw data available and more granular results to demonstrate the overarching principles of the proposed methodology.

A benefit of such methodologies is that they can be applied in preprocessing before resilience studies are performed for specific scenarios, saving computation time, and multidimensional data interpolation is continually improving with increasing power of processors and improving computational techniques.

A comparison between interpolated and non-interpolated data for a representative "extreme wind" scenario (from an extreme wind event in December 2011 [31]) can be observed in Figure 4.8, shown below. These maps were produced using the *python* libraries *Cartopy* [32] and *matplotlib*. Example implementations and documentation can be found online at [33, 34].

Such interpolations can be particularly significant where there are rapid increases in significance at extreme values, which is particularly important in the scenarios being investigated here, given the rapid increase in failure rates on OHL when wind speeds increase from 20 ms^{-1}.

In Section 4.5.5, a method for extrapolating wind power for wind turbines, given the associated wind speed and corrections for different hub heights, is described.

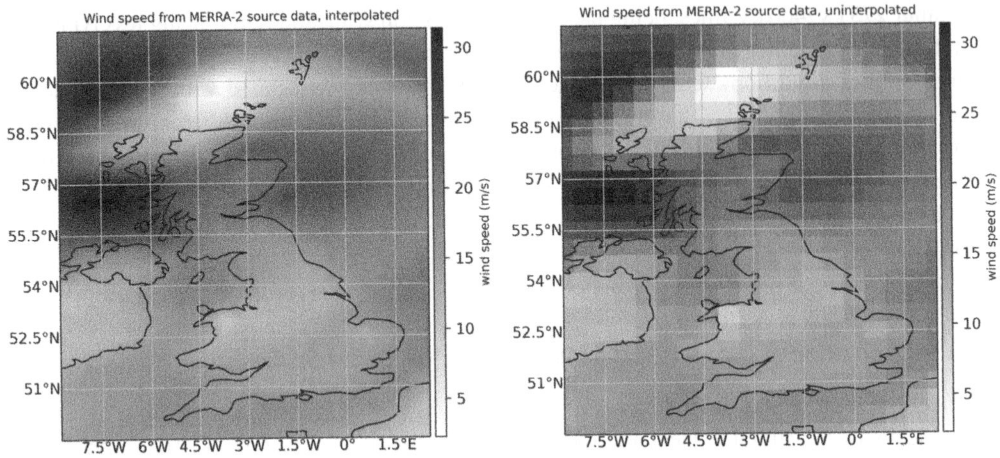

Figure 4.8 Comparison between MERRA-2 resolved wind speed data (three-level); interpolated and non-interpolated data.

4.5.5 A Tractable Method for Calculating Wind Power Outputs in Extreme Wind Events

In extreme weather conditions, wind turbines will engage in self-preservation actions to prevent electromechanical failure, such as deployment of brakes or adjusting angle of turbine blades to reduce the rotational speed of the turbine. This in turn reduces wind turbine output and, when aggregated across wind farm groups or entire wind farms, can cause significant drops in wind power infeed over the scale of minutes or hours, depending on the duration and intensity of the storm. This is called high wind speed shutdown, or HWSS [35].

HWSS can be significant in the context of resilience because such reductions in wind power infeed will occur at times when high wind speeds are also increasing power system risk associated with OHL outages and, in winter, increased life-critical demand from sources like heating. Although an individual turbine will act in a broadly predictable manner with pre-defined manufacturer settings for cut-out and cut-in, the aggregated effects across a wind farm, taking into account local variations in wind conditions and turbine heights, manufacturers, mean aggregating such variation presents a significant challenge; aggregating this to a regional, national, or international scale even more so.

Various approaches have been taken to project wind power outputs given a set of incident weather conditions. Forming detailed analyses for quantifying HWSS is stymied by factors such as the diversity of kinds of wind farms, constantly evolving technologies in wind farms, and sparsity of data for HWSS conditions, and so forming a generalizable curve which can be applied across a system is in itself a rather gross estimation of the wide variety of factors which can affect wind generation (wind direction can generally be considered less of a factor in this instance than in the context of OHL faults, as wind turbines can be designed to "track" the wind direction to optimize output, however, given the impacts of local geography on local wind speeds wind direction may still be a factor).

The approach proposed here, and demonstrated in [18], was to use a simple proxy and amalgamation of myriad methods [29, 36, 37]. A curve representative of a pre-corrected power curve (that is, not considerate of HWSS) was generated using interpretations of data used in [18] interpreted or adapted from [29] and [37]. Formula (4.5) represents the formulation of the curve,

$$y = \frac{a}{1 + e^{(-f(w)(x-w+\delta)}} \tag{4.5}$$

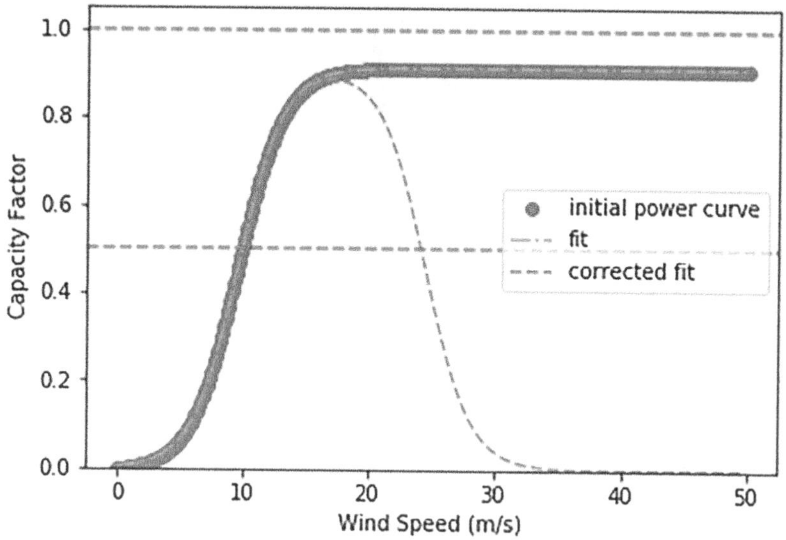

Figure 4.9 Power curve to be used in example. [18] / John Wiley & Sons / CC BY 3.0.

in which a is a normalizing factor set to 0.9128, δ is an x-shift with value -15. $w, f(w)$ represents the power curve as represented as an x, y coordinate set. The *SciPy* function *curve_fit* was used for the regression. This postulated power curve is illustrated in Figure 4.9.

In equivalent studies, a Gaussian filter was applied after some maximum value to force the output to be – or approach – zero. Here, this is not performed; rather, a complementary curve was used to curtail the input at higher wind speeds simply by subtracting this complementary curve from the original, as shown below [18]. This was for simplicity, as the end results were deemed to, by inspection, be adequate to serve as a reasonable approximation that was straightforward to implement.

The scale of challenge presented in producing more representative power curves was, in the context of the examples within this text, deemed out of scope insofar as the exact impact of HWSS itself was not sought to be quantified; rather, an indication of general trends was sought in the context of resilience.

That is, ideally, a more complete curve based fully on empirical data would be preferred, but the aim was to generate a curve that could be productively used to illustrate general trends across a system, and the simple proxy generated was deemed appropriate and proportionate for that purpose. Replacing this curve with data-driven examples, as opposed to proxies as deployed, is well explored in the literature, but not necessarily useful for indicative studies where a more straightforward indicative measure can be used.

With the data corrections and models now complete for the case studies to be conducted, some results can be illustrated to demonstrate the power and potential value of these methods and models when taken together in visualizing overhead line failure probability and how this may relate to wind power projections on a transmission system.

4.5.6 A Case Study Using Demonstrated Methods on the Northern Scottish Transmission System

Using the methods aforementioned, a map of the total exposure across all OHL on the system to extreme wind is shown, with the area of most concern blown up for clarity, shown in Figure 4.10.

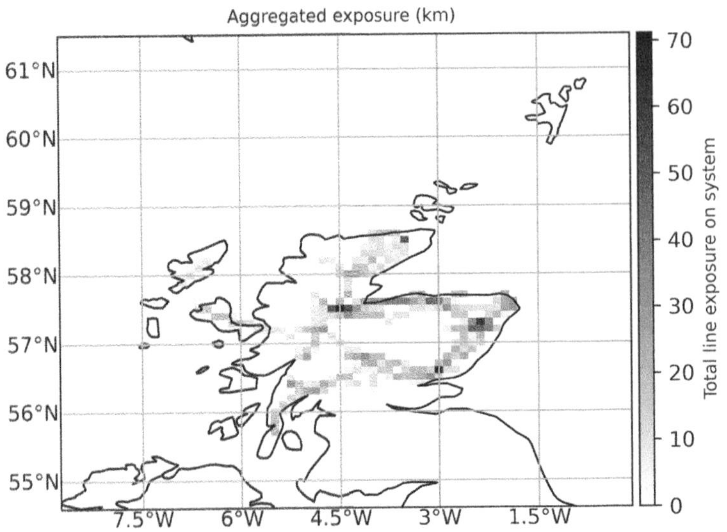

Figure 4.10 Exposure map of northern Scottish OHL network.

For context, this should be compared to a snapshot of the distribution of wind farms on the Scottish grid. As an illustrative example, wind farm capacities were extracted from the Renewable Energy Planning Database (REPD) as of December 2018 [38]. The coordinates were converted to lat-lon pairs using an online Ordnance Survey bulk coordinate conversion tool to get the coordinates into an appropriate format [39]. These could then be scaled and presented on a map. The results are shown in Figure 4.11. Each disc represents a single wind farm, with size and coloration scaled by installed capacity with the network topology synthesized from publicly available sources such as the ETYS Individual companies in GB also now offer significant data portals users can extract datasets for study.[1]

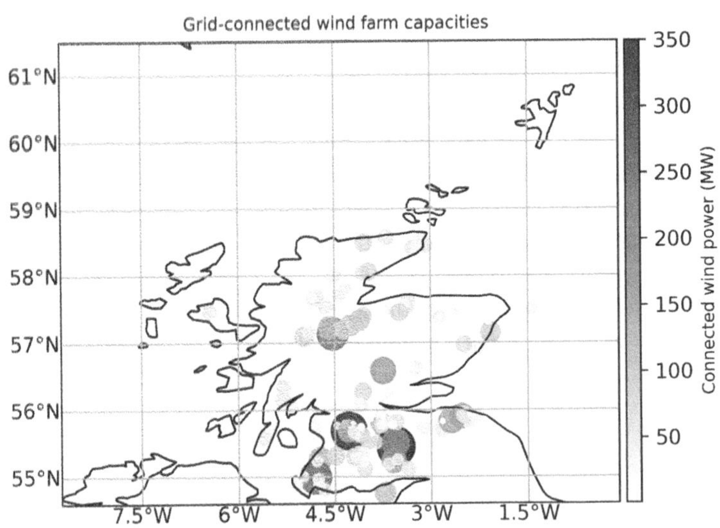

Figure 4.11 Illustrated wind farm capacities in Scotland as of December 2018.

1 For example, SSEN provides a data portal at data.ssen.co.uk with data such as network capacity.

It can be observed that there is significant overlap in regions of SHETL where there is grid capacity and where there is installed wind power capacity. This should be unsurprising – power can only be connected to the grid where a grid exists to which it can be connected.

In Scotland, the majority of demand is in the South-Central region of the country, also known as the Central Belt, just south of SHETL. This is also where the largest onshore windfarms were connected, again in part because there is significant grid capacity to allow connection of wind resources.

A consequence of the overlap between grid capacity installation and wind power installation is that it creates interdependencies between wind power output and system risk in these locations, which is compounded in extreme wind scenarios.

To investigate the failure probability across the whole system is not possible analytically. Therefore, a Monte Carlo simulation must be performed for a sample hour to determine failure probabilities during the time step investigated. This is used to aggregate locational failure probability across the system and produce maps representative of OHL risk across the system. That is, the risk of a failure happening at a given location in a given hour using the models and methodologies formulated thus far. This is done as described in the following pseudocode:

```
For sample in sample size:
    For line in all lines:
        For section of line:
            Generate random number between 0 and 1
            Get failure probability for line section
            If random number < failure probability:
                Fault recorded in block
                Latitude, longitude of fault recorded
```

This simple, but powerful, algorithm records the location of each outage, generated by the fragility curve and the incident weather data. This means that not only can the presence of a fault on a particular line be recorded, but also where that outage has happened, and how many outages on a particular line at a particular time step may have occurred.

This then generates a two-dimensional array with fault numbers that occur in each $\sim 10 \times 10$ km pixel of data recorded. Dividing these counts by the sample size provides an estimate of the overall fault probability by block. This means that a line can "fault" across multiple locations, or multiple faults can be recorded within a given block within a given time step if multiple lines intersect. This provides an estimate, then, of how likely any part of the network is to fault in a given time step.

Individual lines can, of course, be analyzed in this method, as, in a power system simulation, whether a line faults in 10 locations or 1, it is still simply no longer in service. However, the purpose of this exercise is to demonstrate and visualize the overall probability of something breaking in any given location on the network, agnostic of the impacts of that incurred fault. Understanding how many outages have occurred on an individual line may also, of course, significantly impact how much time that line takes to restore.

The average of the latitude and longitude of outage locations within the simulation can then also be used to generate a value representative of the expected location of a fault on the system – the expected fault location (EFL). These results can then be combined with illustrations of overhead line failure probability to illustrate the links between these system risks. EFL is not in and of itself particularly useful on a system scale, as it only marks the "epicenter" of where outages happen. However, across individual lines, it could be useful to indicate where the most likely place for an

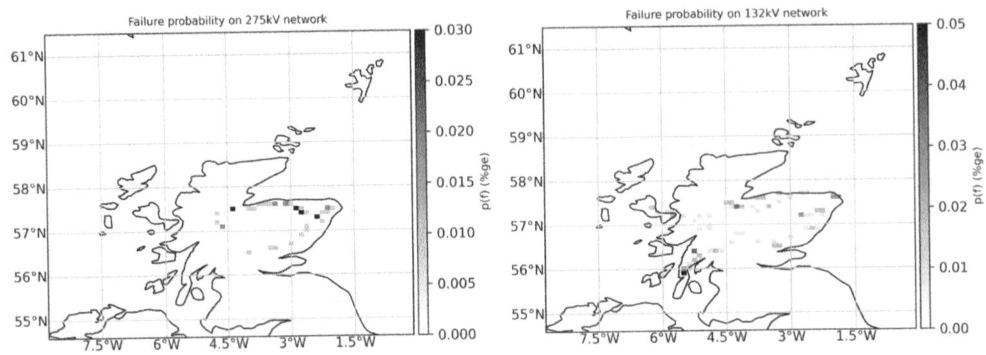

Figure 4.12 Network outage probabilities on Northern Scottish Grid.

outage to occur is, or in smaller systems could still be useful to guide staff to locations where they can optimize restoration efforts on the system.

A sample size of 10 000 is used (but any sample size could be used, dependent on desired precision versus computational and time expense). The case study of Cyclone Friedhelm with the illustrated weather conditions was used. The three-level data (that is, the data with 2 m, 10 m, and 50 m sets extrapolated to relevant asset altitudes) is used first. The results for the estimated OHL fault probability on the 132 kV and 275 kV/400 kV networks are shown in Figure 4.12. In these figures, "p(f)" represents the percentage probability of an outage occurring within the sample hour within that cell.

You may also wish to aggregate outage probability across the entire system. This produces the following output in Figure 4.13.

Using the same weather data and the incident power curve previously mentioned, we can then also visualize the estimated wind generation output across the system as in Figure 4.14.

By inspection, it can be seen that, despite there being significant wind farm capacity installed across the East Coast and Southern regions of the northern Scottish network, the effects of HWSS have reduced significant amounts of infeed to zero due to the extreme wind speeds. That is, the

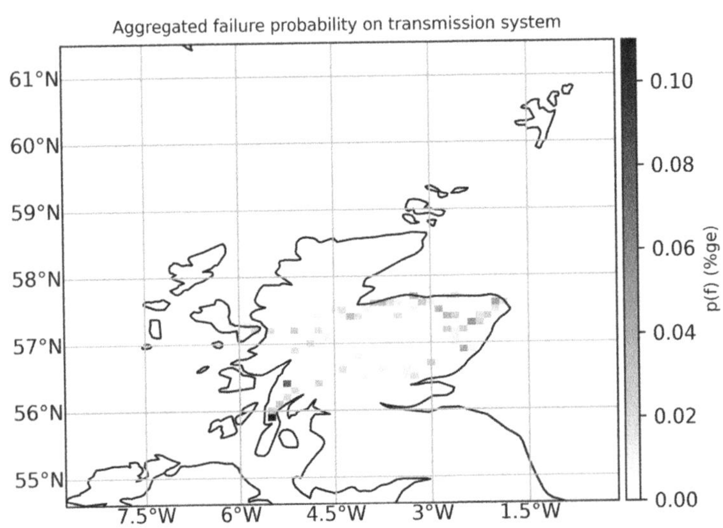

Figure 4.13 Total outage probability across system.

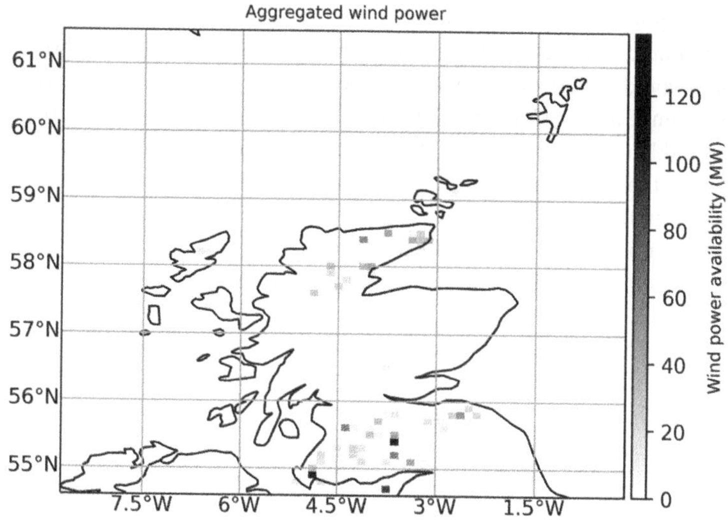

Figure 4.14 Estimated wind farm output across Scotland during storm in hour examined.

areas with highest OHL risk will coincidentally carry significant risk of loss of wind power due to HWSS.

Overlaying the total OHL failure probability with the installed wind farm capacity illustrates this, as can be seen in Figure 4.15. The disc radii have been normalized to the highest capacity single windfarm on the system at the time (350 MW, Clyde Wind Farm, though Whitelee is the largest farm facility at >460 MW it is separated into multiple separate groups within the data itself). This can be seen in Figure 4.15.

The EFL from these results is (57.03, −3.81)°. Visually, this seems reasonable – the most extreme values are in the South, Southwest, and East of the network – the mean value of these locations should reasonably be expected to be somewhere in the middle of these areas. This tells us where a failure is most likely to happen, which can be useful information if attempting to optimize system operation, such as placement of spares or other emergency response equipment or staff

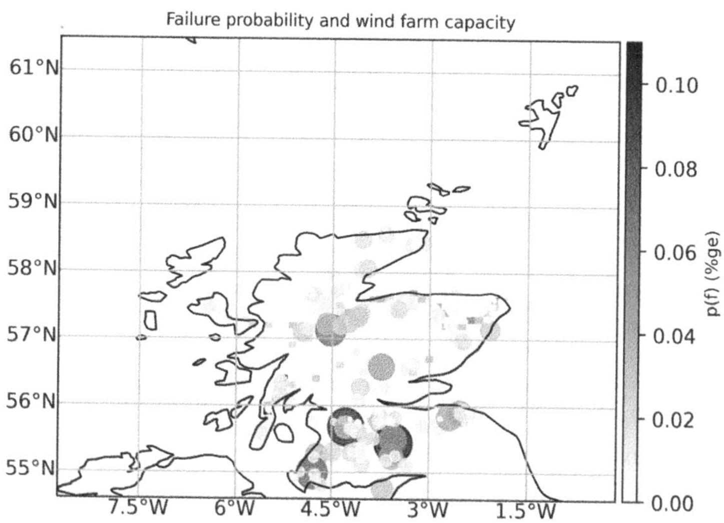

Figure 4.15 Failure probability on OHL overlaid with installed windfarm capacities in SHETL system.

Of course, these values are unweighted by the actual impact of any outages – in reality, an outage on the East coast would be far more consequential than a loss to the relatively small Southwest peninsula in Scotland, and that would weigh significantly on the planning decisions as regards location of repair teams and resilience assets. Conversely, there is significantly more redundancy on the East coast than on the weakly connected fringes of SHETL, meaning any single outage is less likely to lead to disconnections than the weaker segments of the network. This does, however, provide a useful first-order approximation and indication of what the most at-risk areas of the network would reasonably be. This is important to remember: risk is a function of both the *likelihood* of an event and the *magnitude of its impact*. When making planning decisions, it is vital to take into account both these aspects.

Losing electricity supply to an island or rural community, for example, will have a significantly different impact on those in that community than it would, for example, if supply was lost to central London. Island and rural communities, which have long had intermittent or unreliable electricity supply, are likely to have inbuilt resilience – backup generators, alternative heating supplies, alternative cooking supplies, etc., which those in cities will typically lack.

4.6 Exercises

As has been demonstrated throughout this section, the methods themselves are not particularly complicated in isolation, but can be difficult to link together with the calculation of line exposure. That is, for a given fragility curve, a given wind speed, a given height of asset, and a given exposure of line (in km), it is trivial to calculate what the failure rate, and hence probability, should be. However, ascertaining these values is where the difficulty emerges.

Some exercises and questions are proposed here for you to familiarize yourself with the various concepts and thus be ready to implement similar software yourself for comparable purposes and to consider the wider context of the content.

- Concept Questions
  - What factors, other than line exposure and wind speed, may impact the failure rate of a given asset in a storm subject to a given maximum wind gust w? Suggest 3
  - Do the maps of failure probability represent system risk? Why or why not?
  - Assume a storm is moving across a region and you have limited response teams. There are three areas likely to be impacted: a rural village, a large city, and a wind farm with a weak connection to the grid. Why might you decide to locate your response teams in each of these areas? Where is most likely to be the most important place to locate a team optimized to restore downed overhead lines, and why?
- Practical tasks
  - From a source such as Copernicus or MERRA-2, extract a suitable dataset for visualization. These websites provide comprehensive tutorial materials and software wizards to aid the process. Try and choose a nation or region with significant geographic diversity – such as the United Kingdom – to observe how geography can affect weather.
  - Using a dataset such as *Copernicus* or *MERRA*-2, and a plotting suite such as *Cartopy* or *basemap,* replicate the visualizations of the weather data provided here. Experiment with different color schemes and datasets, and with different times. Observe how weather patterns move across regions during storms. Do not interpolate this data.
  - Using the three-level MERRA-2 dataset, produce graphs that show wind speeds at different heights

  - Using the virtual wind farm model as given, extrapolate wind speed to 100 m for each pixel and reproduce a map visualizing the results
  - Using the following pseudocode as a starting guide, write a script to produce a chart indicative of network exposure to wind for a given vector as generated in Figure 4.1:
- Define a set of latitude and longitude increments
- Convert start and end points to lat-lon pairs
- Set a "start" point for the vector and an "end" point
- Define number of iterations (steps along vector) to be taken
- Divide lat/lon distance traversed by the number of iterations to get value of step
- For each step:
  - Add step distance to previous co-ordinate in lat, lon
  - Determine in which pixel spot falls
  - Add to total number of "steps" in pixels
  - Use number of steps in each pixel to determine amount of line present
    - What edge cases may exist? How can these be addressed in your implementation?
    - Consider the differences between the results produced by interpolated data and the original data. What effect might interpolation have on the end results?
- Concept Question Answers :
  - Include but are not limited to:
    - Wind direction relative to line bearing
    - Turbulence of wind
    - Age of assets
    - Condition of assets
  - No, they only represent *probability*. To represent and consider *risk*, there would have to be some consideration of the *impact* of these outages
  - Features to be considered in planning for an extreme storm include (but are not limited to)
    - A rural village will likely have weak lines and be connected radially, meaning they are the most vulnerable of the loads and may be particularly exposed to the most extreme winds, dependent on their location. They may also have limited transportation links, so it may be difficult to access, and may have more vulnerable elderly populations that need rapid sup port should power be lost.
    - In cities themselves, most cables and infrastructure tend to be shielded by buildings, ameliorating the wind speed somewhat, or branches are underground, eliminating the direct risk associated with wind. However, damage to the system in a city will almost certainly mean significant loss of demand and economic damage, meaning the impacts are far more significant than in rural areas.
    - Losing connections to wind farms during extreme events may cause cascading outages, which jeopardize wider system security via the loss of power. However, in high wind situations, the wind farm may actually end up having a low output if wind turbines are "locked out" of service. Therefore, it depends on the resilience of the connecting branches and size of the wind farm. Wind farms may also have their own maintenance teams and plans to respond.
    - Cities in industrialized nations are unlikely to have significant overhead line penetration, particularly near the center, as these will tend to instead rely on underground cables. However, there will be large-scale transmission corridors leading into the city. A planner may sacrifice smaller rural demand centers to ensure system stability by ensuring connections to large supply points remain intact, particularly if the energy resource is grid-forming and

connected to operationally significant branches. During extreme wind events, it is likely wind farm output will be variable or curtailed. Therefore, it is likely that such a team may be deployed to try and minimize the risk from a loss of infeed to a city.

- "Steps" which fall on a boundary between two cells exactly may end up confusing the software as they cannot be assigned to either cell on which the boundary overlaps, so exceptions may need to be scripted in. The software should also be scripted in such a way that the direction in which the data is fed in does not impact the determination of the line exposure.
- More coarse datasets mean that more data points, and more of the system model, fall within the extremes of temperature. The effect of this is that this can significantly overestimate overhead line failure rates for parts of the system. This may be a very conservative assumption, but is also context dependent – if you are security conscious, it may be sensible to be more conservative in your security assessments. There is, therefore, a balance between data granularity, accuracy, and computational expense for these models.

4.7 Correlated Natural Hazards

An issue with modeling resilience and, particularly, wind is that wind-related outages rarely happen in isolation and extracting the causal feature that contributed to a given outage in a given scenario may involve complicated data analysis from datasets which may be incomplete or difficult to ascertain for issues of commercial sensitivity. Research already discussed from Murray et al. [11] and in [40] has looked to investigate features associated with line failure probabilities and what features in particular are useful for building probabilistic relationships for subsequent analyses.

In [11], wind, in combination with snow, sleet, and blizzard, for example, is used to construct cumulative probability distributions representative of the probability of an outage associated with a given wind speed. In Bayesian terms, we might represent this mathematically as $p(o|w)$, that is, the probability of outage o given wind speed w.

Complications emerge, however, because we cannot know for certain whether it was the wind speed itself that caused an outage, or a combination of factors. "Wind" as a parameter has a large number of properties that we may wish to investigate. Due to the properties of large-scale simulations, it is necessary to define clear system boundaries for analysis, and wind *speed* is an intuitively obvious one. However, it is far from the only one. Given are some other features of wind which may impact on the risk associated with outages on a power line.

- Direction
- Turbulence
- Temperature

Other features that occur concurrently with wind may also need to be considered. These include but are obviously not limited to;

- Ambient temperature
- Humidity
- Precipitation
- Lightning
- Wildfires

Therefore, establishing causal relationships in the data between a given parameter and a given outage on a line carries significant complexity at the start of the process. If a line collapses during a storm where there is also frozen water on an exposed conductor, is it the wind that has caused this outage, or the mechanical loading due to a combination of the wind and the added mass of ice? Such an outage can be studied using, for example, computer simulation models to simulate the events ex post, but without on-site inspections of outages and thorough data collection, it can be difficult to be certain.

Further, the capacity to ascertain accurate readings of weather at the time of faults may be significantly variable, particularly in more isolated sections of network, and may rely on reanalysis, downscaling, or interpolation of data, which in turn carries inherent sources of error. These aspects affect the weather, which acts on the system, but it is also important to consider the properties of the system on which the natural hazard is acting.

In [40], different properties of OHL are considered with respect to the causes of outages in a representative power system. In general, it is determined that line length in particular is associated with a high probability of outage. This is intuitively reasonable and should also be considered in the context of where there may be long, exposed OHL and why those lines are long in the first place. That is, connecting distant power systems over areas with little development or few population centers may imply that those areas are inherently inhospitable, or which may be associated with adverse weather conditions.

Building any probabilistic relationship between a parameter and failure probability, therefore, requires implicit assumptions about the parameter under investigation. In "empirical" fragility curves, as defined in [10], for example, to determine if a particular wind speed was associated with a given outage on an overhead line, reanalysis data could be used with geographic data associated with the line outage itself from the relevant asset owner. This is still reliant on the quality of data acquisition of the transmission owner and of the quality of the reanalysis modeling data used.

Further, when using the fragility curves developed on the basis of this data analysis, the appropriateness of the data used to make projections must also be considered. For example, outages caused by turbulent wind gusts, or extreme sudden blasts of wind, will inherently have different mechanics from, for example, sustained high-speed wind gusts which are of a more laminar, stable nature. It may still be useful to use such models with data that is not directly comparable but it should be made clear what compromises are being made and that any judgment of performance based on these models is likely to be most useful in relative terms, rather than a particularly accurate assessment of different scenarios. Examples can be seen in the previous section, which discusses this.

It should also be noted that there are likely to be hidden correlations behind datasets that may not be immediately apparent. In [41], the circumstances behind a catastrophic wildfire that affected North California in 2018 (called "The Camp Fire") are detailed. A brief summary of the event is provided in the following box.

The Camp Fire

Known colloquially as the "Camp Fire," this was a large wildfire event which occurred in Northern California and was traced in origin to significant failures by Pacific Gas and Electricity (PG&E). The Butte County District Attorney's office produced an extensive report detailing the evidence against PG&E and the circumstances leading to the Camp Fire [41], which is publicly accessible and should also be read due to the aspects of resilience that it addresses.

To summarize the consequences of the event:

- At the point of containment, it had burned 153 336 acres of land
- Approximately 18 800 structures were destroyed
- 589 structures were damaged

- 84 people were killed
- Two civilians and one firefighter suffered significant injury

At approximately 0620 on Thursday, November 8,2018 a PG&E Hydro employee reported fire underneath a set of PG&E transmission lines North of the Feather River. As there was no mobile phone signal in the area, he used a radio to signal to a nearby powerhouse to call 911, who in turn reported the fire to the fire service. The fire then travelled more than seven miles in less than 1.5 hours.

Investigations by the authorities determined a hook holding up a conductor had failed, causing the conductor to collide with the steel structure, causing arcing, which sprayed liquid hot metal and sparks onto the dry brush, igniting it. The part that failed was determined to be a "C-hook" suspending the conductor, which showed signs of extensive wear. The line itself was constructed in the aftermath of World War 1. The failed component was "an original 1921 insulator." The line itself was almost 100 years old.

The ground was dangerously dry and in an area prone to wildfires and high winds. The fire continued for 17 days.

The Caribou-Palermo (C-P) line operated by PG&E, which ultimately failed, was built sometime between 1919 and 1921 by the Great Western Power Company. The report's evidence concludes that despite having owned the line since 1930, "PG&E did not catalogue or replace the original conductors, insulators or attachment hardware on many of the towers in the original Caribou-Big Bend section of the transmission line."

The C-hook that broke was determined to be one of the original components installed on the line;thus, it was almost 100 years old. The average age of copper conductors on the PG&E inventory – which the failed Caribou–Palermo line– was 75 years old. 44.4% of their 115 kV structures were installed before 1931. Non-copper conductors averaged 36 years of age. This in particular should be noted due to the fact that most modern power system infrastructure uses aluminum, not copper, as conductors for OHL, and the mean age of the copper conductors was double that of aluminum.

Generally, transmission asset owners will have plans in place to regularly inspect, patrol, and maintain assets to discover flaws in the system that need to be repaired and identify vulnerabilities before they can become problematic. However, visibility and record-keeping was so poor in the system that even lines which *did not exist* were recorded as having had maintenance performed on them – "in 2014 PG&E Transmission Asset Strategy compiled a list of all transmission structures located within the boundaries of a designated wild fire area. Approximately 85 towers on the Caribou-Palermo line between the Butte-Plumas County line and the Big Bend Substation were included on the list. Tower 27/222 for some unknown reason was not on the list, but Towers 22/187 through 23/192 (which did not exist in 2014 because they had collapsed in 2012) were listed".

The Camp Fire fundamentally occurred due to, or was exacerbated by, a combination of factors, according to all available evidence; and could not have occurred at the scale it did without the *combination of these events.*

- Lax maintenance caused by a corporate culture of cost-cutting
- Aging infrastructure
- Lack of system visibility and knowledge
- High winds
- Lack of response time due to inadequate communications
- Failure of forest management to contain the wildfire before it affected such a area of land
- Climate change is exacerbating the drought

These aspects pertain to multiple aspects of resilience – infrastructural, procedural, and climatological. Not all of these can be captured by power system simulations, and in this case, it was a combination of high winds and drought that contributed to the starting of the wildfire itself. Although single extreme events can cause HILP events, so too can a combination of unrelated events be interacting in a destructive manner.

PG&E was punished in court for the extent of their negligence. The final judgment was $50 000 in fines for "starting a fire" and $10 000 fines for each of the 84 people killed through negligence. In the context of a power system, this is an almost trivial sum, but it was the most severe judgment the court could levy in this instance.

At the time of writing, a civil suit remains in progress against the company, and the company itself is now bankrupt as a consequence of this and other incidents detailed in the report, but because of the nature of their failures, no individuals will be punished for the endemic failures that led to it.

Various circumstances had to coincide in order for the wildfire to take place (which resulted in hundreds of casualties and billions of dollars worth of damages). These include, but are not limited to:

- Assets on the network are almost a century old
- Lack of maintenance on transmission infrastructure
- Lack of system visibility to detect faults
- Extended drought in the region
- High winds

The wildfire could only have occurred due to a combination of failures and exogenous threats to the system, paired with systemic failures in terms of operating the transmission system itself. That is, though the end result was a large, high-impact event, the causal event was not a singular failure, but a *concurrent* series of failures across *multiple systems* combining in a catastrophic manner. These may not always be possible to replicate in a power system model, and resilience modeling may not always be able to capture scenarios of this nature because they are so difficult to replicate in their complexity.

While the Camp Fire, and the mass of system outages associated with it, could be attributed to high winds causing the failure of a line due to mechanical stress, clearly, other events on the system exacerbated what should have been a fairly routine and manageable scenario for a power system.

This is important in the context of resilience. Even if every line on the system is constructed in an identical manner and probabilistic relationships are developed based on, for example, voltage level of power lines is assumed, this will mask features such as tower age, condition, and direction. If a singular wind speed is used, this masks features such as bearing of wind on OHL. These could be accounted for with, for example, correction factors or use of multiple different fragility curves for different line features given concurrent weather effects (e.g. by generating fragility curves based on tower construction with different age bands), but this rapidly becomes problematic when data availability becomes scarce. Such relationships are then always a balancing act between ensuring robust and cohesive sets of data from which relationships can be formed, versus the scale of data necessary to ensure adequate precision of this data.

Though this methodology has concentrated on wind-related outages, extensive research has also been conducted into methods for quantifying resilience associated with seismic events. For example, in [42].

4.8 Incorporating Imprecise Fragility Curves in Decision-Making Models

The fragility curves are essential to understand the effects of threat intensity over failure probabilities. In turn, failure probabilities are critical in calculating reliability, resilience and, in general, risk metrics, which are vital for decision-making. In this vein, a decision can be deemed appropriate (or optimal) according to the improvement caused on a given (probabilistic) metric. For example, we can identify the transmission investment portfolio that minimizes the risk level (measured by using a particular metric or combination of metrics) for a given budget constraint [43].

Notably, the fragility curves are built based on assumptions and imperfect information. This is the case even if real historical data is used to build fragility curves. For example, historical data can be used to determine failure probabilities given to a threat intensity value (i.e. a point of the fragility curve). However, estimation of probabilities based on historical data will feature errors, or, more formally, confidence intervals. The width of the confidence interval associated with that probability value will depend on the data volume used to derive it. Unfortunately, data volumes are usually poor in these cases as events that are of importance in resilience analysis are, by definition, rare. To illustrate how (un)reliable the estimation of probabilities can be depending on the volume of observed historical data, we introduce the following example.

Simple example: flipping coin

Let us assume we want to calculate the probability of obtaining heads in an unbiased coin. We do so by throwing the coin several times and dividing the number of times we obtain heads by the total number of times we throw the coin. Notice that if we carry out this calculation after throwing the coin a few times, only twice, for example, twice, there is a large likelihood of estimating a wrong probability value. In fact, after throwing the coin twice, we may get a probability equal to 0 (when we observe heads zero times), 1/2 (if we observe heads once), and 1 (if we observe heads twice). Of course, some of these values are incorrect, as we know the right probability to obtain any side in an unbiased coin is 1/2. The correct value can only be accurately calculated after throwing the coin many times. Still, the calculation of such probability will feature a confidence interval that decreases with the amount of historical data available, in this case, the times we throw the coin. For instance, if we use the Wald approximation to estimate the confidence interval of the success probability $\widehat{p}$ of obtaining heads (assuming a large sample n and that $\widehat{p}$ is not close to 0 or 1), we get:

$$\widehat{p} \pm z\sqrt{\frac{\widehat{p}(1-\widehat{p})}{n}} \tag{4.6}$$

where $\widehat{p} = \frac{n_s}{n}$ with n_s being the number of heads within the sample of size n. Also, $z = 1.96$ for a 95% confidence level. Notice that the larger the n, the smaller the error, and the higher the accuracy of $\widehat{p}$. Interestingly, notice that both coin flipping and equipment failure follow a Bernoulli distribution. In the case of equipment failure, though, $\widehat{p}$ is expected to be close to 0 or 1 for certain values of the threat intensity (for small intensities $\widehat{p} \approx 0$ and for large intensities $\widehat{p} \approx 1$, following the shape of a typical fragility curve). Therefore, other methods for approximating the confidence interval of $\widehat{p}$ should be considered. Despite this difference, it is important to highlight that the estimations of $\widehat{p}$ will be subject to errors that may be significant if the datasets are poor.

Moreover, fragility curves feature other problems. For example, they are calculated for assets that, in practice, may greatly vary due to real characteristics (for example, the type of soil on which a tower or substation is constructed is of great importance for earthquake analysis). Thus, applying the same fragility curve to assets that are not strictly the same may be good on average but not in every particular case. Another problem we face is the presence of non-stationary processes, which

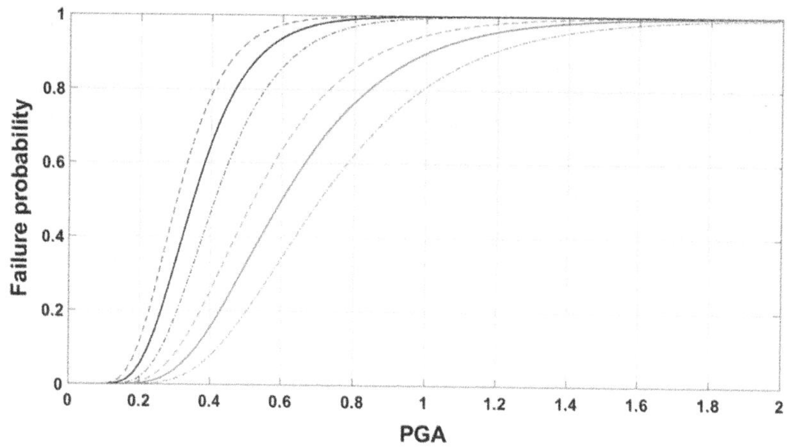

Figure 4.16 Two central fragility curves and their upper and lower bounds.

are becoming increasingly important due to climate change. In general, we assume that we can explain future occurrences based on past observations. This may not be the case if, for instance, the number of adverse weather events increases in intensity and frequency without following the historical pattern.

Interestingly, the fact that fragility curves are imprecise can be incorporated in optimization decision-making models such as in [44]. In fact, the work in [44] incorporates confident intervals associated with a central fragility curve as shown in Figure 4.16. The central fragility curve represents the expected value of the failure probabilities for a given threat intensity level, while the confident intervals can be used to build upper and lower bounds against the decision-making model must remain robust. An interesting finding of [44] is that decisions can be justified to hedge not only against uncertainty in natural hazards and their effects but also against the imprecise representation of fragility curves. The particular technique used here is based on distributionally robust optimization and interested readers are advised to study [44] for further details.

4.9 Summary

An approach for visualizing overhead line failure probability on a power system associated with extreme wind has been presented. Examples and pseudocode are provided to demonstrate the fundamental principles and the concepts surrounding fragility curves are described to aid understanding of the statistical models underpinning the approach used, and similar approaches deployed elsewhere.

The demonstrated methods are used for quantification of *probability* as opposed to *risk*, as they do not consider the *impacts* of the outages analyzed. However, the quantification of impacts can readily be applied to the modeling of probability as in the models used the assessment of probability is ambivalent to the state of the assets being modeled. That is, the failure probability is normally unaffected by features such as load flow across a given line based on the assumptions made. In reality, for certain natural hazards, this may not always be the case – for example, if considering line icing, there may be a direct relationship between line loading associated with power transmission, icing on the line, and mechanical loading caused by the mass of the ice accumulating on the conductor.

Such analysis is more complex and requires modeling of the underlying thermodynamic processes at play and is not considered here.

Uncertainty and the granularity also feature as points to consider. The datasets that underlie fragility curves may at times be sparse and limited, and this may have an impact on the usefulness of the fragility curves generated in terms of using them for future projections and modeling scenarios. Ensuring the data used in simulations based on the fragility curves used are from comparable sources is important for ensuring meaningful results – if curves derived from *gust* data are used with *average speed* data, this may, for example, underestimate the actual speed which lines experience. Data corrections may therefore have to be applied as necessary.

The granularity of data used can also be consequential – data with a higher resolution may allow more precise representation of individual branches' failure risks, for example, but over-interpolation may overestimate how precise the data actually is, especially at highly local scales where geographic features may begin to have significant impacts. Overly coarse, data, however, may result in highly conservative results wherein the effects of high wind speed areas concentrate fault risk and may not represent a reasonable assessment of the actual risk experienced by lines in that region.

Acknowledgments

The authors would like to thank Jacob Kelly for his contributions and insights, which supported the research underlying [18].

References

1 C. Grigg *et al.*, "The IEEE reliability test system-1996. A report prepared by the reliability test system task force of the application of probability methods subcommittee," *IEEE Transactions on Power Systems*, vol. 14, no. 3, pp. 1010–1020, 1999, doi: 10.1109/59.780914.

2 R. Billinton, C. Wu, and G. Singh "Extreme adverse weather modeling in transmission and distribution system reliability evaluation," in *Power Systems Computation Conf*. Spain: *(PSCC)*, Seville, 2002.

3 M. Panteli, P. Mancarella, D. N. Trakas, E. Kyriakides, and N. D. Hatziargyriou, "Metrics and quantification of operational and infrastructure resilience in power systems," *IEEE Transactions on Power Systems*, vol. 32, no. 6, pp. 4732–4742, 2017, doi: 10.1109/TPWRS.2017.2664141.

4 N. Balijepalli, S. S. Venkata, C. W. Richter, R. D. Christie, and V. J. Longo, "Distribution system reliability assessment due to lightning storms," *IEEE Transactions on Power Delivery*, vol. 20, no. 3, pp. 2153–2159, 2005, doi: 10.1109/tpwrd.2005.848724.

5 M. Jamieson, G. Strbac, S. Tindemans, and K. Bell, "A simulation framework to analyse dependent weather-induced faults," in *IET International Conference on Resilience of Transmission and Distribution Networks (RTDN 2017)*, Birmingham, UK, 26–28 Sept 2017 2017, pp. 1–6, doi: 10.1049/cp.2017.0346.

6 NERC. "Transmission Availability Data System (TADS)." North American Electric Reliability Corporation. http://www.nerc.com/pa/RAPA/tads/Pages/default.aspx (accessed 6 July 2020).

7 World Nuclear Association. "Three Mile Island Incident." https://www.world-nuclear.org/information-library/safety-and-security/safety-of-plants/three-mile-island-accident.aspx (accessed 16 July 2019).

8 T. Lagos *et al.*, "Identifying optimal portfolios of resilient network investments against natural hazards, with applications to earthquakes," *IEEE Transactions on Power Systems*, vol. 35, no. 2, pp. 1411–1421, 2020, doi: 10.1109/TPWRS.2019.2945316.

9 S. Dunn, S. Wilkinson, D. Alderson, H. Fowler, and C. Galasso, "Fragility curves for assessing the resilience of electricity networks constructed from an extensive fault database," *Natural Hazards Review*, vol. 19, no. 1, 04017019, 2017.

10 A. Serrano-Fontova *et al.*, "A comprehensive review and comparison of the fragility curves used for resilience assessments in power systems," *IEEE Access*, vol. 11, pp. 108050–108067, 2023, doi: 10.1109/ACCESS.2023.3320579.

11 K. Murray and K. R. W. Bell, "Wind related faults on the GB transmission network," *Probabilistic Methods Applied to Power Systems (PMAPS), 2014 International Conference on, Durham, UK, 7–10 July 2014*, pp. 1–6, 2014, doi: 10.1109/pmaps.2014.6960641.

12 M. Panteli and P. Mancarella, "Modeling and Evaluating the Resilience of Critical Electrical Power Infrastructure to Extreme Weather Events," *IEEE Systems Journal*, vol. 99, pp. 1–10, 2015, doi: 10.1109/jsyst.2015.2389272.

13 D. Clements and P. Mancarella, "Fragility curve based storm modelling of distribution networks with staff constraints," *IET International Conference on Resilience of Transmission and Distribution Networks (RTDN 2017), Birmingham, UK, 26–28 Sept 2017*, pp. 1–5, 2017, doi: 10.1049/cp.2017.0342.

14 E. Ciapessoni, D. Cirio, A. Pitto, and M. Sforna, "A quantitative methodology to assess the process of service and infrastructure recovery in power systems," *Electric Power Systems Research*, vol. 189, p. 106735, 2020.

15 Office of Gas and Elecricity Markets. "Investigation into 9 August 2019 Power Outage." Ofgem,. https://www.ofgem.gov.uk/publications-and-updates/investigation-9-august-2019-power-outage (accessed 16 August 2021).

16 'Black system south australia 28 september 2016 - final report', https://www.aemo.com.au/-/media/Files/Electricity/NEM/Market_Notices_and_Events/Power_System_Incident_Reports/2017/Integrated-Final-Report-SA-Black-System-28-September-2016.pdf, accessed 9 January 2023

17 NASA. "Modern-era retrospective analysis for research and applications, version 2." NASA. https://gmao.gsfc.nasa.gov/reanalysis/MERRA-2/ (accessed 20 July 2021).

18 M. R. Jamieson, G. Strbac, and K. R. Bell, "Quantification and visualisation of extreme wind effects on transmission network outage probability and wind generation output," *IET Smart Grid*, vol. 3, no. 2, pp. 112–122, 2019.

19 M. Koenig, S. Sagareli, M. Vaiman, and M. Vaiman, "Node-breaker topology representation of Con Edison's Stations for planning studies," in *2015 IEEE Power & Energy Society General Meeting*, 26–30 July 2015 2015, pp. 1–5, doi: 10.1109/PESGM.2015.7286158.

20 'QGis - A free and open source geographic information system', https://www.qgis.org/en/site/, accessed 16 January 2023

21 'Wget - GNU Project', https://www.gnu.org/software/wget/, accessed 16 January 2023

22 ECMWF (2022). "Welcome to the Climate Data Store." Retrieved 15 February 2022, from https://cds.climate.copernicus.eu/cdsapp#!/home.

23 NESO, 'Electricity Ten Year Statement (ETYS) Appendix A' (National Grid ESO, 2018), pp. 2–8

24 E. A. Morris, K. R. Bell, and I. M. Elders "Spatial and temporal clustering of fault events on the GB transmission network," in *In 2016 International Conference on Probabilistic Methods Applied to Power Systems (PMAPS)*. IEEE, 2016, October, pp. 1–9.

25 Sharp, J., M. Miligan, and H. Bloomfield. "*Weather Dataset Needs for Planning and Analyzing Modern Power Systems.*" ESIG, October 2023 (2023).

26 'MERRA-2 tavg1_2d_flx_Nx: 2d, 1-Hourly, Time-Averaged, Single-Level, Assimilation, Surface Flux Diagnostics V5.12.4', https://disc.gsfc.nasa.gov/datasets/M2T1NXFLX_V5.12.4/summary, accessed 22 February 2022

27 'MERRA-2 inst1_2d_asm_Nx: 2d, 1-Hourly, Instantaneous, Single-Level, Assimilation,Single-Level Diagnostics V5.12.4', https://disc.gsfc.nasa.gov/datasets/M2I1NXASM_V5.12.4/summary, accessed 22 February 2022

28 L. C. Dawkins, *Weather and Climate Related Sensitivities and Risks in a Highly Renewable UK Energy System: A Literature Review.* United Kingdom: The Met Office, 2019.

29 I. Staffell and S. Pfenninger, "Using bias-corrected reanalysis to simulate current and future wind power output," *Energy*, vol. 114, pp. 1224–1239, 2016.

30 P. Virtanen, R. Gommers, T. E. Oliphant *et al.*, "SciPy 1.0: fundamental algorithms for scientific computing in Python," *Nat Methods*, vol. 17, pp. 261–272, 2020, doi: 10.1038/s41592-019-0686-2.

31 'Scotland Hit By 165mph Winds As Twitter Dubs Storm 'Hurricane Bawbag' (Photos) (Video)', https://www.huffingtonpost.co.uk/2011/12/08/scottish-winds-hurricane-bawbag_n_1136460.html, accessed 22 February 2022

32 'Cartopy', https://scitools.org.uk/cartopy/docs/latest/, accessed 13 July 2022

33 'The Matplotlib Basemap Toolkit User's Guide',https://matplotlib.org/basemap/users/index.html, accessed 22 February 2022

34 J. D. Hunter, "Matplotlib: A 2D graphics environment," *Computing in science & engineering*, vol. 9, no. 3, p. 90, 2007.

35 H. Macdonald, G. Hawker, and K. Bell "'Analysis of wide-area availability of wind generators during storm events," in *2014 International Conference on Probabilistic Methods Applied to Power Systems (PMAPS).* UK: Durham, 2014.

36 D. J. Brayshaw, A. Troccoli, R. Fordham *et al.*, "The impact of large scale atmospheric circulation patterns on wind power generation and its potential predictability: A case study over the UK," *Renewable Energy*, vol. 36, no. 8, pp. 2087–2096, 2011.

37 D. J. Cannon, D. J. Brayshaw, J. Methven *et al.*, "Using reanalysis data to quantify extreme wind power generation statistics: A 33 year case study in Great Britain," *Renewable Energy*, vol. 75, pp. 767–778, 2015.

38 'Renewable Energy Planning Database quarterly extract,' https://www.gov.uk/government/publications/renewable-energy-planning-database-monthly-extract, accessed 23 March 2022

39 'Batch coordinate transformation tool,' https://www.ordnancesurvey.co.uk/gps/transformation/batch#, accessed 23 March 2022

40 S. Ekisheva, M. Papic, M. G. Lauby, and M. Elkins, "Assessment of impact of AC circuit attributes to outage frequency in the WECC system," *IEEE Transactions on Power Delivery*, vol. 36, no. 2, pp. 1159–1172, 2020.

41 Ramsey, M. L., et al. (2020). The Camp Fire Public Report - A Summary Of The Camp Fire Investigation.

42 Villamarín-Jácome, A., & Moreno, R. (2022, June). Seismic Resilience Assessment of Electric Power Systems Using a Substation Bay-level Model. In 2022 17th International Conference on Probabilistic Methods Applied to Power Systems (PMAPS) (pp. 1–6). IEEE.

43 T. Lagos *et al.*, "Identifying optimal portfolios of resilient network investments against natural hazards, with applications to earthquakes," *IEEE Transactions on Power Systems*, vol. 35, no. 2, pp. 1411–1421, 2020.

44 D. Alvarado, R. Moreno, A. Street, M. Panteli, P. Mancarella, and G. Strbac, *Co-Optimizing Substation Hardening and Transmission Expansion Against Earthquakes: A Decision-Dependent Probability Approach.* Early Access: IEEE Transactions on Power Systems, 2022.

5

Resilient Investment Planning

5.1 Network Investment Decisions: Unraveling the Planner's Trilemma

5.1.1 Expanding and Hardening Solutions

Natural hazards threaten the integrity of power systems across the globe, with catastrophic consequences to the electricity consumers and the overall society. To protect power systems against natural hazards such as earthquakes, tsunamis, and hurricanes, there is a rich set of investment options that can be used to effectively enhance system resilience, going beyond those traditionally considered in reliability analysis that fundamentally enhance the redundancy levels of the system. Indeed, while the economic and reliability performance of power networks have been classically improved by expanding the network (installing more network infrastructure, e.g. new powerlines), resilience can also be tackled by hardening existing infrastructure as indicated in [1, 2]. Such hardening decisions feature the effect of decreasing failure probabilities in those assets being hardened when a natural disaster occurs. Hence, rather than adding more assets to increase capacity redundancy, hardening focuses on changing the fragility features of the existing network. Examples of hardening decisions include [1, 3–5]:

- undergrounding lines to deal with high wind speeds and adverse weather conditions,
- upgrading poles with stronger, more robust materials that can withstand high winds and other stressors,
- elevating substation equipment or building substation defenses to deal with flooding events,
- anchoring substation equipment to better deal with earthquakes, and
- reconductoring with covered conductor solutions.

In case of seismic hazards, for instance, substation outages pose an important risk to the continuity of supply. As noted by [6] in the context of the 8.8 Mw-magnitude earthquake that struck Chile in 2010, a quarter of the substations of the main transmission grid was affected. Figure 5.1 shows the main damages of the 2010 Chilean earthquake on the main grid's substations. Therefore, in order to design networks that are resilient against earthquakes, it is of great importance to consider substation hardening along with traditional investment alternatives focused on network expansion (i.e. new lines and transformers).

In another case, Pacific Gas & Electric (PG&E) has implemented several measures to harden its electricity grid against wildfires, focusing on upgrading infrastructure in high fire-threat areas. This includes system hardening with stronger poles (fire-resistant ones made of materials like steel or composite) and covered powerlines to prevent fire ignition. These measures are part of

Fundamentals of Power System Resilience: Disruptions by Natural Causes, First Edition.
Mathaios Panteli, Rodrigo Moreno, Dimitris Trakas, Magnus Jamieson, Pierluigi Mancarella, Goran Strbac, and Nikos Hatziargyriou.

Figure 5.1 Damage to substation due to earthquake, February 27, 2010, Chile.

an overarching strategy that also encompasses tree trimming and removal, inspections, repairs, and improved situational awareness [7].

5.1.2 Digitalized, Non-Wire and Smart Grid Solutions

Apart from these network-based solutions associated with grid expansion and hardening, there are other non-network (or non-wires) and innovative solutions based on the concept of the so-called smart grid (as also explained in Chapters 6 and 7), aiming to improve system flexibility and responsiveness. In this context, several advancements are envisaged that could fundamentally transform the way networks are controlled and operated, making them more flexible and adaptable to the evolving system conditions, including high-impact low-probability (HILP) events, and thus more resilient. For instance, as explained in [8], distribution networks are transitioning from mostly passive networks, where the control problem is resolved at the planning stage by designing for the worst-case peak-demand scenarios, to active systems, where new Information and Communications Technologies (ICT) and controllable distributed energy resources (DER), including storage and flexible demand, can provide real-time services associated with reliability and resilience. DERs can also be controlled through the concept of microgrids (as explained in Chapter 7), which, in a decentralized fashion, can manage local DER to keep essential loads being supplied and thus provide resilience after a natural hazard. Additionally, the transmission system will become much more active through (i) new controllable and flexible technologies such as Flexible AC Transmission Systems (FACTS) [9] and high-voltage DC (HVDC) systems [10] that can control power flows and voltage levels through the network without changing power injections/withdrawals, (ii) System Integrity Protection Schemes (SIPS) that can enforce rapid increase/reduction of power in importing/exporting areas after a network outage occurs by, for instance, curtailing generation and/or demand [11], and (iii) various Wide Area Monitoring and Control (WAMC) equipment supported by ICT technologies that will increase the capability of system operators to monitor and control electricity assets in real time and throughout wider areas through increased communication [12, 13]. Furthermore, demand will be controllable and end users will become active participants in system and market operation, thereby opening up opportunities for aggregating and coordinating consumers and system needs (e.g. for system balancing and congestion management), taking advantage of flexibility from smart appliances (e.g. dishwashers, tumble dryers, etc.), electric heaters, batteries, etc. [14, 15]. Also, electrification of other energy demands from the heating/cooling (e.g. electric heat pumps and heating ventilation and air conditioning equipment) and transport (e.g. electric vehicles) sectors can further expand the opportunities to control electricity loads and coordinate them with system needs owing to their intrinsic virtual storage capabilities [16, 17].

In a similar vein, the digitalization of energy systems will provide unique opportunities for much smarter management of the electricity system during extreme events. For instance, this smart management can include switching off non-essential demand when the network is stressed while supplying essential demand. The supply of essential loads during emergencies will also be enabled by DERs and virtual storage capabilities from demand (e.g. battery units from electric vehicles). This would significantly enhance the resilience of supply delivered as energy consumers will have their essential load supplied during high-impact events [18].

5.1.3 The Resilience Enhancement Trilemma

Hence, finding the correct portfolio of resilience enhancement measures can be a difficult task since there is a rich variety of options with synergies and conflicts that can be used and combined. Additionally, the appeal of various investment options is case-dependent. For example, while undergrounding might be an effective strategy for mitigating adverse weather impacts, it may not be as suitable for addressing earthquake-related risks. To streamline the complexity inherent in this extensive range of options, we have categorized these investment alternatives into three broad groups, as illustrated in Figure 5.2: "Stronger," "Bigger," and "Smarter" [2]. This classification aids in synthesizing the wide array of choices available for enhancing resilience and helps in better understanding their respective impacts.

In the coming sections, we discuss some important topics that can enlighten planners to select the right portfolio of resilient enhancement measures, with a focus on investment.

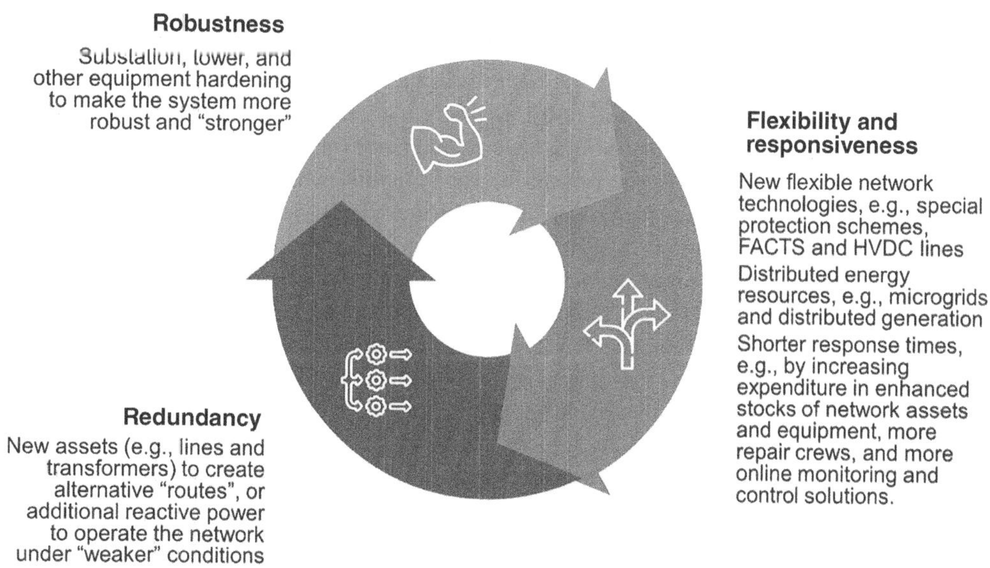

Figure 5.2 The resilience enhancement trilemma. *Source:* [19] / CIGRE.

5.2 Resilience and Risk Metrics for Risk-Averse Decision-Making

Historically, (deterministic) network reliability standards have typically ignored any contingency beyond "credible" ones, e.g. *N*–1/*N*–2. This has resulted in a bias toward building more and more infrastructure, mainly to provide redundancy in the efforts to deal effectively with any outage that might threaten the uninterrupted power supply. However, as elaborated in [20], experience with extreme events clearly shows that this reliability-driven approach of making the infrastructure "bigger and stronger" through redundancy and reinforcements may not be capable of hedging against multiple simultaneous outages or ones occurring in rapid succession. In fact, very extreme events cause outages well beyond credible ones, potentially affecting hundreds of network components. In other words, extreme events typically lead to *N*–*X* outages, with *X* greater or much greater than 1 or 2 and even in the order of 100 seconds or more! Indeed, here lies one of the fundamental differences between reliability and resilience, at least in the context of network planning. Planners may then intuitively realize that, while improving reliability, more redundancy may not necessarily improve system resilience to extreme events. Alternatively, flexible ("smarter") solutions could provide more viable options to enhance resilience by helping to withstand the initial adverse impacts of extreme events as well as by supporting the efficient response and prompt recovery of the system.

In the new context outlined above, could "hybrid" solutions (where "hybrid" refers to both infrastructure/network, i.e. to provide redundancy and robustness, and non-infrastructure/non-network or smart operational solutions, i.e. to provide flexibility) constitute the optimal portfolio to boost resilience to extreme events? While this is still an open research point and likely to be case-specific, if we want to keep the lights on or at least pursue an acceptable level of system operation under a large array of circumstances (beyond the so-called credible contingencies), it is evident that current planning standards need to be modified to allow HILP events to be accounted for within the network design and expansion decision making process. However, the key question here is *how*? While no straightforward answer exists, below we discuss a few possible approaches, also recognizing that it is technically unrealistic (and economically not viable) to consider targets of 100% reliable supply after extreme events and, at the same time, the system should meet classical (deterministic) reliability standards that consider only credible outages.

A first approach to incorporate HILP events within network investment planning could adopt probabilistic (or, in mathematical programming terminology, "stochastic") models that explicitly consider the associated probabilities and resulting impacts of many states of the system, including the intact system and simultaneous outage scenarios. These impacts are usually measured in terms of energy not supplied (ENS) and valued through economic metrics such as the value of lost load (VoLL). More specifically, in the probabilistic reliability assessment pioneered by Billinton and Allan [21], the resulting estimated costs from ENS are averaged (weighted by probability) across all modeled scenarios and optimized against additional investment and operational costs. Network investments are thus well justified as a trade-off between economics and security of supply and, if we neglect for simplicity operational costs (e.g. congestion costs, losses, etc.), they should be carried out up to the point where the marginal cost of additional investment equalizes the marginal benefit of enhanced reliability. This is graphically illustrated in Figure 5.3, where reliability cost is measured as expected energy not supplied (EENS) times VoLL. This probabilistic approach, however, presents a fundamental limitation to properly address HILP events and inform appropriate investment decisions as a hedge against them, as we discuss below.

HILP events are, by definition, very rare, and their impact on average indicators such as EENS is therefore very limited. For example, our own analysis shows that, on average, it would be economically optimal to suffer the consequences of very large earthquakes in Chile occurring approximately

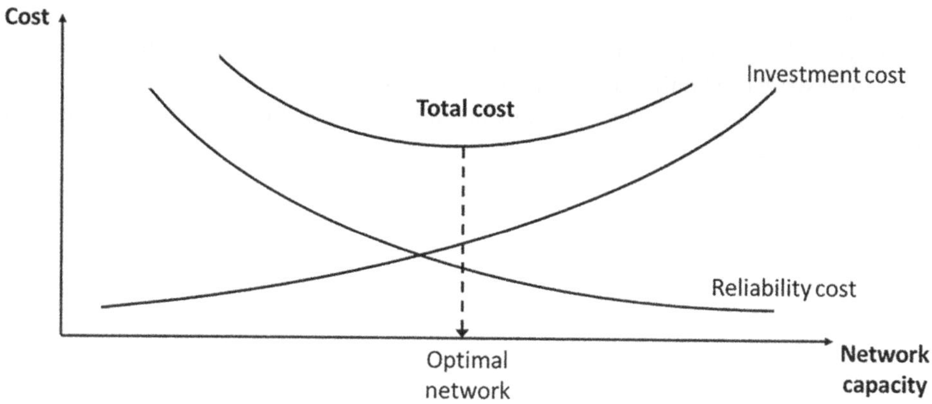

Figure 5.3 Optimal balance between investment and reliability costs.

every 15 years, rather than invest in further assets to reinforce and harden the power system. It is therefore worth asking ourselves why we should be concerned about events that, on average, have a relatively small effect. We argue that the answer to this question may be with the *risk attitude* of electricity consumers and policy makers (and therefore network planners too, who somehow "serve" both). In fact, as suggested by empirical evidence, customers and policy makers generally dislike risks associated with the highly adverse consequences often associated with HILP events, and would thus like to reduce them as much as possible. But how can we model this risk attitude, and what is the underlying risk attitude assumption in the probabilistic assessment above?

In risk analysis, the attitudes toward risk are usually classified into three categories, which are exemplified as follows. Consider an electricity consumer who is given the choice between the following two options. In the first option, the consumer pays \$90 for a network service that hardly ever fails and, when it does, small amounts of energy are unserved, totalizing an associated expected cost of ENS equal to \$10. In the second option, the consumer pays \$50 for a network service that fails more often and with larger amounts of ENS each time, totaling an associated expected cost equal to \$50. Note that, in both cases, the total amount paid by the consumer for network and EENS is the same. In this context, the consumer is said to be:

- *Risk neutral*: if the consumer is indifferent between these two options.
- *Risk averse*: if the consumer prefers the first option over the second one.
- *Risk seeking*: if the consumer prefers the second option over the first one.

The reader can easily deduce that risk neutrality is intrinsically part of traditional probabilistic analysis because the above two options seem equally attractive. In reality, however, consumers are typically risk-averse and arguably prefer a more stable and predictable outcome from the electricity network, even if this may (slightly) increase its costs, as mentioned earlier. In fact, as in many other industries, consumers might be willing to pay the price of insurance policies that eliminate (or at least mitigate) the losses associated with some rare but high-impact scenarios that may occur. In this context, risk-averse electricity consumers (which we argue comprise the majority) might prefer to be hedged against the consequences of HILP events on their electricity supply and pay for the corresponding cost increase, even if these events occur rarely or might not happen at all. Apart from really extreme cases such as earthquakes, which may be life-threatening and for which higher risk aversion may be justified for different reasons in any case, evidence of such consumer attitudes can be seen more and more often even for relatively smaller impact events. For example, heat waves in

Australia in January 2019 led to sporadic rolling load shedding events in several areas, including central Melbourne. The relatively short outages were considered outrageous by many consumers, even though they experienced only a minor overall adverse impact.

In addition to the consumer attitude, governments must consider the welfare of their citizens and, understandably, might want to take a risk-averse approach in dealing with HILP events for political reasons, irrespectively of the classical economics associated with traditional power system planning methodologies. That is why specific regulatory and market mechanism responses were called for following the South Australia black system event of September 2016, and more is expected in response to the bushfires that again occurred throughout Australia in January 2020.

In any case, once again, the main noteworthy message here is that risk-averse approaches and metrics that should be contemplated for HILP events are typically not present in current system planning practices.

5.2.1 A Probabilistic Risk Metric

Hence, to incorporate an appropriate risk metric in resilience assessment, we need to move away from purely considering indicators like the expected value. In this vein, the *Conditional Value at Risk (CVaR)* measure can be used so as to consider only those scenarios representing the worst cases. More specifically, CVaR (also referred to as CVaRα, although in this chapter we use CVaR without the subscript to simplify the notation) represents the expected value across a predetermined set of worst cases. An illustrative example of probability distribution function (PDF) and relevant parameters, with the ENS used as the resilience metric, is shown in Figure 5.4. It should be noted that the value of the parameter α also provides an explicit indication of risk attitude, since $1-\alpha$ indicates the size of the considered set of worst cases. A reliability assessment would use the measure of EENS by sampling across *all* credible sets of outages. In the averaging, the impact of non-credible worst-case outages would normally be outweighed by the much more frequent credible outages. In the context of our framework, then, this means that in first approximation the mean value (i.e. the EENS) and the CVaR of the PDF shown in Figure 5.4 could be used as the reliability assessment and the resilience assessment measures, respectively. Interestingly, although optimizing the mean value or CVaR is useful to identify decisions from either a reliability or resilience

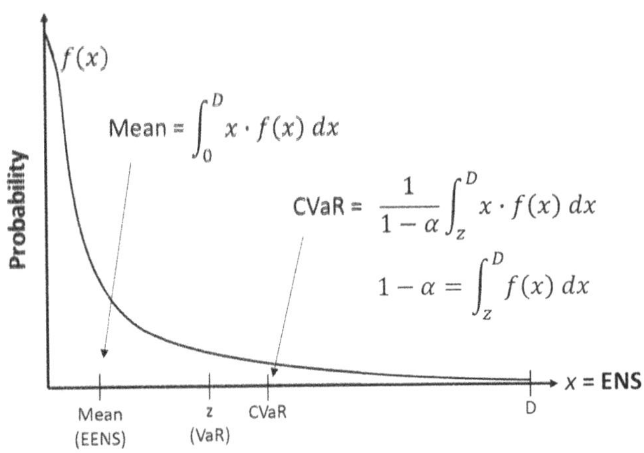

Figure 5.4 CVaR concept for risk-averse resilience assessment, where $1-\alpha$ indicates the size of the considered set of worst cases, VaR (or z) is Value at Risk, CVaR is Conditional VaR, D is maximum demand, x is energy not supplied (ENS), and the Mean value is expected energy not supplied (EENS).

perspective, respectively, in practical settings, where planners need to consider both criteria to make reasonable decisions, optimizing a linear combination of both will be more suitable (we will expand on this in Section 5.4.5).

5.2.2 Risk Versus Resilience Metrics

Here, it is important to distinguish between risk and resilience metrics. The risk metric measures the size of the relevant tail of a PDF. If the analysis focuses on cost, the relevant tail will be that on the right-hand side (i.e. right tail with the higher costs). If the analysis focuses on revenues or profits, the relevant tail will be that on the left-hand side (i.e. left tail with the lower profits). The main idea is to capture the "worst" cases. Hence, VaR, CvaR, and even the standard deviation serve as risk metrics, although the latter measures positive and negative deviations with respect to the mean, making it less attractive for worst-scenario analysis. We choose to illustrate CVaR as the risk metric due to its attractive mathematical properties, particularly in terms of its straightforward incorporation within network expansion problems tackled through linear mathematical programs. More details about the properties of CVaR can be found in [22].

The resilience metric, on the other hand, parametrizes a characteristic of the unsupplied demand curve during a particular outage scenario. Hence, ENS can be used as well as any other metric from the FLEP metric system explained previously in Chapter 3. For example, Figure 5.4 illustrates a PDF associated with the ENS. In this case, CVaR is used as the risk metric, and ENS corresponds to the resilience metric. Rather than minimizing ENS as a conditional expected value across the worst cases (i.e. minimize CVaR of ENS), we can alternatively choose, for instance, to maximize the recovery demand connection rate after the outage occurs as a conditional expected value across the worst cases (i.e. maximize CVaR of the recovery rate). Reference [1] shows the optional portfolio of decision investments found to enhance transmission network resilience when optimizing different resilience metrics.

5.2.3 Dependencies and Common Mode

Importantly, outage dependencies and common mode failures need to be properly acknowledged when constructing PDFs [23]. Furthermore, risks due to natural hazards are mainly encapsulated in the tails of the PDFs, which greatly depend on outage dependencies. For example, the current literature usually models simultaneous outages like a series of outages that cascade very rapidly across the system. This is the case of models such as those explained in [24–27], in which several power flow simulations are undertaken after triggering an initial network outage that can overload other parts of the network and so cascade into a major event. A stochastic optimization model for network investment, however, requires encapsulation of the aforementioned series of outages into a scenario tree [28] (Figure 5.5). Such scenario tree should describe the final state of the system after outages occur. In this vein, a series of cascading outages can be represented by a single contingent state of the system where multiple elements failed simultaneously. As failures in a cascading event are clearly not independent, dependent probabilities must be used to correlate the outage probabilities of multiple elements after a given triggering event has happened. This triggering event can also be exogenous. For example, an earthquake that accelerates the ground above acceptable levels will trigger the failure of various components simultaneously. Clearly, the probability of multiple simultaneous outages becomes much higher in this case. These dependencies ultimately mean that the usual formula (5.1) to calculate the probability of outage scenarios cannot be utilized, as (5.1) only works when failures are independent from each other. Hence, the construction of PDFs based

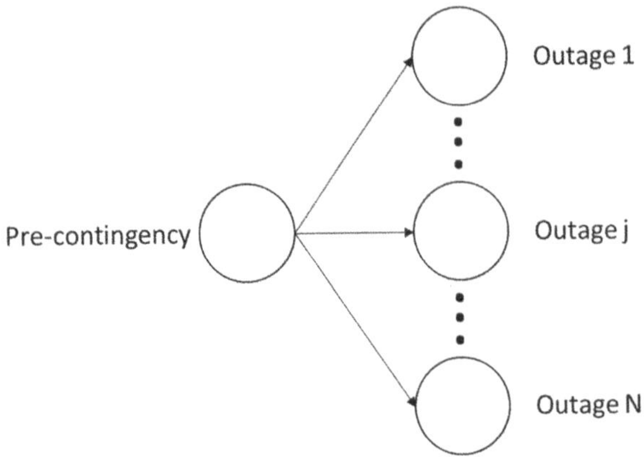

Figure 5.5 Scenario tree of outages.

on simulations, for example, becomes a complex task where outage dependencies have to be clearly captured. Reference [23] discusses this topic in-depth.

$$P\,(\text{Outage}\,j) = \prod_{k \in A(j)} P_k \prod_{k \in U(j)} (1 - P_k) \tag{5.1}$$

In the multiplication rule for independent events (5.1), $P\,(\text{Outage}\,j)$ denotes the probability of the specific outage scenario j occurring. P_k represents the failure probability of each individual element k. The set $A(j)$ contains the elements that are operational in outage scenario j, while $U(j)$ comprises the elements that have failed or are not operational in that scenario. The total number of possible scenarios is 2^M, where M is the total number of elements. Among these scenarios, one represents all elements being operational (no outages), and another represents all elements having failed.

5.2.4 Example of Probability Capacity Tables Accounting for Dependencies in a Double Circuit

Table 5.1 illustrates the probabilities of outage states for a double circuit within an area with natural hazard potential. For illustration purposes, let us assume that the natural hazard is a wildfire that happens, on average, every five years in a specific area that contains a double circuit powerline. The table shows probabilities with and without the common-mode failure of both circuits due to wildfires (the common event being the wildfire, and when it occurs, both circuit fails). Column A shows the conditional probability assuming independent failures and ignoring common mode failures. These probabilities are commonly used in reliability analysis and were calculated assuming independence (equation (5.1) was applied) and using the available statistics under normal conditions shown at the beginning of Table 5.1. Column B presents the probability of likely events that will happen if a wildfire occurs. In this case, we assume that the wildfire will cause a double outage with a probability equal to 1 (meaning that the double outage will occur with a probability equal to 1 if the wildfire occurs; note that this probability can be changed). Column C shows the marginal probabilities with dependencies, including both conditional probabilities, with and without a wildfire that causes a double outage (calculated through equation (5.2)). These probabilities are those to be used in the resilience assessment. Notice the significant increase in the probability of state four in Column C with respect to that in Column A.

Furthermore, in the case where dependency is considered (Column C), the probability of a double outage is larger than that of single outages, demonstrating the importance of this consideration.

Table 5.1 Statistics and failure probabilities of a double circuit.

Network failure statistics (normal conditions)

One circuit's failure rate (λ) = 1 occ/year

Mean repair time = 1 days

Repair rate (μ) = 365 occ/year

One circuit's failure probability = 0.0027

Wildfire statistics

Wildfire frequency = 0.2 occ/year

Wildfire duration (T) = 7 days

$1/T$ = 52.14 occ/year

Wildfire probability = 0.0038

State	Availability		(A) Conditional probability (no wildfire)	(B) Conditional probability (wildfire)	(C) Probability (marginal)
	Circuit A	Circuit B			
1	1	1	0.99454	0.00000	0.99074
2	1	0	0.00272	0.00000	0.00271
3	0	1	0.00272	0.00000	0.00271
4	0	0	0.00001	1.00000	0.00383

While previous equation (5.1) shows the expression used to calculate the state probabilities in Col umn A, equation (5.2), that represents the law of total probability, is the formula used to calculate the marginal (or total) probability of Column C.

$$P(j) = \sum_{k \in K} P(j \mid k)P(k) \tag{5.2}$$

In the law of total probability (equation (5.2)), $P(j)$ denotes the marginal (or total) probability of state j, $P(k \mid j)$ the probability of state j conditional to event k (where the set K = {no wildfire, wildfire}), and $P(k)$ the probability of event k.

5.2.5 Deterministic or "Robust" Approach to Risk

Two major challenges in probabilistic resilience analysis in practice are the difficulty (even feasibility) of calculating probabilities and the very large number of scenarios to be analyzed (theoretically, all of them). A natural solution would be to use other approaches that do not depend on probabilities and, still, can deal with HILP events, and can limit the number of scenarios such as robust optimization. Under this approach, the decision-maker plans against the worst cases (this is why it is called "robust"). One of the main challenges associated with this setting is the need to carefully choose the scenarios against which the decision-maker wants to be protected. Such scenarios must preserve some statistical features of the actual uncertainty being faced and represent appropriate risk-aversion levels, neither too conservative nor too risky. Notice that this is similar to deciding the right value of the parameter α under the probabilistic CVaR approach that defines the set of worst cases on which the conditional expectation should be calculated. In both probabilistic and deterministic or robust approaches, we seek to find investment solutions to optimize the system performance with a special focus on worst cases. The interested reader is referred to [29, 30] for applications of advanced robust optimization models applied to system outage scenarios.

5.2.6 Distributionally Robust Approach to Risk

The distributionally robust approach to risk is used to tackle uncertainty in probability distributions. It combines probabilistic/stochastic and robust approaches explained earlier. This focuses on planning against the worst-case scenario within a set of possible probability distributions. Unlike traditional planning studies, which often assume a specific known probability distribution (as in probabilistic analysis) or ignore probability distributions entirely in favor of worst-case scenarios (as in robust analysis), the distributionally robust analysis considers a range of potential distributions. This approach seeks planning solutions that are robust not just against variations in data but against variations in the underlying distributions themselves (for more details, see Chapter 4). The interested reader is addressed to [29, 30] for applications of advanced distributionally robust optimization models applied to resilient network planning against earthquakes.

5.3 Mathematical Frameworks to Plan Resilient Grids

5.3.1 Possible Theoretical Formulations: From Robust to Stochastic Network Planning

5.3.1.1 Two-Stage Framework

Finding the right solution to the trilemma problem of resilience (invest in "Stronger," "Bigger," or "Smarter" solutions) is a non-trivial task. We need to find a balanced portfolio of investment and operational solutions that can provide the best possible hedge against an array of HILP events. Hence, we present a modeling framework based on mathematical programming concepts, which can capture the importance of risk/uncertainty due to natural hazards when planning new investments as well as the effects of relevant operational solutions. To do so, the proposed framework considers decision variables in two different timescales (investment and operation; see operational measures in detail in Chapter 6), and the presence of uncertainty to reflect the changing landscape faced by system operators, especially in terms of simultaneous outages driven by natural hazards. In this framework, optimal investment decisions are made before the realization of uncertain scenarios (e.g. investment cost of a new technology), whereas operational decisions aim to run the resulting infrastructure after each possible uncertain scenario realizes (system outages triggered by a natural hazard) and for a number of operating conditions (e.g. demand levels, wind power outputs, etc.). The optimal investment decisions minimize a given metric or objective function (e.g. expected total cost, maximum cost, expected total cost plus CVaR, and maximum regret among all scenarios) and comply with a set of constraints that ensure feasibility under an array of physical laws and market rules as well as certain reliability levels (e.g. system reserves requirements). The mathematical framework is introduced in [8] and shown in equations (5.3)–(5.5), where functional $f_{\varepsilon}\{\cdot\}$ represents the chosen utility function to select the optimal investment vector x. $C^{I}(\cdot)$ and $C^{O}(\cdot,\cdot)$ respectively represent the investment and operational cost functions, $y(\cdot)$ is the matrix of operational decisions (per component and per operating condition or time period), and ε represents a given scenario of the uncertainty set E, representing the set of outages triggered by natural hazards. X and $Y(\cdot, \cdot)$ are the feasibility sets of investment and operational decisions, respectively, where the latter depends on the optimal investment vector x.

$$\min_{x,y(\cdot)} f_\varepsilon\{C^I(x) + C^O(y(\varepsilon), \varepsilon)\} \tag{5.3}$$

$$\text{s.t. } x \in X \tag{5.4}$$

$$y(\varepsilon) \in Y(x, \varepsilon); \forall \varepsilon \in E \tag{5.5}$$

The proposed framework is presented as a two-stage optimization program, where investment decisions are taken in the first stage, and operational decisions are made per outage scenario and taken in a second stage. Functional $f_\varepsilon\{\cdot\}$ defines whether the model corresponds to a stochastic (with and without risk aversion) [28, 31] or robust optimization program [32]. Set $Y(\cdot, \cdot)$, that aims to capture how the resulting infrastructure is operated and its corresponding costs, contains the physical (and market) laws associated with power system operation and may include: first and second Kirchhoff's laws and power flow equations, generation and network capacity constraints (including minimum and maximum limits), inter-temporal constraints including minimum up and down times and ramping limitations, system reserves requirements, etc. [33].

5.3.1.2 Multi-Stage Framework with Two Layers of Uncertainty

We can add another, more realistic level of complexity to our framework if we consider that system operation has to be planned ahead of real time (as well as investment has to be planned ahead of system operation). Under this new consideration, two layers of uncertainty (typically called long-term and short-term uncertainties) are explicitly recognized, and network operations have to be differentiated between pre- and post-fault actions. While the short-term uncertainty represents system failures driven by natural hazards (as in the previous two-stage framework), the long-term uncertainty can represent fundamental changes in the future trend of climate parameters that can affect the frequency and intensity of natural hazards. In this context, new investments should be able to cope not only with long-term uncertainties (e.g. climate states), but also with uncertain changes that may occur more rapidly in real time (e.g. natural hazards driving outages).

To that end, the aforementioned framework can be extended as shown in equations (5.6)–(5.9), where $f_{\varepsilon,\xi}\{\cdot\}$ models the objective function being minimized (e.g. stochastic, expected costs plus CVaR, or robust optimization), whereas $y(\cdot)$ and $z(\cdot, \cdot)$ represent the operational decisions undertaken in the scheduling or operations planning timescale (e.g. day-ahead) and in real time, respectively. Set $Z(\cdot, \cdot, \cdot)$ represents the operational decisions that can be made in real time (e.g. actual deployment of generation reserves, demand curtailments, etc.), which clearly are constrained by prior decisions made during the investment timescale (vector $x(\cdot)$) and operations planning timescale (matrix $y(\cdot)$). ξ represents a given scenario of the uncertainty set Ξ that contains a number of realizations of the long-term uncertainty such as global temperature or any other climate parameter that can be considered relevant for resilience analysis.

$$\min_{x(\cdot),y(\cdot),z(\cdot,\cdot)} f_{\xi,\varepsilon}\{C^I(x(\xi), \xi) + C^O(y(\xi), z(\xi, \varepsilon), \xi, \varepsilon)\} \tag{5.6}$$

$$\text{s.t. } x(\xi) \in X(\xi); \forall \xi \in \Xi \tag{5.7}$$

$$y(\xi) \in Y(x(\xi), \xi, \varepsilon); \forall \xi \in \Xi, \forall \varepsilon \in E \tag{5.8}$$

$$z(\xi, \varepsilon) \in Z(x(\xi), y(\xi), \xi, \varepsilon); \forall \xi \in \Xi, \forall \varepsilon \in E \tag{5.9}$$

Both frameworks (equations (5.3)–(5.5) and equations (5.6)–(5.9)) pose important challenges at the computational and algorithmic levels that need to be carefully addressed. Under these frameworks, we can classify several methods of the current state of the art in network investment resilience. For example, references [1, 30] belong under framework in (5.3)–(5.5), while references [34–36] belong under framework in (5.6)–(5.9).

One plausible implementation of the two-stage stochastic optimization framework in (5.3)–(5.5) can be observed in [1], where Optimization via Simulation (OvS) is used (presented in Section 5.3.3).

5.3.2 CVaR Formulation

In mathematical programming, CVaR is a practical risk measure because it can be formulated using linear equations, allowing for straightforward incorporation into stochastic investment planning models, more specifically, probabilistic security-constrained optimal power flow (OPF)-based network expansion planning problems like those presented in [37]. The linear constraints that model CVaR, initially pioneered in [22], are detailed in equations (5.10)–(5.12), where $\overline{CVaR}$ is the upper bound of CVaR, representing in this case the maximum system demand that can be lost on average in the worst $1-\alpha$ events, z is the minimum amount of system lost load associated with the worst $1-\alpha$ events (also namely Value at Risk, VaR); and δ_k is the (positive) amount of system lost load in event k with respect to z (note that $\delta_k = 0$ if system lost load in event k is lower than z). Note that $\overline{CVaR}$ can be either a fixed parameter, as seen in CVaR-constrained models, or a variable. In cases where it is variable, $\overline{CVaR}$ needs to be included as part of a minimization-type objective function.

$$\delta_k \geq 0;\ k = 1 \dots M \tag{5.10}$$

$$\sum_{n=1}^{N} ENS_{k,n} - z \leq \delta_k;\ k = 1 \dots M \tag{5.11}$$

$$z + \frac{1}{1-\alpha} \sum_{k=1}^{M} p_k\, \delta_k \leq \overline{CVaR} \tag{5.12}$$

5.3.3 Optimization via Simulation Approach: A Two-Stage Stochastic Model

There are several ways to apply the above frameworks, especially to implement the two-stage model presented in (5.3)–(5.5). As explained in [20], we adopt an Optimization via Simulation (OvS) approach, which determines the (nearly) optimal portfolio of network enhancements based on a series of simulations. In this setup, the optimizer corresponds to the first stage, while the simulator represents the second stage and is treated as a black-box model with no explicit mathematical formulation. One of the key advantages of the OvS approach is that it allows inclusion of a great deal of operational details in the simulation stage, e.g. minimum stable generation levels, ramp rate limits, minimum startup and shutdown times, etc., which require a non-convex formulation and are complex and hard to manage in closed form.

Hence, we introduce a two-stage model (Figure 5.6) that intelligently selects, in the first stage, specific network investments out of a set of candidate options, which are, in the second stage, tested through quantification of their resilience benefits in probabilistic outage scenarios originated by stochastic simulation of natural hazards. As a result, the optimal portfolio of network investment decisions with the highest impact on a given resilience metric (measured across various scenarios) is identified.

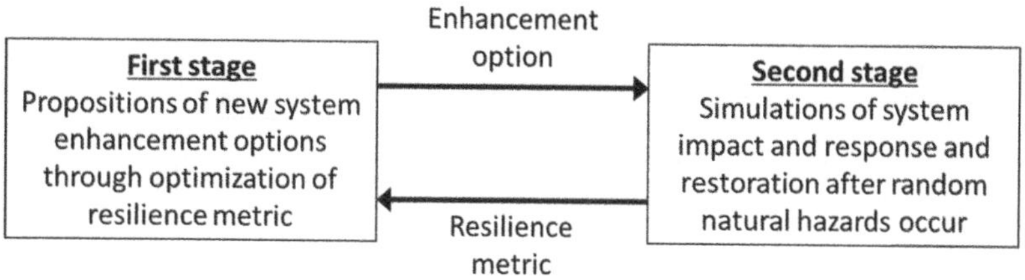

Figure 5.6 Quantitative approach to identify optimal resilient system enhancement options.

The stochastic generation and assessment of hazard and outage scenarios is carried out through the following simulation-based steps:

1) *Hazard characterization:* In the first step, we generate various hazards with random intensities and locations (this can be done by respecting historical patterns). Additionally, spatiotemporal profiles may be necessary to model hazards that change position and intensity dynamically (e.g. storms), spread (e.g. bushfires), or attenuate their magnitude with distance (e.g. earthquakes).
2) *Vulnerability assessment of system components:* By using fragility curves (illustrated in Chapter 3) which are assumed to be known for various natural hazards and system components, either through historical data or on the basis of engineering modeling), we determine both (i) the hazard-dependent failure probabilities of every network component (e.g. towers, lines, as well as substation and generation equipment) and (ii) the outage scenarios across the system which are generated from these probabilities.
3) *System response:* This is the step where we simulate, for each outage scenario identified above, the potential system cascading from automatic power flow rerouting, load/generation disconnection, and post-contingency redispatch (once cascading has ended). We then assess the spatially resolved volumes of ENS. Importantly, prior to the outage, we assume a normal operation of the system by running an economic dispatch problem, in which the system infrastructure is intact.
4) *System restoration:* We simulate both (i) the reconnection of failed/damaged network components once these have been repaired (whose reconnection times are determined probabilistically, assuming that the reconnection events are exponentially distributed) and (ii) the reconnection of load/generation, which is obtained by a post-contingency dispatch model.

Figure 5.7 illustrates the aforementioned process, and Figure 5.8 shows a typical curve for the supplied demand which results from the simulation of the post-fault events associated with a scenario that follows this sequence: (i) random generation of hazard; (ii) random generation of network outage; and (iii) random generation of equipment repairs. Note that the shape of the curve in Figure 5.8 can vary and can be hazard- and system-specific (note this is the more realistic version of the generic resilience trapezoid presented in Chapter 3). This power supply curve is generated several thousand times (e.g. 10,000) for different hazard-outage-repair sequences by using Monte Carlo simulation. We consider a large number of simulations so as to adequately capture the potential consequences of a comprehensive set of natural hazards on the power network.

After the simulations have been performed under several outage scenarios originated by natural hazards, a suitable resilience metric can be selected and calculated for every scenario. The metric could, for example, be the ENS, or any other taken from the *FLEP* resilience metric system (see details of the Fast Low Extensive Promptly [FLEP] metric system in Chapter 3). After quantifying the effect of each scenario with the selected metric, the optimization problem minimizes risk exposure subject to a budget constraint (representing the total amount available to be invested in

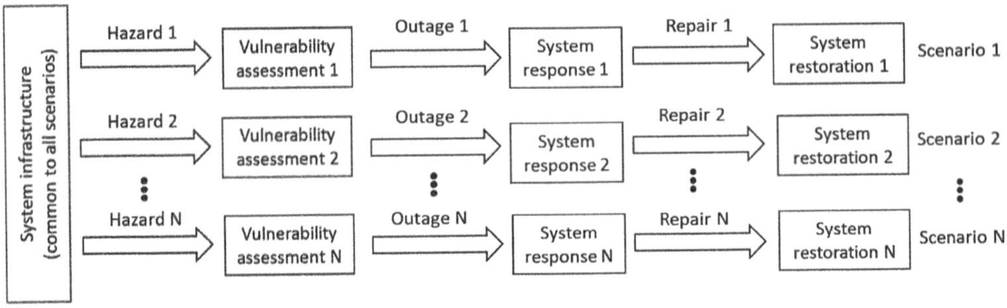

Figure 5.7 Simulation of hazard-outage-repair scenarios for the system infrastructure under study.

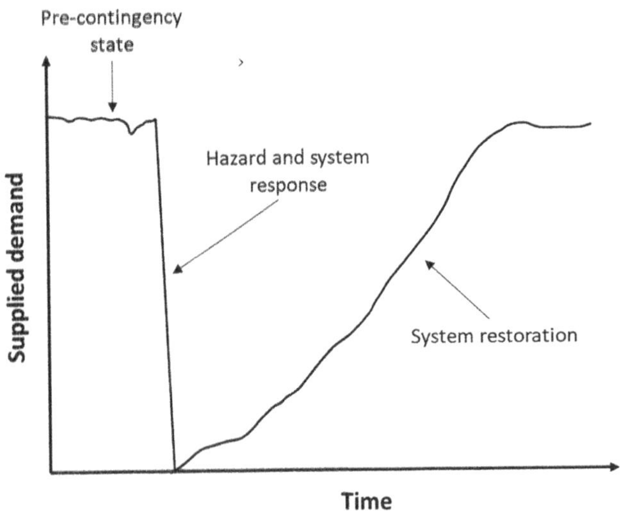

Figure 5.8 Example of supplied demand curve resulting from each scenario simulated.

resilient network enhancement), thus identifying the optimal portfolio of investment decisions that provides the best hedge against the HILP events simulated. The risk measure being minimized can be determined by calculating the expected value of the selected resilience metric (e.g. ENS) across an appropriately selected *worst* set of scenarios resulting from the stochastic simulation.

To do so, the *CVaR* measure can be used to consider only those scenarios representing the worst cases, as explained earlier in Section 5.3.2.

Note that we have applied the two-stage formulation from Section 5.3.1.1 with the following considerations: The operational cost is now solely represented by the demand curtailment cost, as this is typically the most significant cost component following a contingency in practical scenarios. Additionally, the investment cost has been removed from the objective function and is instead incorporated into the constraints, with an upper bound defined by the available budget to invest in resilience enhancements. Furthermore, the functional $f_{\varepsilon}\{\cdot\}$ now only considers the average across a selected set of worst-case scenarios, specifically focusing on the CVaR.

The proposed mathematical framework can be used to select a wide-ranging portfolio of resilient network enhancement options depending on the specific decision variables considered in the problem, such as hardening infrastructure, installation of new assets, provision of better response times for repairing damaged infrastructure, and restoring power. This addresses the so-called *resilience trilemma*, that is, the need for balancing the portfolio solution among several options to make the system *stronger*, *bigger*, and/or *smarter* in a resilient and cost-efficient manner. In particular, among the set of investment decisions to enhance system resilience, the following could be considered:

- New lines and transformers, to create alternative "routes" to transfer power and provide redundancy, or additional reactive power to operate the network under "weaker" conditions when several network assets are outages due to HILP events ("bigger").
- Substation, tower, and other equipment hardening, to make the system more robust and "stronger" against HILP events.
- Shorten response times by increasing expenditure in enhanced stocks of network assets and equipment, more repair crews, and more online monitoring and control solutions ("smarter").
- Installation of new flexible network technologies such as special protection schemes, energy storage units, FACTS, and HVDC, so as to make the system more flexible to adapt to different conditions post-fault, helping to mitigate the consequences of HILP events.
- Installation of DER (such as microgrids and distributed generation), to provide localized energy solutions when the main system fails.

5.3.4 Advanced Resilient Investment Planning Models

To properly quantify risks, the aforementioned methodologies are probabilistic. Furthermore, they are based on stochastic optimization. One of the main challenges associated with stochastic optimization approaches is their tractability (or "difficulty") when applying them to large-scale power networks. Hence, advanced decomposition techniques and heuristics (as the OvS approaches presented) will become critical to identify and study resilience enhancements against natural threats in real-world power systems.

Another major challenge of stochastic models for resilience decision-making is the difficulty (or even feasibility) of calculating probabilities. A natural solution would be to use other approaches that do not depend on probabilities and, still, can deal with uncertainties, such as robust optimization (the advanced reader can note that, depending on functional $f_\varepsilon\{\cdot\}$, the frameworks in (5.3)–(5.5) and (5.6)–(5.9) can be stochastic or robust). Under this approach, the decision-maker plans against the worst cases. A notable advantage of robust optimization is that the worst cases can be found using highly efficient algorithms, improving the execution time (tractability) compared with stochastic optimization. One of the main challenges associated with this setting is the need to carefully choose the scenarios against which the decision-maker wants to be protected. Such scenarios must preserve some statistical features of the actual uncertainty being faced and represent appropriate risk-aversion levels, neither too conservative nor too risky.

Finally, there are other types of models to make decisions under uncertainty that combine the features of stochastic and robust optimization. Under this hybrid approach, probabilities are assumed to be partially known (or "ambiguous"), which can be more suitable in reality if some information regarding the PDFs of rare events is available. Remarkably, these models may offer a straightforward way to solve problems with decision-dependent probabilities (i.e. probabilities that are not constant and depend on the decisions being optimized), critical in resilience modeling to incorporate hardening decisions aiming to reduce the fragility or failure probability of the existing assets against given threats (e.g. replacing overhead lines with underground cables reduces the risk exposure of an existing link rather than expands the network). References [29, 30] propose such an approach to plan resilient power networks.

5.4 Merging Resilience and Reliability Criteria for Practical Decisions

In practice, policymakers and network planners have to consider various criteria to make sensible investment decisions. The following simple example illustrates the different concepts and criteria that can be used to make investment decisions, identifying resilient enhancement options against HILP events in network planning and, crucially, how these differ from other decisions that are more reliability-oriented. At the end, we present a common framework that combines these criteria for investment purposes.

5.4.1 Illustrative Case Study

The two-busbar network in Figure 5.9, introduced in [20], features one 500 MW generating unit in node 1, one load in node 2 with a constant demand of 500 MW, and a transmission link between the two nodes. Depending on the configuration (i.e. number of circuits and their capacities) and reliability characteristics of this link, and assuming perfect reliability for the generator, this power network can be adequate, secure and/or resilient. As adequacy and security have been historically part of reliability analysis, we will consider here that a network is reliable if it is both adequate and secure. We also use the DC power flow approximation (and no network losses) for the sake of simplicity.

5.4.2 Reliability I: Adequacy

Considering that adequacy is the ability of a power system (including generation and network capacities) to supply the aggregate electricity demand at all times, the power network in Figure 5.9 is adequate if the total (thermal) capacity of the link is at least 500 MW. This can be achieved by a number of network configurations, including the option of one single circuit of 500 MW transferring power from node 1 to node 2. This particular option, however, is evidently not N–1 secure. In this case, the N–1 criterion requires the presence of, at least, another circuit between the nodes.

5.4.3 Reliability II: Security

From our previous work, we distinguish between two types of security: deterministic N–k security and probabilistic security. To comply with a deterministic N–1 security standard for network design, we need two circuits, each of (at least) 500 MW, to link nodes 1 and 2. A probabilistic standard, instead, demands an appropriate balance between the cost of improving reliability, here in the form of investment costs, and the associated savings in reliability operational costs, measured

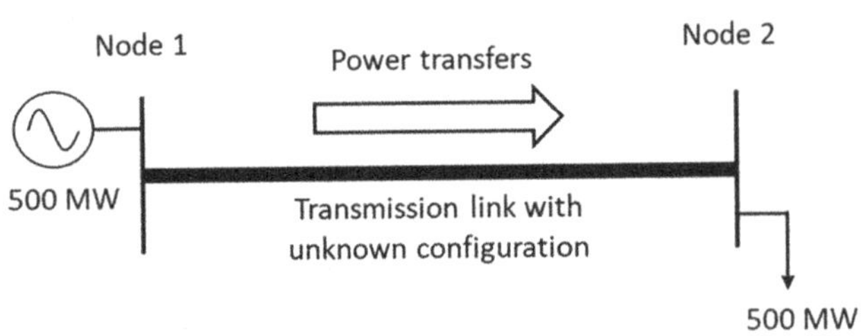

Figure 5.9 Illustrative two-busbar system.

Table 5.2 Reliability and cost information associated with two alternative network design options.

Option N-0

Cost of investment calculation

Unit cost of investment [$/MW.km.yr]	100
Lenght [km]	200
Capacity per circuit [MW]	250
No. Circuits	2
Cost of investment [$]	**10,000,000**

Cost of EENS calculation

VoLL [$/MWh]	1000
Expected power not supplied [MW]	0.06148
Time horizon [h]	8760
Cost of EENS [$]	**538,532**

Failure states

Circuit 1	Circuit 2	State probability	Available capacity [MW]	Power not supplied [MW]
Available	Available	0.9997629	500	0
Unavailable	Available	0.0001142	250	250
Available	Unavailable	0.0001142	250	250
Unavailable	Unavailable	8.782E-06	0	500

Total cost calculation

Total cost [$]	**10,538,532**

Option N-1

Cost of investment calculation

Unit cost of investment [$/MW.km.yr]	100
Lenght [km]	200
Capacity per circuit [MW]	500
No. Circuits	2
Cost of investment [$]	**20,000,000**

Cost of EENS calculation

VoLL [$/MWh]	1000
Expected power not supplied [MW]	0.00439
Time horizon [h]	8760
Cost of EENS [$]	**38,464**

Failure states

Circuit 1	Circuit 2	State probability	Available capacity [MW]	Power not supplied [MW]
Available	Available	0.9997629	1000	0
Unavailable	Available	0.0001142	500	0
Available	Unavailable	0.0001142	500	0
Unavailable	Unavailable	8.782E-06	0	500

Total cost calculation

Total cost [$]	**20,038,464**

through the improvement in EENS (resulting from the new investment) times the VoLL. Let us now assume that there are only two possible "secure" configurations (i.e. number of circuits and their capacities) for the transmission link in question, and let us evaluate the cost and benefits of each of them to identify the appropriate, optimal solution. Table 5.2 provides the basic reliability information and costs of the following two alternatives:

- *N*–0 *option*: where 2 circuits of 250 MW are installed
- *N*–1 *option*: where 2 circuits of 500 MW are installed

In both cases, the network is 99.976% of the time with full available capacity (further reliability information associated with outage and repair rates is presented later in Table 5.3).

In this case, the *N*–0 solution demonstrates to be more economically efficient and therefore should be the one selected under a probabilistic security approach. However, this solution is sensitive to an array of economic and reliability parameters. For example, if VoLL is increased from 1000 $/MWh to 30,000 $/MWh, then the secure option changes from the *N*–0 to the *N*–1 design solution.

5.4.4 Resilience

One of the key characteristics of the probabilistic security analysis carried out above is its focus on EENS, which, as discussed, is not suitable for resilience studies dealing with HILP events. For the

Table 5.3 Reliability data and probabilities of failure under fair and adverse weather and marginal probabilities of failure for the four states considered.

	Fair weather	Adverse weather
Failure rate [occ/yr]	0.2	20
Repair rate [occ/yr]	2190	52
Unavailability	9.1E-05	0.27723
Availability	0.99991	0.72277
Duration [h]	8759	1

Failure states: Circuit 1	Circuit 2	State probability (fair weather)	State probability (adverse weather)	State probability (marginal)
Available	Available	0.999817	0.522400	0.999763
Unavailable	Available	9.13E-05	0.200373	0.000114
Available	Unavailable	9.13E-05	0.200373	0.000114
Unavailable	Unavailable	8.34E-09	0.076855	8.78E-06

purposes of assessing resilience, then, let us assume that the (marginal) probabilities of the four states previously evaluated (a no-fault state, two single-fault states, and a double-fault state) originate from the conditional probabilities displayed in Table 5.3. More specifically, Table 5.3 shows the probability of the four states under two different weather conditions, namely, fair and adverse weather, as well as the marginal probability of the four states considering *both* weather conditions.

For illustrative purposes, we assume that the failure rate is 100 times higher for adverse weather and that repair times increase from four hours under fair weather conditions (i.e. repair rate of 2190 occurrences per year or occ/yr) to seven days under adverse weather conditions (i.e. repair rate of 52 occ/yr). Also, adverse weather, in this example, is limited to one hour per year (thus conventionally representing a HILP event), while fair weather conditions occur during the remaining time (i.e. 8759 hours).

In order to provide a hedge against such an HILP event, we analyze the following three options within the concept of resilience trilemma (assuming that the initial condition is the same *N*–0 network configuration selected under the probabilistic security approach described in the previous section):

- *N*–1 *design*: where we re-evaluate the option to install two circuits of 500 MW, i.e. making the network "bigger" by adding redundancy;
- *N*–0 *with shorter response times under extreme events*: where we evaluate a contract with other network companies to use their repair crews under extreme events, which reduces from 7 to 3 days the repair times (i.e. making the network "smarter");
- *N*–0 *with underground cables*: where we evaluate the option to underground the current double circuits, each of 250 MW of capacity, thus halving the failure rate under both weather conditions, respectively, and assuming at the same time, for simplicity, that the repair rate stays the same. This is equivalent to a "stronger" system option.

Table 5.4 shows the impact of these new options on various average and risk indicators, including:

- *EENS*: the expected value of ENS across all scenarios (economically valued at VoLL).
- *CVaR*: the conditional value at risk of ENS, that is, the average ENS across the 0.001% worst cases (economically valued at VoLL).
- Probability of double outage under adverse weather conditions, which occurs one hour per year only.

As it could be expected, changing the network design from *N*–0 to *N*–1 provides the best results in terms of EENS cost, reducing it by 93% from approximately 539 k$/yr to 38 k$/yr. However, this decision provides a very limited hedge against HILP events, reducing the CVaR by 6% only from

Table 5.4 Average and risk indicators of the four considered network design options.

Metric	*N*–0 base case	*N*–1	*N*–0 shorter repair time	*N*–0 underground
VoLL × EENS ($)	538,532	38,464	470,506	280,428
VoLL × CVaR ($)	4,113,206,199	3,846,412,398	2,690,095,838	2,837,833,988
Probability of double outage under adverse weather (%)	7.7%	7.7%	2.0%	2.6%

4113 k$/yr to 3846 k$/yr. (Note that in this case and in the following ones, the probability of double outage and CVaR follow a similar pattern, so, for the sake of simplicity, we just focus on CVaR.) Interestingly, reducing the repair time under adverse weather reduces CVaR significantly (by 35%, from 4113 k$/yr to 2690 k$/yr) while the corresponding EENS reduction is more limited (only 13%). This exemplifies the fact that one enhancement solution may be preferred from a reliability perspective, while, from a resilience perspective, other options may be more attractive.

5.4.5 Combining Reliability and Resilience

Interestingly, Table 5.4 shows that the option in which lines are undergrounded features an attractive compromise between reliability and resilience indicators, reducing EENS cost by 48% (which is not as good as the 93% reduction related to the *N*–1 solution, but not as bad as the 13% reduction related to the *N*–0 case with shorter response times) and reducing CVaR by 31% (which is not as good as the 35% reduction related to the *N*–0 case with shorter response time, but not as bad as the 6% reduction associated with the *N*–1 solution). This is particularly important in practice in the presence of budget constraints and under the need for undertaking *both* reliable and resilient enhancements in power networks.

As, in practice, network investment decisions should consider both resilience and reliability implications, we can create a new function equal to the weighted sum of the energy not supplied in reliability (EENS) and resilience (CVaR). That is, for given weights $\lambda_1, \lambda_2 \geq 0$, consider the objective in (5.13).

$$\lambda_1 f_{\text{EENS}}(x) + \lambda_2 f_{\text{CVaR}}(x) \tag{5.13}$$

This multi-objective function, proposed in [1], is used to obtain the convex envelope of the Pareto frontier (of efficient/non-dominated solutions) by changing the weights λ_1, λ_2. Hence, the ultimate solution will depend on the preferences of the decision maker between reliability and resilience objectives.

It is important to highlight that this example demonstrates the fundamental differences between investment propositions that aim to hedge against average and more risky scenarios.

5.5 Realistic Application to Earthquakes in Chile

In order to demonstrate the applicability of the proposed resilience planning framework to the real world, particularly the OvS approach presented in Section 5.3.3, the following case study, introduced in [20], is used to identify resilience enhancement decisions among an array of multiple candidate solutions to protect against earthquakes in the Chilean transmission system.

5.5.1 Case Study Description

The Chilean transmission system, which covers more than 3200 km from Arica to Chiloe, is modeled through an equivalent network composed of 40 nodes/substations and 56 transmission corridors (shown in Figure 5.10), representing the infrastructure in 2018. For that year, electricity demand was approximately 76 TWh and generation supply included mainly hydro (23 TWh, 30%), coal (30 TWh, 39%), and gas (11 TWh, 15%) units, with minor participation from wind (4 TWh, 5%) and solar resources (5 TWh, 7%) too. The total installed generation capacity was 24 GW.

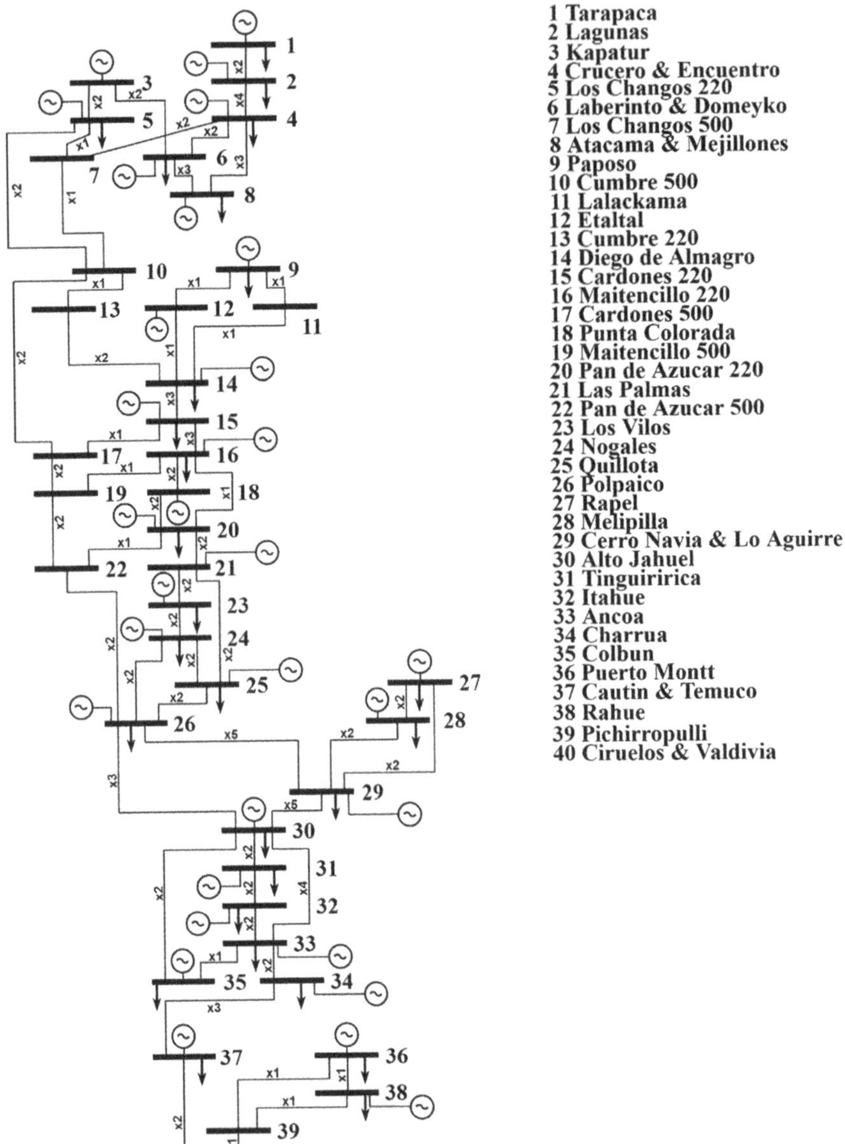

Figure 5.10 Chilean transmission network model used in the case study, indicating topology, generation and demand nodes, and number of circuits per corridor.

For the purposes of modeling the potential failure of system infrastructure during an earthquake, we have used the fragility curves of towers, generation units, and substations adopted by the Federal Emergency Management Agency, USA (Hazus-MH2.1), which relates the probability of failure of these system components with the peak ground acceleration (PGA) at their particular locations.

For the calculation of the PGA in different locations following an earthquake with a given epicenter, we have used validated models that could characterize the strong ground motion attributes observed in the 2010 earthquake in Chile.

We then randomly generated a comprehensive set of scenarios (e.g. 10,000) that follows the sequence:

1) *Random generation of earthquakes and PGA calculation*: With random location and fixed magnitude equal to 8.5 Mw, equalizing the conditions of the latest 2010 earthquake, which is one of the worst earthquakes experienced in Chile, the PGA is determined at the location of each system equipment.
2) *Random generation of network outage*: Once the probabilities of outages are obtained from the fragility curves, outages are simulated through a Monte Carlo process (as explained in Chapter 3).
3) *Random generation of equipment repairs*: Once pieces of equipment fail, they are recovered following a random process (as explained in Chapter 3).

The system dispatch before, during, and after the earthquake was obtained by simulating five days, where the earthquake occurs in the first hour, in order to capture the system collapse as well as the system recovery. The analysis captures key features of a resilient power grid.

5.5.2 Results: Portfolio Solutions for Resilience Enhancement

Figure 5.11 shows the Pareto frontier between the risk measure and the budget used to improve resilience. Here, the risk measure is the Conditional EENS (CEENS), where ENS is averaged across worst scenarios, e.g. all scenarios originated by very large earthquakes with a magnitude of 8.8 Mw (in practice we can assume that CEENS $\approx$ CVaR by an appropriate selection of the value of the parameter α), and the "budget" indicates the number of available investment options allowed as

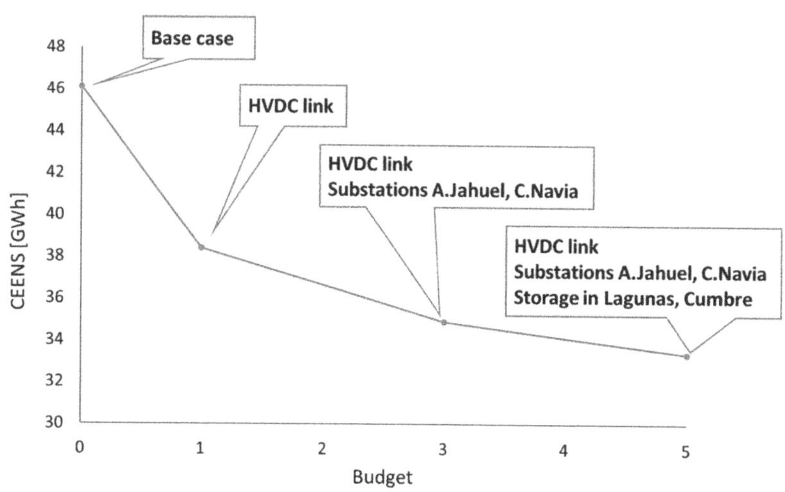

Figure 5.11 Optimal portfolio solutions for resilience enhancement for different budgets.

a proxy for their total cost (e.g. budgets equal to 1, 3, and 5 mean that up to 1, 3, and 5 investment decisions can be made, respectively). The figure shows that the most economical option is to invest in an HVDC link connected point to point between the Atacama desert (node 4) and Santiago (node 29). (This is observed under a budget of 1, reflecting our assumption that each investment proposition carries the same cost, thereby allowing the budget to essentially serve as a count of the actions implemented.) This enhancement option, in fact, features two main characteristics. First, pre- and post-fault power flows are fully dispatchable due to power electronics equipment that allows system operators to have a higher level of controllability of the power flows, co-optimizing them with other operational measures, e.g. generation (re-)dispatch, exercise of reserve services, etc. The second characteristic is the unique topological connection of this candidate cable, which bypasses most substations between the north and the center of the country, reducing in this way its exposure to earthquakes (note that the main impact from earthquakes is on substation equipment rather than towers and transmission lines). In fact, empirical evidence from past events in Chile, strongly suggests that substations (rather than lines) feature the most severe problems during earthquakes. For example, in the 8.8 Mw earthquake that occurred in Chile in 2010, only 3 towers failed, while 12 (out of 46) substations in the high-voltage transmission system showed some level of damage. For this very same reason, the best complements for this HVDC link are the hardening of substations supplying Santiago, i.e. Alto Jahuel and Cerro Navia (budget = 3).

Interestingly, for a budget equal to 5, two storage facilities, each being a pumped-hydro storage unit of 300 MW, are added to the optimal portfolio. In this case, the storage facilities located in Cumbre and Lagunas allow the system operator to more efficiently manage the large amounts of renewable generation located in the north (under any outage conditions, with and without earthquakes), whose production is transferred through the HVDC link to the central area of the country in Santiago, which is in turn supplied by Cerro Navia and Alto Jahuel substations (among others). All of the aforementioned assets are part of the selected portfolio solution, clearly indicating the value of coordinated investment. These five enhancement options, selected as a unified portfolio, thus act as one synergic multi-asset enhancement, providing the best possible insurance to the main system load center against the occurrence of large earthquakes.

5.6 Main Grid Resilience Investments Versus DER

It is envisaged that network reliability and resilience services, which have been historically delivered through asset redundancy, should evolve and hence be provided by emerging and innovative non-network technologies, especially those at the demand side, in order to release latent network capacity and thus make more efficient use of the existing assets. In this vein, advanced technologies for post-contingency control, which can take advantage of a range of DER, can effectively provide security services and thus displace the need for redundant network capacity. The set of DER includes distributed generation (DG), backup generation, an array of storage technologies and demand itself (by utilizing the inherent demand-side flexibility, particularly from non-essential loads and demand associated with the heat and transport sectors).

5.6.1 A Suitable Methodology to Design DER Portfolios Against Wildfires

The following methodology, derived from (5.3)–(5.5) seeks optimal investments in DER portfolios to efficiently hedge a distribution system against HILP events such as wildfires. The methodology, introduced in [18], is based on an optimization model and, importantly, also allows capturing the substitution effect between DER and network investments since it co-optimizes both. In operational timescales, the model determines the optimal operation of the distribution system with DERs, including purchases from the main grid and topology control. Hence, the optimization model can identify optimal preventive and corrective measures to hedge the distribution system against potential outage scenarios originating from wildfires. The model optimizes the following set of decisions:

- *Preventive measures*: Investments in DER equipment such as storage plants, backup generation, and network investments. The model also finds the optimal volume of demand response contracted. These measures are made upfront, pre-contingency, and thus are present in all scenarios.
- *Corrective measures*: These measures depend on the specific contingency and are scenario dependent. We model two types of corrective measures, fast and slow:
  - *Fast*: Refers to the distribution system operation itself, including demand curtailments and a (smart) operation of system assets (topology control and dispatchable DER). These actions can occur right after a contingency occurs.
  - *Slow*: Installing and dispatching mobile DER. These actions feature a lag associated with the arrival of mobile equipment.

The proposed optimization model is probabilistic with the scenario tree considered in Figure 5.12, and minimizes expected costs (including the cost of investment, operation, and ENS), considering risk aversion. Risk aversion can be incorporated by adding to the optimization model either an extra constraint to limit a given risk measure (like in [38]) or an extra term in the

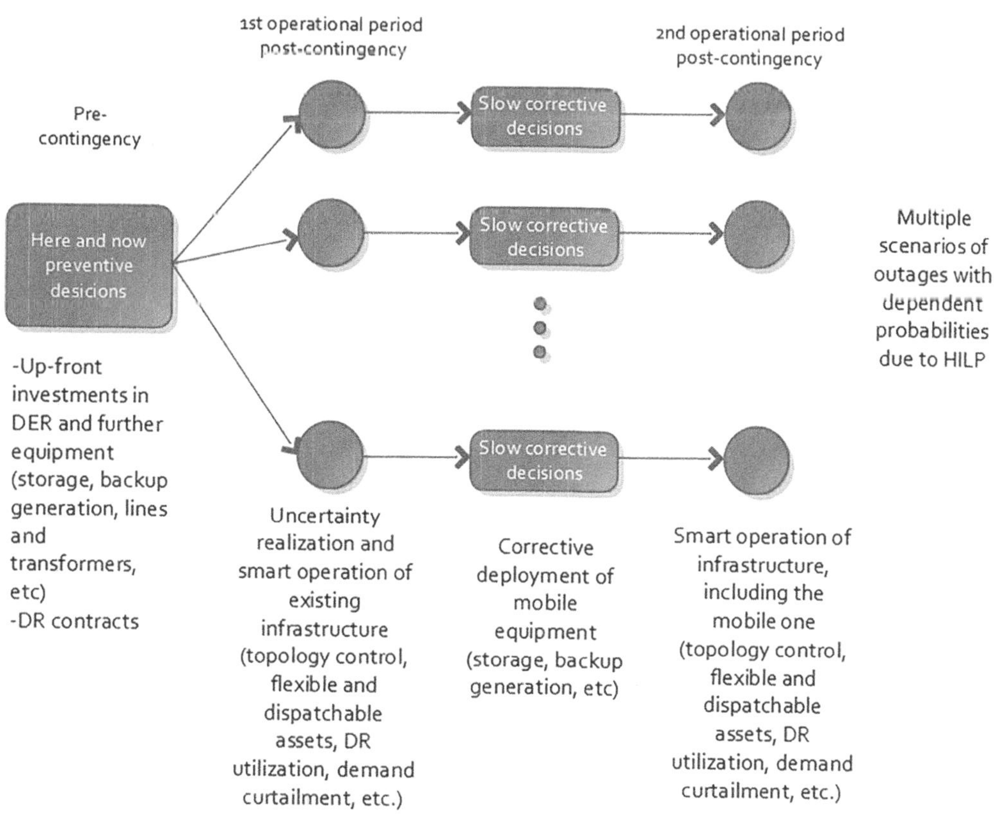

Figure 5.12 Scenario tree of the proposed methodology.

objective function to penalize the given risk measure [39]. The CVaR is a commonly used risk measure that can be straightforwardly incorporated into any optimization model. Importantly, in the event of a wildfire, the probability of simultaneous outages becomes high since a single fire event can affect various pieces of system equipment. Here, ignition probability maps, like those in [18], can inform about the places with the highest risks of wildfires.

5.6.2 Illustrative Case Study Example

The model above was applied to the textbook-like system design example displayed in Figure 5.13 and introduced in [18]. This example was used in the "Review of Distribution Network Security Standards" in the United Kingdom to illustrate, from a fundamental viewpoint, the problems of the current network standards and the potential solutions going forward. (Note that, as the risk of wildfire is not a concern in the UK, the original example focused on the risk of flooding; for detailed information, refer to [40]. In this chapter, we adapted the example to wildfires.) This example seeks to determine, in a greenfield fashion, the optimal system design to supply Areas A and B. The figure shows all candidate assets (i.e. investment propositions) to supply the constant loads in the two distribution networks (25 MW in Area A and 50 MW in Area B). The set of candidate assets includes six power lines and distributed PV and battery systems in Area A. In case of network outages, the choices of renting mobile generation units and exercising DR contracts in Areas A and B as corrective actions could also be considered as part of the system design. Importantly, the failure rates of lines 5 and 6 are affected by the risk of wildfires, which increases the probabilities of failures in that network corridor and originates dependencies of line failure probabilities between lines 5 and 6.

Under normal conditions, the outage rate of lines is 1 occ/yr with a mean time to repair of one day. Under catastrophic wildfire conditions in the neighboring area of lines 5 and 6 (1 occurrence in 10 years), we assume these lines will fail (simultaneously) and that their mean time to repair

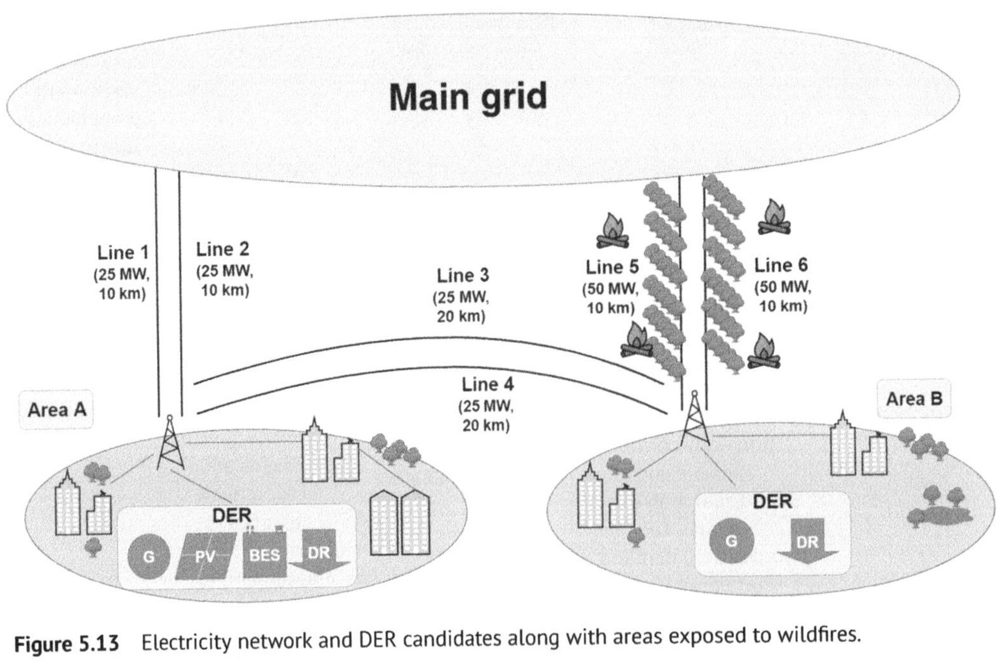

Figure 5.13 Electricity network and DER candidates along with areas exposed to wildfires.

Table 5.5 Scenario probabilities without and with the risk of wildfires. Dependent probabilities have been marginalized (using the fraction of the time to which the network is exposed to a wildfire).

	Probabilities		
	Independent	**Dependent**	**Variation**
Intact system	9.84E−01	9.76E−01	−1%
All N–1 failures	1.62E−02	1.60E−02	−1%
All N–2 failures but L5–6	1.03E−04	1.03E−04	−1%
Double failure of L5–6	7.38E−06	8.16E−03	110,404%
All failures beyond N–2	4.05E−07	4.02E−07	−1%

increases to 30 days. Other relevant input data include an energy price of \$50/MWh (use to buy energy from the main grid), VoLL equal to \$10,000/MWh, and the investment costs of lines, PV, and storage of \$75/MW.km.yr, \$500/kW, and \$200/kWh, respectively.

Regarding the mobile generation units in both areas, the system operator can rent them at an hourly cost of \$68.5/MW and operate them with a \$200/MWh fuel cost. For simplicity, DR features the same costs as those of mobile units. We also assume that the system operator takes an average of 2.4 hours and 3 days, respectively, to install the mobile units under normal and wildfire conditions. Mobile units and DR measures can cope with up to half the power demand in each area.

We consider 22 scenarios, including one intact system (with no outages), six N–1 line outages, and 15 N–2 line outages. The probabilities of these scenarios are calculated assuming independence and dependence under normal and wildfire conditions, respectively. Hence, once a wildfire occurs (every 10 years on average), a common mode failure arises, affecting lines 5 and 6 simultaneously. The probabilities with and without the risk of wildfires are shown in Table 5.5 (using equations (5.1, 5.2), respectively). Notably, the probability of a double contingency affecting lines 5 and 6 increases by more than 110,000% when we assume a wildfire every 10 years (on average).

Note that knowing the actual PDFs of rare events such as wildfires may be difficult. Nevertheless, the probabilistic framework may still be beneficial for deriving good estimates when certain probability values are assumed. Alternatively, other optimization frameworks (like the robust and distributionally robust optimization framework discussed in Sections 5.2.5 and 5.2.6) that do not rely on (exact) probability values may be promising for resilience analysis.

Table 5.6 shows the results found for three cases, namely, Case A, Case A Re-evaluated, and Case B. Case A corresponds to the results when the risk of wildfires in the neighboring area of lines 5 and 6 is neglected (i.e. independent probabilities used in the optimization). Case A Re-evaluated features the same infrastructure as Case A, where the costs of operation and unserved energy have been re-evaluated, including the risk of wildfires (i.e. independent probabilities used in the optimization and dependent probabilities used in the re-evaluation). Case B corresponds to the optimal design when the risk of wildfires is appropriately considered (i.e. dependent rather than independent probabilities have been considered in the optimization).

5.6.3 The Classic *N*–1 Design

Suppose we neglect the risk of wildfires in the neighboring area of lines 5 and 6. In that case, the optimal system design found by the probabilistic model will be the classic N–1 configuration,

Table 5.6 Results with costs in thousand dollars (k$) per year. L, MG, DR, PV, and BES refer to line, mobile generator, demand response, PV panels, and battery energy storage, respectively.

	Case A	Case A (Re-evaluated)	Case B
Assets and measures	L1, L2, L5, L6, MG, DR	L1, L2, L5, L6, MG, DR	L1, L2, L3, L4, L5, PV, BES, MG, DR
PV + BES investment cost	—	—	11,500
Line investment cost	113	113	150
Operational cost	32,850	33,115	21,901
Lost load cost	27	19,665	6
Total cost	32,990	52,893	33,558

installing lines 1 and 2 (25 MW each) and lines 5 and 6 (50 MW each) without installing up-front DER capacity (Case A in Table 5.6). Interestingly, mobile generators and DR in both areas can be used as corrective actions to mitigate the impacts of N–2 line outages and thus supply part of the load.

5.6.4 The Resilient Design

Suppose we consider the risk of wildfires in the area in question. In that case, the optimal system design will include a richer set of DERs, combining up-front investments in PVs and battery systems with corrective actions in the form of mobile generating units and DR (Case B in Table 5.6). Remarkably, the consideration of this richer set of DERs will reduce line investments through the risky network corridor (investment in line 6 is dropped) and increase investments in lines to transfer power between areas A and B (lines 3 and 4). This resilient design solution will limit the lost-load cost to only $6000 per year (which is significantly smaller than the lost-load cost of the other solutions), including the unserved demand during wildfires. Another remarkable result is that the PVs and battery systems, triggered by increasing the risk of wildfire from no occurrence to 1 occurrence in 10 years, are also used to reduce the power imported from the main grid under normal conditions, decreasing operational costs. These results demonstrate the multiple benefits of DERs and the importance of capturing these in the cost–benefit analysis to justify investments.

Note that, in this analysis, CVaR has been omitted, as we have adopted a risk-neutral stochastic approach (functional $f_\varepsilon\{\cdot\}$, introduced in Section 5.3.1, considers purely the expected costs). This indicates that employing a risk-neutral framework can still yield more resilient investment propositions when appropriate probabilities are utilized. These probabilities should accurately reflect the likelihood of simultaneous outages and the dependencies between faults, as well as between faults and external natural hazards.

5.6.5 The Actual Risks Associated with the Current Security Standards

Table 5.6 conveys another lesson. The actual risk to which the system is exposed due to applying the current security standards is significant. Indeed, the expected unsupplied energy cost of Case A Re-evaluated is 700 times higher than the initial evaluation (see differences between Case A and Case A Re-evaluated), increasing the total cost of the system design to almost twice the first estimate. Instead, suppose wildfire risks were appropriately recognized from the beginning to plan the system accordingly (as in Case B). In that case, the increase in total costs of the resilient design will be almost negligible compared with the initial cost estimations of the classical N–1 design (see differences in total costs between Case A and Case B). This result demonstrates the importance of appropriate resilient planning.

5.7 Investment Versus Operational Measures

The application of corrective control and other innovative operational measures based on non-wires technologies can deliver latent network capacity to network users, displacing, totally or in part, the need for investment in network reinforcements as explained in [37]. This is so since the use of new ICT and new flexible network technologies can maximize the utilization of existing electricity infrastructure. These new technologies could reduce network redundancy in providing security of supply and resilience by enabling the application of a range of advanced, technically effective and economically efficient corrective (or post-fault) actions that can release latent network capacity of the existing system.

For example, under adverse weather conditions (when the probability of simultaneous line outages increases), rather than increasing network redundancy (by reducing power transfers or investing in network capacity), higher volumes of generation reserves can be held to mitigate the effects of simultaneous line failures. In fact, based on numerical examples run over the Great Britain (GB) system, we demonstrate that, under an appropriate risk-averse probabilistic framework, the network can be operated more robustly against HILP events and, at the same time, more efficiently than N–2 deterministic solution, typically used in GB. In Table 5.7, risk-averse probabilistic solution suggests higher commitments of reserves (by 200 MW), rather than constraining transfers as suggested by deterministic standards (which drives the need for network investments), as a measure

Table 5.7 N–K deterministic against risk-neutral and risk-averse probabilistic solutions. Constraints cost refers to the cost of congestion. X refers to the expected cost of the unsupplied demand. O refers to the expected cost of operation.

	Constraints cost (£/30 min)	Reserve holding cost (£/30 min)	Reserve and intertrip utilization cost (£/30 min)	X (£/30 min)	O + X (£/30 min)
Deterministic N–1	1597	9,000	702	188	11,487
Probabilistic risk-neutral	1480	9,000	696	154	11,331
Deterministic N–2	3995	9,000	612	7	13,615
Probabilistic risk-averse	2248	10,500	621	4	13,373
	EENS (MWh/h)	**CVaR-ENS (MWh/h)**	**Generation reserves (MW)**	**Scotland-England transfer (MW)**	**Overall network utilization (%)**
Deterministic N–1	1E–02	1999	1200	3949	33
Probabilistic risk-neutral	1E–02	1600	1200	4420	34
Deterministic N–2	5E–04	378	1200	2401	26
Probabilistic risk-averse	2E–04	189	1100	3619	31

to limit impacts of extreme events (e.g. simultaneous multiple outages of both generation and network assets). This is particularly beneficial in the context of increased generation reserves due to growing penetration of variable renewable generation that can, apart from dealing with renewable generation uncertainty, support higher power transfers across the network.

5.7.1 A Comment on Resilience and CVaR Effects on Portfolio Diversification

Generally, it is noted that the greater the resilience level of an investment portfolio, the more diversified it tends to be. This observation is significant as we progress toward a broader array of technologies (a detailed list of technologies can be found at [41]) that facilitate operational measures and DER to complement asset-heavy investments. This also aligns with the principles of risk management and CVaR, which advocate for distributing solutions across a diverse set of options to mitigate the impact of tail risks. For instance, reference [37] demonstrates how incorporating CVaR considerations in the investment planning problem can shift from a solution focused solely on redundancy to a more comprehensive portfolio approach that includes DER and operational measures in the design of a substation.

5.8 Fairness Considerations in Resilience

Network planning methodologies typically employ deterministic N–k security criteria. While the same level of redundancy may be applied uniformly across the network, the resulting reliability and resilience levels can vary significantly in different areas. Furthermore, it can be demonstrated that the majority of outages mostly affect very specific locations in a network. This uneven distribution of reliability and resilience is increasingly concerning policymakers and regulators, and it may affect the acceptability of transmission plans.

In the GB transmission network, for instance, we demonstrate in [37] that the security performance experienced by customers in different locations can vary significantly, e.g. consumers in England will experience more outages than those in Scotland, and hence current planning (deterministic) methodologies do not provide equal reliability and resilience to all network users. Demand shedding profiles in Figure 5.14, which illustrate its uneven distribution across the country, were obtained by running several simulations of outages (including simultaneous outages under both fair and adverse weather conditions) on the 29-busbar representation of the GB transmission network presented in [37].

As an attempt to improve reliability and resilience fairness, further metrics can be considered in the planning methodologies. For example, reference [42] study the impacts of new investment propositions on a Gini coefficient that is calculated based on nodal reliability levels (the Gini coefficient is usually a measure of income inequality within a population; it ranges from 0 to 1, where 0 represents perfect equality, everyone has the same income, and 1 represents perfect inequality, one person has all the income and everyone else has none. Based on this concept, reference [42] proposed using the Gini coefficient to measure reliability and resilience fairness). Expanding on this idea, we have developed a set of constraints to incorporate the Gini coefficient of EENS in the network expansion planning problem. First, we present the inequality ratio r_n defined as (5.14).

$$r_n = \frac{\frac{\mathrm{EENS}_n}{\sum_{i=1}^{N} \mathrm{EENS}_i}}{\frac{\mathrm{Dem}_n}{\sum_{i=1}^{N} \mathrm{Dem}_i}} \tag{5.14}$$

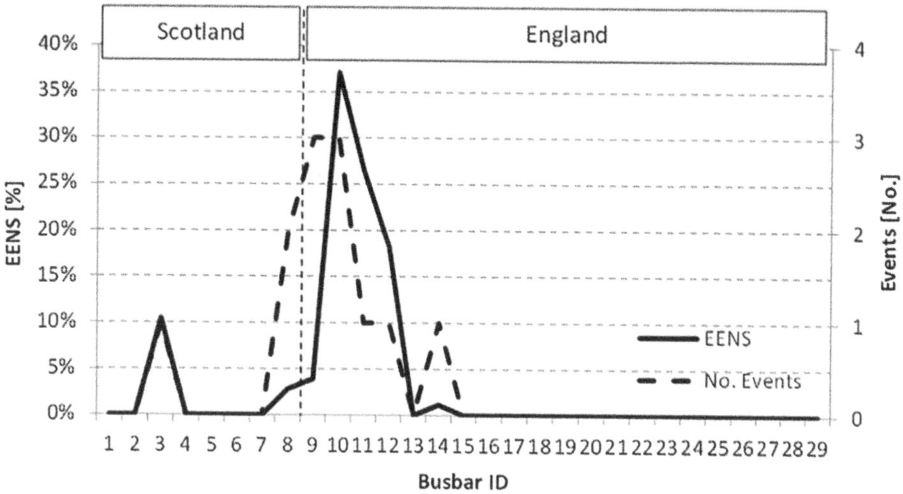

Figure 5.14 EENS distribution across the GB network.

This ratio becomes equal to one if the proportion of EENS of node n (EENS_n) with respect to the system-wide level of EENS ($\sum_{i=1}^{N} \text{EENS}_i$) is the same as the proportion of demand of node n (Dem_n) with respect to the system-wide demand ($\sum_{i=1}^{N} \text{Dem}_i$). Then, we write the mathematical constraint for the Gini coefficient as (5.15), where the left-hand side is the Gini coefficient and the right-hand side is a parameter representing an upper bound that can take values between 0 and 1. Here, a Gini coefficient of 0 represents perfect equality (i.e. everyone has $r_n = 1$), while a Gini coefficient of 1 indicates perfect inequality (i.e. one person has all EENS, and everyone else has none).

$$\frac{\sum_{i=1}^{N} \sum_{j=1}^{N} |r_i - r_j|}{2N \sum_{i=1}^{N} r_i} \leq \hat{G} \tag{5.15}$$

Note that the term $\sum_{i=1}^{N} \text{Dem}_i / \sum_{i=1}^{N} \text{EENS}_i$ is present in the numerator and denominator of (5.14) and thus can be cancelled. Thus, we can re-write the previous expression by using $r'_n = \frac{\text{EENS}_n}{\text{Dem}_n}$, originating (5.16). Note that r_n is non-linear, whereas r'_n is linear.

$$\sum_{i=1}^{N} \sum_{j=1}^{N} \left| r'_i - r'_j \right| \leq 2N\hat{G} \sum_{i=1}^{N} r'_i \tag{5.16}$$

Finally, we eliminate the absolute value of the previous expressions by introducing auxiliary variables $Q_{i,j}$, creating a set of linear constraints that can be introduced in a network expansion planning problem.

$$r'_i - r'_j \leq Q_{i,j}; \forall i \in \{1..N\}, j \in \{1..N\} \tag{5.17}$$

$$r'_i - r'_j \geq -Q_{i,j}; \forall i \in \{1..N\}, j \in \{1..N\} \tag{5.18}$$

$$\sum_{i=1}^{N} \sum_{j=1}^{N} Q_{i,j} \leq 2N\hat{G} \sum_{i=1}^{N} r'_i \tag{5.19}$$

We applied the previously mentioned constraints, incorporating equations (5.17)–(5.19), to a two-stage stochastic network expansion planning problem, specifically a probabilistic security-constrained OPF-based network expansion planning problem. In this case, the costs of

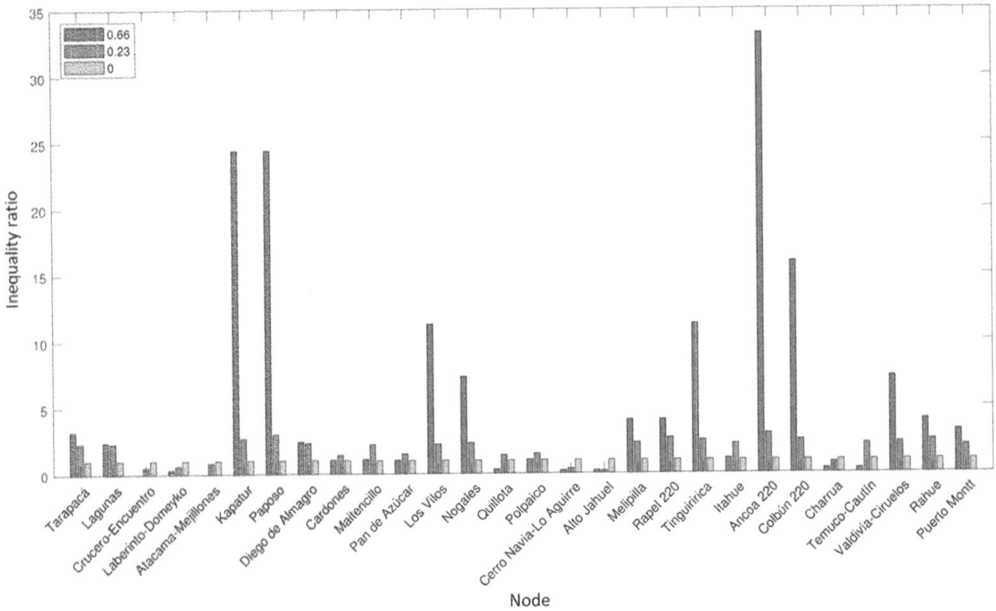

Figure 5.15 Inequality ratio distribution across the Chilean network for three solutions of the investment planning and operation problem with $\hat{\mathbf{G}}$ unbounded (0.66), 0.23, and 0.

new investments are balanced against the benefits, which include cost savings during normal operation and contingencies. The latter savings encompass demand curtailment costs, as detailed in [37]. In various case studies where $\hat{G}$ is adjusted from its unbounded value (typically near 1) down to zero, there is an observed increase in the cost function (due to the presence of more stringent constraints in the network expansion planning problem). This increase is attributable to additional investments and operational measures aimed at enhancing the equitable distribution of EENS across the network. In fact, these extra transmission investments can enable more balanced demand curtailment schemes (where $r_n = 1$ for all n), preventing network overloads post-fault. Figure 5.15 displays the inequality ratios r_n for all nodes within the Chilean network, as presented in Section 5.5.1. These ratios are obtained for $\hat{G}$ values 0.66, 0.23, and 0. Notably, the cost increase in these cases is very marginal, each falling below 1% in terms of the objective function.

5.9 Beyond Networks: Long-Term Duration Energy Storage to Enhance Future Energy System Resilience Against Prolonged Low Output of RES

As more than 80% of electricity demand would be met by renewable energy sources (RES) to achieve net-zero emissions, there is a growing concern about the energy system resilience against prolonged low output of renewables (understood here as HILP events), particularly wind. The availability of wind varies in time, and there will be periods with prolonged low wind output during cold-spell conditions (peak demand periods), which can lead to energy supply scarcity if it is not anticipated in advance [43, 44]. The United Kingdom (UK) meteorological studies [45] demonstrate a correlation between extreme cold conditions with low wind output. The duration of such events can be from several days to several weeks. When cold conditions, which drive up energy demand, particularly for heating, occur with high pressure over a large area (e.g. Northern Europe) for many days, these conditions bring still air with little or no wind. In these extreme events, wind turbines, which will be the backbone of the future energy supply, will produce near-zero output. These conditions were observed in the past: seven days of low wind between December 16 and 22, 2021 and 10 sequential days of low wind in March 2022 across Great Britain [46].

The unavailability of RES during peak demand conditions causes several system challenges that must be solved:

1) Capacity adequacy issue – as wind and solar power are not firm energy sources and their capacity contribution to peak demand is low or none, the system must have a sufficient dispatchable peaking generation capacity to meet the peak demand. A few possible solutions for this issue include the deployment of low-carbon peaking power generation technologies such as combined or open cycle hydrogen gas turbine (CCHGT or OCHGT), electricity storage such as pumped hydro, compressed air, or battery energy storage systems, or reducing the peak of electricity demand using demand side management. A well-designed system should have sufficient dispatchable capacity to meet the peak demand, considering the risk of having prolonged low RES output.
2) Even if the system has sufficient power capacity, alternative energy sources must be available for generating electricity during low RES periods. This requires a substantial volume of energy storage, depending on how long the system needs to deal with low RES outputs.

There are a few promising long-duration storage technologies; they include:

1) Pumped hydro, which moves water from a low to a high reservoir when electricity is cheap and uses the gravitational potential energy stored in the water to rotate the turbine when needed. However, most pumped hydro was designed to have 4–6 hours of storage. Some exceptions exist in regions with mountainous terrain where larger reservoirs (> 100 h) are possible.
2) Compressed air energy storage, which stores pressurized air in liquid and releases the energy stored when needed.
3) Seasonal thermal storage (heat/cold) for several months using underground thermal storage or phase-change material (PCM).
4) Hydrogen storage, which is produced during high RES output and stored in large underground hydrogen storage to be used during low RES output.

However, storing energy incurs losses and costs associated with the investment and operation of energy storage. The bigger the storage requirements, the higher the costs. Finding the optimal portfolio of long-duration energy storage technologies becomes important to minimize the overall system costs.

An illustrative example of the volume of hydrogen reserve needed as a function of the number of weeks with low wind output is shown in Figure 5.16. The analysis considers the efficiency of CCGHT (58%) in producing electricity from hydrogen.

We quantify the range (min–max) of wind power energy that needs to be compensated if a prolonged low wind output event occurs. The volume of reserve increases to deal with longer low-wind periods. The requirements vary in a large range, indicating high variability of wind

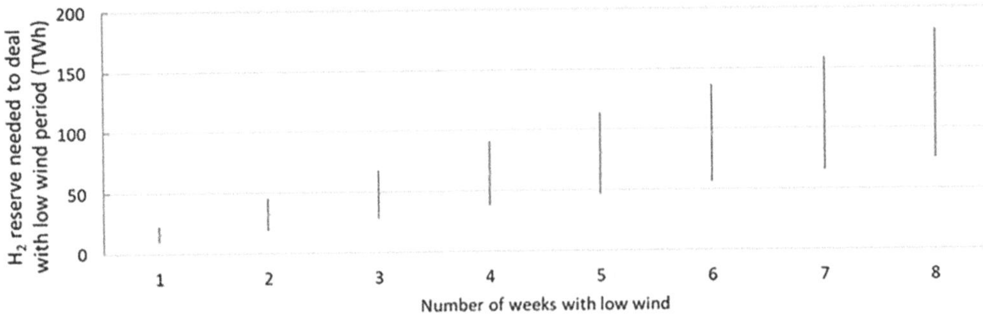

Figure 5.16 Range of hydrogen reserve needed to deal with prolonged low-wind periods.

energy availability. For a week with low wind, the median of hydrogen reserve required is about 10 TWh and increases when the system is designed to cope with a longer period of low wind. To put it into context, in 2022, the natural gas storage in the UK was around 16 TWh. If all storage were converted to hydrogen, the energy system could only deal with less than two weeks of low wind.

If the system needs to deal with three weeks of low wind, the hydrogen reserve needed is around 68 TWh (if we select the highest requirement). The annuitized fixed cost of hydrogen storage is £0.23 million per year per GWh storage [47]. Therefore, the cost of having 68 TWh of hydrogen storage is £15.6 billion/year. If we assume the extreme three weeks of no wind occur once in 10 years, the hydrogen energy that needs to be reserved every year is 6.8 TWh. The annual cost of producing blue hydrogen from methane reforming is relatively low (i.e. £164 million/year[1]) compared to the cost of hydrogen storage. It is assumed that producing blue hydrogen for reserve will not increase hydrogen production capacity, as the reserve can be produced when the capacity is idle.

If 6.8 TWh hydrogen reserve is produced by wind and electrolyzer, the cost is larger. Assuming 80% electrolyzer's efficiency, additional electricity required from wind is 8.5 TWh/year. If the LCoE of offshore wind is £40/MWh, the additional investment cost in wind generation is £340 million/year for 1.62 GW of additional offshore wind capacity (with 60% load factor). Assuming the annualized cost of electrolyzer is around £60/kW per year, the annual cost of additional electrolyzer is £97.2 million/year. The total cost of producing 6.8 TWh of green hydrogen reserve is £437.2 million/year.

It is clear that the cost of storing very large amount of hydrogen is significantly larger (£15.6billion/year) than the cost of producing hydrogen (between £164 million/year and £437.2 million/year).

Further investigations on the impact of climate change and weather variations and their associated risks to the reliability of energy supply are needed to understand better how to develop a low-carbon, cost-efficient, and resilient energy system.

It is important to note that the risk-aware concepts and models introduced in this chapter, initially applied to network design, can also be utilized to determine the optimal portfolio of long-duration energy storage technologies. Our research indicates that risk-averse investment planning models, such as those employing CVaR, can justify larger amounts of energy storage. This increase in storage capacity is to mitigate tail risks associated with unserved energy and spikes in operational costs, which may arise from price shocks in international fossil fuel markets [39, 48].

---

1 Assuming the cost of natural gas £21.6/MWh.

References

1 T. Lagos *et al.*, "Identifying optimal portfolios of resilient network investments against natural hazards, with applications to earthquakes," *IEEE Trans. Power Syst.*, vol. 35, no. 2, pp. 1411–1421, 2020.

2 M. Panteli and P. Mancarella, "The grid: Stronger, bigger, smarter?: Presenting a conceptual framework of power system resilience," *IEEE Power Energy Mag.*, vol. 13, no. 3, pp. 58–66, 2015.

3 Z. Bie, Y. Lin, G. Li, and F. Li, "Battling the extreme: a study on the power system resilience," *Proc. IEEE*, vol. 105, no. 7, pp. 1253–1266, 2017.

4 N. Abi-Samra and W. Henry, "Actions before... and after a flood," *IEEE Power Energy Mag.*, vol. 9, no. 2, pp. 52–58, 2011.

5 B. Chiu, R. Roy, and T. Tran, "Wildfire resiliency: California case for change," *IEEE Power Energy Mag.*, vol. 20, no. 1, pp. 28–37, 2022.

6 J. Araneda, H. Rudnick, S. Mocarquer and P. Miquel, "Lessons from the 2010 Chilean earthquake and its impact on electricity supply," *Proceedings of the 2010 International Conference on Power System Technology (POWERCON)*, Hangzhou, China, 24–28 October 2010. IEEE.

7 Pacific Gas and Electric Company, "Investor PGE Corp," Mar. 2023. [Online]. Available: https://investor.pgecorp.com/news-events/press-releases/press-release-details/2023/PGE-Outlines-Continued-Risk-Informed-Improvements-to-Wildfire-Safety-Efforts-Through-2025/default.aspx#:~:text=PG%26E%27s%20WMP%20outlines%20critical%20layers,repairs%3B%2. (accessed November 2023).

8 R. Moreno, A. Street, J. M. Arroyo, and P. Mancarella, "Planning low-carbon electricity systems under uncertainty considering operational flexibility and smart grid technologies," *Phil. Trans. R. Soc. A*, vol. 375, 20160305, 2017.

9 X.-P. Zhang, C. Rehtanz, and B. Pal, *Flexible AC Transmission Systems: Modelling and Control*. Heidelberg, Germany: Springer, 2012.

10 C.-K. Kim, V. K. Sood, G.-S. Jang, S.-J. Lim, S.-J. Lee, and H. V. D. C. Transmission, *Power Conversion Applications in Power Systems*. Hoboken, NJ: John Wiley & Sons, 2009.

11 V. Madani *et al.*, "IEEE PSRC report on global industry experiences with system integrity protection schemes (SIPS)," *IEEE Trans. Power Delivery*, vol. 25, no. 4, pp. 2143–2155, 2010.

12 M. Zima, M. Larsson, P. Korba, C. Rehtanz, and G. Andersson, "Design aspects for wide-area monitoring and control systems," *Proc. IEEE*, vol. 93, no. 5, pp. 980–996, 2005.

13 K. Zhu, M. Chenine, and L. Nordstrom, "ICT architecture impact on wide area monitoring and control systems' reliability," *IEEE Trans. Power Delivery*, vol. 26, no. 4, pp. 2801–2808, 2011.

14 G. Strbac, "Demand side management: benefits and challenges," *Energy Policy*, vol. 36, no. 12, pp. 4419–4426, 2008.

15 P. Siano, "Demand response and smart grids—a survey," *Renewable Sustainable Energy Rev.*, vol. 30, pp. 461–478, 2014.

16 D. Papadaskalopoulos, G. Strbac, P. Mancarella, M. Aunedi, and V. Stanojevic, "Decentralized participation of flexible demand in electricity markets—part II: application with electric vehicles and heat pump systems," *IEEE Trans. Power Syst.*, vol. 28, no. 4, pp. 3667–3674, 2013.

17 P. Mancarella, "MES (multi-energy systems): an overview of concepts and evaluation models," *Energy*, vol. 65, pp. 1–17, 2014.

18 R. Moreno *et al.*, "Microgrids against wildfires: distributed energy resources enhance system resilience," *IEEE Power Energy Mag.*, vol. 20, no. 1, pp. 78–89, 2022.

19 M. Panteli, R. Moreno, D. Trakas, M. Jamieson, P. Mancarella, G. Strbac and N. Hatziargyriou, "Enhancing the infrastructure and operational resilience of power systems against wildfires,"

ELECTRA CIGRE's Digital Magazine, 2022, 323. [Online]. Available: https://electra.cigre.org/323-august-2022/technology-e2e/enhancing-the-infrastructure-and-operational-resilience-of-power-systems-against-wildfires.html.

20 R. Moreno *et al.*, "From reliability to resilience: planning the grid against the extremes," *IEEE Power Energy Mag.*, vol. 18, pp. 41–53, 2020.

21 R. Billinton and R. N. Allan, *Reliability Evaluation of Power Systems*, 2nd ed. New York, NY: Springer, 1996.

22 R. Rockafellar and S. Uryasev, "Optimization of conditional value-at risk," *J. Risk*, vol. 2, no. 3, pp. 21–41, 2000.

23 J. Barrera, P. Beaupuits, E. Moreno, R. Moreno, and F. D. Muñoz, "Planning resilient networks against natural hazards: understanding the importance of correlated failures and the value of flexible transmission assets," *Electr. Power Syst. Res.*, vol. 197, 107280, 2021.

24 A. Bernstein, D. Bienstock, D. Hay, M. Uzunoglu and G. Zussman, "Power grid vulnerability to geographically correlated failures — analysis and control implications," *IEEE INFOCOM 2014 - IEEE Conference on Computer Communications,* Toronto, ON, Canada, April 27–May 02, 2014, pp. 2634–2642, 2014.

25 I. Dobson, J. Chen, J. Thorp, B. Carreras and D. Newman, "Examining criticality of blackouts in power system models with cascading events," *Proceedings of the 35th Annual Hawaii International Conference on System Sciences,* Big Island, HI, USA, Jan. 10, 2002, vol. 2, p. 63, 2022.

26 M. Papic *et al.*, "Survey of tools for risk assessment of cascading outages," *2011 IEEE Power and Energy Society General Meeting*, Detroit, MI, USA, July 24–28, 2011, pp. 1–9, 2011.

27 A. Villamarin-Jacome, R. Moreno, M. Panteli, M. Noebels, E. A. Martinez-Cesena, and R. Preece, "Optimizing network expansion and battery storage to mitigate cascading outages via detailed system failures simulations," *IEEE Access*, vol. 13, pp. 81462–81473, 2025.

28 J. R. Birge and F. Louveaux, *Introduction to Stochastic Programming*. Springer Science & Business Media, 2011.

29 D. Alvarado, R. Moreno, A. Street, M. Panteli, P. Mancarella, and G. Strbac, "Co-optimizing substation hardening and transmission expansion against earthquakes: a decision-dependent probability approach," *IEEE Trans. Power Sys.*, vol. 38, no. 3, pp. 2058–2070, 2023.

30 D. Alvarado, A. Moreira, R. Moreno, and G. Strbac, "Transmission network investment with distributed energy resources and distributionally robust security," *IEEE Trans. Power Syst.*, vol. 34, no. 6, pp. 5157–5168, 2019.

31 A. Conejo, M. Carrión, and J. Morales, *Decision Making Under Uncertainty in Electricity Markets*. New York, NY: Springer, 2010.

32 D. Bertsimas, D. B. Brown, and C. Caramanis, "Theory and applications of robust optimization," *SIAM*, vol. 53, no. 3, pp. 464–501, 2011.

33 M. Carrion and J. Arroyo, "A computationally efficient mixed-integer linear formulation for the thermal unit commitment problem," *IEEE Trans. Power Syst.*, vol. 21, no. 3, pp. 1371–1378, 2006.

34 A. Moreira, G. Strbac, R. Moreno, A. Street, and I. Konstantelos, "A five-level MILP model for flexible transmission network planning under uncertainty: a min–max regret approach," *IEEE Trans. Power Syst.*, vol. 33, pp. 486–501, 2018.

35 X. Zhang and A. J. Conejo, "Robust transmission expansion planning representing long- and short-term uncertainty," *IEEE Trans. Power Syst.*, vol. 33, no. 2, pp. 1329–1338, 2018.

36 A. Moreira, D. Pozo, A. Street, E. Sauma, and G. Strbac, "Climate-aware generation and transmission expansion planning: a three-stage robust optimization approach," *Eur. J. Oper. Res.*, vol. 295, no. 3, pp. 1099–1118, 2021.

37 G. Strbac, D. Kirschen, and R. Moreno, "Reliability standards for the operation and planning of future electricity networks," *Found. Trends® Electr. Energy Syst.*, vol. 1, pp. 143–219, 2016.

38 A. Inzunza, R. Moreno, A. Bernales, and H. Rudnick, "CVaR constrained planning of renewable generation with consideration of system inertial response, reserve services and demand participation," *Energy Econ.*, vol. 59, pp. 104–117, 2016.

39 A. Inzunza, F. D. Muñoz, and R. Moreno, "Measuring the effects of environmental policies on electricity markets risk," *Energy Econ.*, vol. 102, 105470, 2021.

40 G. Strbac, P. Djapic, R. Moreno, I. Konstantelos, D. Papadaskalopoulos, J. Calvo, D. Pudjianto, S. Tindemans, S. Kairudeen, Y. Yang, H. Karimi, E. Ortega and M. Aunedi, "Review of distribution network security standards – extended report," The Distribution Code of Licensed Distribution Network Operators of Great Britain - DCODE, 2015.

41 L. Meegahapola, P. Mancarella, D. Flynn, and R. Moreno, "Power system stability in the transition to a low carbon grid: a techno-economic perspective on challenges and opportunities," *Wiley Interdiscip. Rev.: Energy Environ.*, vol. 10, no. 5, e399, 2021.

42 E. Heylen, M. Ovaere, S. Proost, G. Deconinck, and D. Van Hertem, "Fairness and inequality in power system reliability: summarizing indices," *Electr. Power Syst. Res.*, vol. 168, pp. 313–323, 2019.

43 L. C. Dawkins, "Weather and climate related sensitivities and risks in a highly renewable UK energy system: a literature review," Met Office, 2019.

44 J. Oswald, M. Raine, and H. Ashraf-Ball, "Will British weather provide reliable electricity?" *Energy Policy*, vol. 36, no. 8, pp. 3212–3225, 2008.

45 National Infrastructure Commission - NIC, "Weather and climate change risks in a highly renewable UK energy system," [Online]. Available: https://www.preventionweb.net/publication/weather-and-climate-change-risks-highly-renewable-uk-energy-system-literature-review. (accessed November 2021).

46 C. Best, "Wind Lulls," Mar. 2022. [Online]. Available: https://clivebest.com/blog/?p=10242.

47 Element Energy, "Hydrogen supply chain evidence base," a report for Department for Business, Energy & Industrial Strategy, 2018.

48 G. Diaz, A. Inzunza, and R. Moreno, "The importance of time resolution, operational flexibility and risk aversion in quantifying the value of energy storage in long-term energy planning studies," *Renewable Sustain. Energy Rev.*, vol. 112, pp. 797–812, 2019.

6

Operational Resilience Planning

Nomenclature

Indices, sets, functions, and symbols

g	Indices for generators.
i, j	Indices for buses.
(i, j)	Index for line from bus i to bus j.
m	Index for monitored lines.
$\tilde{l}$	Index for line interconnector.
o, o'	Indices for outaged lines.
s	Index for scenarios.
t	Index for time intervals.
β, γ	Indices for areas.
C^{UC}	Unit commitment cost function.
C^{OC}	Operational cost function.
$\mathbb{U}$	Feasible unit commitment set.
$\mathbb{W}$	Uncertainty set of damaged lines.
$\mathbb{O}$	Feasible system operation set.
$\mathcal{G}$	Set of generators.
$\mathcal{L}$	Set of transmission lines.
$\mathcal{M}$	Set of monitored lines.
$\mathcal{O}$	Set of line outages.
$\mathcal{S}$	Set of scenarios under consideration.
$\mathcal{T}$	Set of time intervals for the examined period.
,	Symbols for upper and lower bounds.

Parameters

c^G	Generation cost.
D	Composite load-damping factor of area.
H	Total kinetic energy.
M_1	Sufficiently big positive number.
P^L	Load in area.
pr	Scenario probability.

Fundamentals of Power System Resilience: Disruptions by Natural Causes, First Edition.
Mathaios Panteli, Rodrigo Moreno, Dimitris Trakas, Magnus Jamieson, Pierluigi Mancarella, Goran Strbac, and Nikos Hatziargyriou.

X	Reactance of line.
$VoLL$	Value of lost load.
w	Weight of edge.
π	Line failure probability.
π_{thres}	Predefined threshold to consider system degraded states with probability over this threshold.
ρ	Inverse of droop parameter.
$\boldsymbol{PTDF}$	Power transfer distribution factor matrix.
$\boldsymbol{W}$	Weighted adjacency matrix.

Variables

f	Angular frequency.
FC	Flow canceling transaction.
p^{inj}	Active power injection to the network.
k	Binary variable indicating line damage (1 in time interval of line damage, 0 otherwise).
p^{G}	Active power generation.
p^{shed}	Active power load shedding.
pf	Line active power flow.
z	Line status.
ϑ	Bus voltage angle.
$\boldsymbol{h}$	Matrix of unit commitment decision variables.
$\boldsymbol{z}$	Matrix of line damage decision variables.
$\boldsymbol{o}$	Matrix of system operation decision variables.
$\boldsymbol{P}$	Net nodal injected power vector.

6.1 Introduction

This chapter aims to describe the operational measures applied to enhance power system resilience. As mentioned in Chapter 2, operational measures refer to "smart" control-based actions taken to provide the assets with control capability and resources to effectively deal with the emergency as it unfolds [1]. *In the event of an external threat, such as a natural hazard, operational measures should be implemented to prevent or mitigate the impact of the threat on power system. These measures are not able to make the system "bigger" or "stronger"*, as explained in Chapter 5, and thus avert the damages of system components provoked by the external threat, but they can reduce the power outages and prevent a partial or total black out due to the damages. In general, the aim of the operational resilience planning is to minimize load curtailment under external threats, while respecting all system constraints.

In comparison with system secure operation under normal conditions, when usually the $N-k$ criterion is adopted, where k is mainly considered equal to 1 and rarely equal to 2, the operational resilience planning aims to mitigate the impact of several outages, which is not the case under normal conditions. In contrast, dozens to hundreds or even thousands of components may be damaged in the face of an extreme event, significantly degrading the performance of the system and its capacity to meet the demand [1]. These damages are detected within the region hit by the extreme event, during which the network topology and generation capacity are constantly changing due to the damages, and therefore the operational measures should be adapted to this dynamic behavior [2].

The operational resilience planning aims at the proper use of system assets, taking into account the damages to minimize load shedding until the system is fully restored and can be divided into three responses [2, 3]:

- ***Preventive response*** refers to the measures applied to prevent potentially dangerous situations when the predictions warn about an extreme event threatening the normal operation of the system. They are applied a few hours or days before the event hits the system and aim to increase power system robustness to withstand the initial shock, as well as the effectiveness of emergency and restorative responses, under explained. For instance, an appropriate unit commitment (UC) and dispatch of the generators could be determined in the event of an imminent extreme weather, taking into account potential damages. In contrast, it is not possible to predict an earthquake, and therefore, preventive measures are not applied in that case. A proper UC could have a large impact on the effectiveness of emergency response, as committed generators must be dispatched to meet demand, taking into account the continuous failures during the event. In addition, the preparation of the necessary repair crews and their preventive allocation have a significant impact on how fast the system can be restored and thus on the effectiveness of the restorative response.
- ***Emergency response*** refers to the real-time measures applied during the extreme event to mitigate its impact on system operation. As the event unfolds, the operational measures are adjusted according to the actual impact of the event on the power system and the current system state. The strategy comprising emergency measures depends to a large extent on the preventive measures. Generation re-dispatch depends heavily on the UC determined in the preventive response. The implementation of an emergency response depends on the duration of the external shock. An earthquake lasts from seconds to a few minutes. Hence, emergency measures are not applied in the event of an earthquake. On the contrary, an extreme weather event can last for hours, and the implementation of the appropriate emergency measures is critical.
- ***Restorative response*** refers to the measures applied immediately after the end of the extreme event until the system is fully restored. The restorative measures aim for the rapid repair of the damages provoked by the event and the minimization of the impact (e.g. minimization of energy not supplied) in the aftermath. For example, a proper schedule of the repair crews could accelerate the repair of the damages, minimizing the load curtailment. In addition, the schedule and dispatch of the generators should be determined, taking into account the damaged components and the schedule of the repair crews. The development of a proper restorative strategy requires a good knowledge of the damaged components and system state.

Figure 6.1 shows the three responses as a function of time with respect to an extreme event. All responses can be planned before the external event hits the system, given that it has been predicted, and adapted to the prevailing conditions when they applied. For more details on Figure 6.1, see Chapter 2.

The description of the responses makes clear that acquiring adequate situational awareness is particularly critical. The effective preparation of all the responses depends on the prediction of extreme events. An effective assessment of the vulnerable components is useful for determining the preventive response and when it should be implemented, as well as for developing an efficient emergency and restoration response. Moreover, the implementation of the emergency and

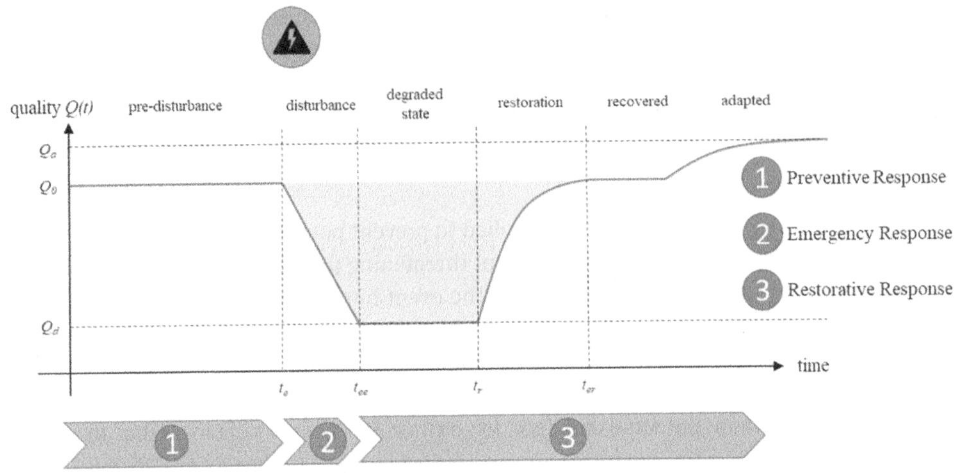

Figure 6.1 Operational planning response as a function of time with respect to the external event.

restorative responses depends heavily on the observability of the system during and after the event. Knowledge of the damages and system state is critical for the proper implementation of these responses. Hence, the information from a situational awareness system supports the effective and timely decision-making in all the phases that a power system might reside in the event of an external threat. An adequate situational awareness system at the early stages of an extreme event enables the effective and fast reaction, preventing cascading outages and averting uncontrollable situations [4]. To achieve this, *the situational awareness system should provide information not only on the system state and the status of power system components, as it is required in normal conditions, but also on the characteristics of the extreme event when it can be predicted or monitored.* The knowledge of these characteristics assists in assessing the spatiotemporal impact of the threat on the system, which helps to project the future behavior of the power system.

For instance, when the intensity and the path of an upcoming extreme weather event can be predicted, using the spatial data of an endangered system, e.g. by a geographic information system (GIS), potential damages can be assessed, and a suitable preventive response can be developed. Additionally, the assessment of the damaged components is vital to proactively allocate the repair crews throughout the region where the power system is extended before the event hits the system in order to derive the optimal repair sequence, initiate the restoration rapidly and minimize the travel time of the repair crews [5]. Also, looking at a wildfire case, once it is detected or reported, its propagation can be assessed and the distance between the wildfire and system components can be estimated during the wildfire's progression. Based on that, the wildfire's impact on them can be assessed to determine the preventive response [6].

Furthermore, real-time measurements of key power system parameters and system estimation are critical for developing an efficient emergency and restorative response. All system constraints should be satisfied during the extreme event and the restoration phase to avoid uncontrollable situations that might lead to partial or total blackouts. The identification of system components' status is also very important to schedule their repair efficiently. To achieve a reliable observability of the power system, the functionality of the hardware and software components applications of the situation awareness system, including telecommunications, needs to be ensured under extreme conditions. Potential failures due to the external event will affect the observability of the system, affecting the effectiveness of the emergency and restorative responses.

6.2 Smart Operational Measures

A list of resilience operational measures and a brief description of each measure are provided below [7]. Later in this chapter, some of these measures will be discussed in more detail.

- Preventive UC
- Adaptive wide-area protection and control schemes (e.g. defensive islanding)
- Network reconfiguration
- Intentional power shut-offs
- Decentralized control
- Advanced and accurate weather forecast
- Advanced visualization and situation awareness systems
- Advanced and adaptive restoration
- Dispatch of the repair crews
- Distributed energy systems (distributed generation and storage)
- Demand side management
- Microgrids

Smart preventive actions, such as preventive UC, and adaptive wide-area protection and control schemes, including defensive islanding, are effective solutions to enhance operational resilience [7, 8]. Protection and control schemes can be critical [9, 10], particularly in situations where the cascading phenomena are developing so quickly that it is impossible for the human operator to react. *Cascading events are the main reason for large blackouts* [11] *and during extreme weather conditions cascading faults may happen more than five times per day* [12]. Preventive UC aims to determine the schedule of the generators so that the proper dispatch of the generation to minimize load shedding despite the multiple line failures can be achieved. In addition, preventive re-dispatch, as well as network reconfiguration, close to the occurrence of the extreme event can reduce the power flow of the vulnerable lines, reducing in that way the consequences of losing a line that transfers a high amount of power and the risk of severe dynamic disturbances that might lead to uncontrollable situations. Intentional power shut-offs are applied during very warm days, especially when combined with high winds, to avoid wildfires sparked by electric power infrastructures. The de-energization of power lines results in the elimination of wildfire ignitions by the power system infrastructure, but also in power outages since the system capacity is reduced. Therefore, the development of a proper power shut-off strategy during extreme conditions is of high importance.

Decentralized control can improve the power system resilience when communication faults or the failure of the central control do not allow the centralized control of the system, or a part of the system does not communicate properly with the central control. As explained earlier, the advanced and accurate weather forecast integrated into an advanced visualization and situation awareness system [4, 13] is critically important for monitoring the evolving system and weather conditions. The development of adequate situation awareness would enable the effective and timely decision-making by the system operator, in order to assess the damage, set priorities, and proceed to the appropriate actions for mitigating the impacts and restoring the system to its pre-event state. Advanced restoration techniques that are adaptive to widespread outages can enhance resilience as an additional operational solution. After a natural disaster, the occurrence of multiple damages is very probable, and thus, repair crews play a critical role during restoration phase. The efficient management and allocation of repair crews reduces the restoration time and customer disconnections, increasing power system resilience.

The rapid installation of distributed energy resources (DERs) provides an additional solution to improve power system resilience. Decentralized energy systems with the large deployment of DERs (mainly distributed generation and energy storage devices (ESS)) [14], in conjunction with demand side management, can play a key role in the resilience boosting efforts. Generating, storing, and controlling energy locally without the need of relying on long transmission lines that might be prone to different threats can make the network less vulnerable, as well as improve the response and restoration times. Also, microgrids have proven to be one of the most effective ways to improve resilience [15]. When the capacity of the upward system is not enough to meet the total demand, microgrids can islanded and supply, in a controlled and coordinated way, all or part of their customers. In addition, microgrids can act as a source, providing power to the upward system, when they have the necessary resources to achieve this. More details on the contribution of microgrids can be found in Chapter 7.

This chapter focuses on preventive response. Emergency response follows a similar approach to preventive response. At each time interval during the event, the emergency response could be updated, taking into account the current degraded system state due to the damages and the prediction of the event progress. As mentioned before, the knowledge of the system state during an extreme event requires adequate situational awareness. Restorative response could be modeled similarly to preventive response, incorporating damage repair modeling that affects the status of the damaged components. Restorative response could be updated at regular intervals based on the progress of the restoration process.

6.3 Preventive Unit Commitment to Enhance Power System Resilience Under Extreme Weather Events

As mentioned in Section 6.2, preventive UC aims to decide the status of the generators to enable the generation dispatch that minimizes the operation cost, taking into account potential component failures due to an imminent extreme event. When the system operational limits and the impact of potential contingencies, such as generator or transmission line outages, are taken into account, the problem is called security-constrained unit commitment (SCUC). While UC, even in its simplest form, is difficult to solve, it becomes more challenging when additional uncertainties are considered, especially when network constraints are also considered. This is the case when a severe weather condition is predicted to hit the grid and cause damages to the transmission network [16].

Stochastic UC (SUC) is an efficient method for addressing uncertainties, such as uncertain damages, in the UC problem. Solving the SUC problem with explicit modeling of these uncertainties, even with simplifying techniques, is challenging [16]. A set of scenarios over the uncertain future is defined and used to model the stochastic nature of damages as a set of discrete futures. To achieve quality results, scenario generation, reduction, and aggregation must be properly performed [17]. Generally, the higher the number of scenarios, the better the quality of the solution. This, however, comes at the cost of heavier computation [18]. Scenario-based SUC has the primary advantage of explicitly modeling the uncertainties into the problem, thus offering a reliable solution [19]. However, its main disadvantage that limits its applications is the demanding calculation time. While modeling new uncertainties into the SUC adds to the complexity of the problem, studies such as [20] claim that by using high-performance computing systems and utilizing parallelization, it is possible to solve the problem within acceptable time for online and offline applications with uncertainties related to generation, transmission failures, and demand. However, given the need for modern power systems to consider more and more uncertainties, it is desirable to develop SUC computational approaches further [21].

Enhancement of modeling the SUC problem is necessary when the scope of the target uncertainties is beyond the operation of the system under normal conditions. An example of such an application is when a severe weather event, such as a hurricane, is predicted to impact the network [5, 18]. Hurricanes can cause tens or even hundreds of transmission outages over the course of their impact, often within a few hours. Although these outages can be predicted in advance, the predictions include uncertainties [22]. While scenario selection for preventive stochastic SCUC (SSCUC) during hurricanes is difficult, solving the problem within a sufficient time can be challenging as well [16].

Despite having different well-known techniques for scenario generation, reduction, and aggregation with applications in power systems, such as [19, 23–25], they may not be fit for extreme cases, such as power system operation during a hurricane. This is due to the probabilistic nature of line failures and that the set of possible network topologies might be extremely large [18]. An efficient method to generate scenarios, called multidimensional scenario selection (MDSS), is introduced in [18], where multiple aspects of information regarding each uncertainty are used to produce the preferred number of scenarios. In [26, 27], an enhanced formulation capable of handling multiple line outages is introduced to model the problem as an SSCUC problem. While a combination of the MDSS and the formulation introduced in [26, 27] delivers a high-quality solution, the overall required time to solve the SSCUC can still be very long for large networks [16].

Robust programming can be used to deal with the uncertainties of component failures while keeping the computational time within acceptable limits. As explained in Chapter 5, robust optimization considers an uncertainty set where the binding worst-case scenarios can be efficiently identified. Robust optimization has been used for solving the SCUC problem in [28–30]. These studies consider the $N-k$ criterion, where the k components are usually lines and/or generators, to deal with the SCUC problems. Conventional $N-k$ criterion ignores the component failure probabilities, which are taken into account in [31–34]. However, other methods, such as adjustable robust, stochastic, distributionally robust, and chance constrained optimization [31–34], can be applied to solve the problem, taking into account the failure probabilities.

Machine learning (ML) algorithms can also be used to reduce the calculation time of the SUC problems. ML techniques have been successfully used for more than a decade to solve SCUC problems in order to determine dynamic security constraints for system stability, e.g. [35], or to solve dynamic security problems, such as in [36, 37]. Refs [38, 39] claim that it is possible to estimate the final solution when planned outages are evaluated by utilizing ML with long offline training time. However, with a lower quality of the final solution in the SUC. In [40], the authors suggest that instead of solving the complete form of SCUC, it is possible to use an ML model to learn about congestion in the network and reduce the unnecessary constraints from the problem. In the mentioned study, it is assumed that the network topology remains unchanged, and the congestion patterns do not change over the study horizon of their data-driven method. In [41], the authors model the UC problem as a Markov decision process and solve it by approximating the results through reinforcement learning [16].

The following sections present two methods to solve the SCUC when the system is threatened by an upcoming extreme weather event and component failures are uncertain. *First, a robust optimization problem characterized by an uncertainty set comprising scenarios of system degraded states due to line failures with a probability exceeding a predefined threshold is presented* (Section 6.3.1). Particularly, only the degraded system states due to line outages with probability over a predefined threshold are considered. Also, the proper selection of this predefined threshold is discussed.

Then, the contribution of ML techniques to the improvement of the solution time of an SSCUC problem without sacrificing accuracy and without the need for historical data of network operation to train the model is presented (Section 6.3.2). In particular, a trained model that predicts the constraints that can be removed from the original problem, thus speeding up the solver, is presented. The model determines the lines that could be congested. These lines are included in the set of the monitored transmission lines and their constraints are included in the SSCUC. On the contrary, the constraints of the lines that are not expected to reach their capacity limits are removed from the problem to speed up the solution. Additionally, the sensitivity of a feasible solution to the errors in predictions by the ML is investigated, when an imperfect ML model is trained and used.

The aim is to examine how these methods can be used to determine operational measures to improve power system resilience and how potential system component failures could be incorporated into the problem to make sound decisions. In both methods, it is assumed that the intensity of the weather event, as well as the region of the system that will be affected by this event, can be forecasted. This is necessary to estimate the line failure probabilities. This estimation can be done using the fragility curves of the vulnerable lines (see Chapters 3 and 4).

6.3.1 Resilience Preventive Unit Commitment Using Robust Optimization

This section presents a resilience SCUC problem that aims to minimize the total cost, taking into account the failure probabilities of the transmission lines due to an upcoming extreme weather. The problem is formulated as a robust optimization problem and the aim is to minimize the total cost under the binding worst-case scenarios with a probability over a predefined threshold [42]. As the worst-case scenarios, the scenarios of transmission line failures that lead to the maximum total cost are considered. During an emergency, the primary goal is to minimize load shedding. Hence, the total cost could only consist of load shedding cost. However, additional operational costs, such as generation cost and UC cost, could also be considered in the total cost. UC cost usually consists of start-up and shut-down costs of generators. When the generation and UC cost are considered, by setting the cost of load curtailment (e.g. Value of Lost Load (*VoLL*)) high enough in relation with the other costs, the minimization of load shedding is prioritized.

6.3.1.1 Mathematical Formulation

The SCUC problem can be formulated as a tri-level two-stage robust optimization problem with the following objective function [42]:

$$\min_{\boldsymbol{h}\in\mathbb{U}} \left\{ C^{UC}(\boldsymbol{h}) + \max_{\boldsymbol{z}\in\mathbb{W}} \min_{\boldsymbol{o}\in\mathbb{O}(\boldsymbol{z},\boldsymbol{h})} C^{O}(\boldsymbol{o}) \right\} \tag{6.1}$$

The objective function (6.1) aims to minimize the UC cost (C^{UC}) and operational costs (C^{O}) under the worst-damage scenarios, while respecting operating limits of the system. $\boldsymbol{h}, \boldsymbol{z}, \boldsymbol{o}$ are the matrices of UC, line damage and system operation decision variables, respectively. The first stage refers to the preventive response (decision of UC), whereas the system operation under the worst-damage scenario is aggregated into the second stage. Operational costs could be expressed as:

$$C^{O}(\boldsymbol{o}) = \sum_{t\in\mathcal{T}}\sum_{i\in\mathcal{B}} VoLL_i p_{i,t}^{shed} + \sum_{t\in\mathcal{T}}\sum_{g\in\mathcal{G}} c_g^G p_{g,t}^G \tag{6.2}$$

The first term of (6.2) expresses the load shedding cost, while the generation cost is expressed by the second term.

The problem is formulated as a tri-level problem, in which the upper level determines the UC, the middle level identifies the worst-damage scenario given the first stage decision, and the lower

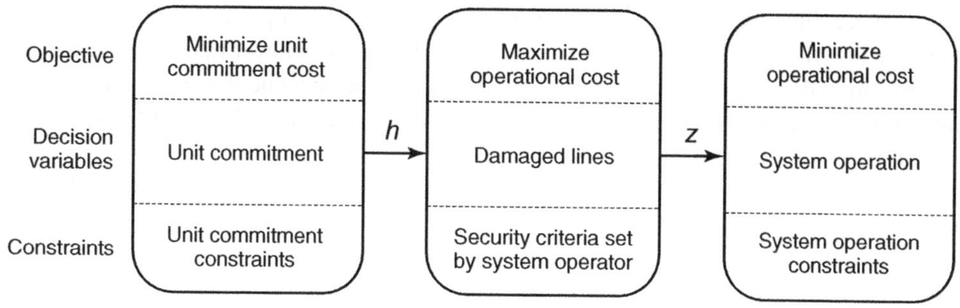

Figure 6.2 Framework of the robust optimization problem.

level determines power system operation given the determined UC (first stage decision) and under the worst-damage scenario. The decisions in the second stage depend on those that have been made in the first stage. In other words, worst-damage scenario and system operation in the second stage depend on the UC determined in the first stage. The framework of the robust problem previously mentioned is shown in Figure 6.2.

Several constraints can be added at each level, which can be expressed mathematically in different ways. Some "typical" constraints that can be included at each level are discussed, while the mathematical expression of some critical constraints is presented.

Some constraints that can be included at the first level are [42]:

- minimum up/down limits of generators to guarantee that a generator will remain on-line or off-line for the required period;
- reserve requirements.

As mentioned before, the second level maximizes the operational cost due to the line damages caused by the extreme weather. Therefore, the constraints of this level are used to determine the status of the transmission lines under the worst damage scenario based on the determined UC in the first level. These constraints can be formulated as follows to take into account the failure probabilities of the lines [42]:

$$\prod_{(i,j)\in\mathcal{L}} \left(\pi_{ij,t}{}^{k_{ij,t}}(1-\pi_{ij,t})^{z_{ij,t}}\right) \geq \pi_{thres}, \forall t \in \mathcal{T} \tag{6.3}$$

$$z_{ij,t} \leq z_{ij,t-1}, \forall (i,j) \in \mathcal{L}, t \in \mathcal{T} \tag{6.4}$$

$$k_{ij,t} = z_{ij,t-1} - z_{ij,t}, \forall (i,j) \in \mathcal{L}, t \in \mathcal{T} \tag{6.5}$$

Assuming that line failures are independent from each other, constraint (6.3) is used to consider the degraded system states due to line outages only with probability over a predefined threshold π_{thres} at each time interval of the examined time horizon. This threshold should be defined by the system operator and determines the security level to be respected. If a line (i, j) is not damaged at time t, only the probability $(1 - \pi_{ij,t})$ holds; otherwise, if the line gets damaged, only the probability $\pi_{ij,t}$ holds. When a line (i, j) is already damaged at time t, the binary variables $k_{ij,t}$ and $z_{ij,t}$ are zero, and therefore none of the probabilities are taken into account in (6.3). This constraint can be linearized by taking the logs of both sides [42]. Constraint (6.4) guarantees that if a line is outaged due to extreme weather, it will remain off-line until the end of the extreme event, as it is assumed that restoration is not possible during the event, due to safety reasons [43]. Using constraint (6.5), the binary variable $k_{ij,t}$ becomes equal to 1, only in the time interval t the line (i, j) is damaged. The value of π_{thres} affects the feasibility of the optimization problem. For example, if, from the set of the possible system states at time t, the probability of each system state is lower than the predefined π_{thres}, the problem will be infeasible. Therefore, there is a maximum value that π_{thres} can obtain. This maximum value depends on the probabilities of the lines to fail at each time interval of the examined horizon [42].

Equations (6.3–6.5) are an example of the constraints that could be added in a robust optimization problem to take into account the failure probabilities. A different mathematical formulation could be used.

The third level minimizes operational cost based on the proposed UC and the line damages determined in the second level and the constraints that should be included at this level are:

- power balance at each bus;
- technical limits of the generators, such as maximum/minimum output and ramp up/down limits;
- power flow equations;
- line/transformer capacity limits;
- load shedding limits;
- voltage magnitude (when AC power flow equations are used) and angles permissible limits.

Regarding load shedding limits, the total load of each bus should be allowed to be shed in order to allow the problem to detect the worst-damage scenarios that lead to the highest load shedding.

Direct current (DC) power flow equations [44], instead of alternating current (AC) ones, could be considered to simplify the problem. In that case, only the active power balance at each bus and the active output limits of the generators are considered.

The set of constraints that allow power flow through the lines, only when they are not damaged, and represent DC power flow equations, can be expressed as follows [42]:

$$z_{ij,t}\underline{pf_{ij}} \leq pf_{ij,t} \leq \overline{pf_{ij}} z_{ij,t}, \forall (i,j) \in \mathcal{L}, t \in \mathcal{T}, \tag{6.6}$$

$$-(1 - z_{ij,t})M_1 \leq pf_{ij,t} - \frac{(\vartheta_i - \vartheta_j)}{X_{ij}} \leq (1 - z_{ij,t})M_1, \forall (i,j) \in \mathcal{L}, t \in \mathcal{T} \tag{6.7}$$

Equations (6.6) and (6.7) link the status of the lines, determined in the second level, with the decision of the system operation in the third level. Constraint (6.6) allows power flow through the line, only when the line is functional. Constraint (6.7) is the big-M representation of DC power flow equations. The equation is converted to equality when the line is functional and relates phase angles to power flow. M_1 is activated when the lines are outaged and is suitably selected to allow the calculation of the phase angle of the buses at the ends of lines set out of order [42].

6.3.1.2 Problem Reformulation and Solution Algorithm

The problem reformulation and the solution algorithm that is used to solve a tri-level problem are described in Appendix 6.6. In short, the problem is formulated into a bi-level problem, considering the dual formulation of the third level. Then, the bi-level problem is decoupled into a master problem and a subproblem. The master problem is a mixed integer linear program (MILP) that selects optimal UC, and the subproblem is an MILP that identifies the line failures that lead to the maximum operational cost [42].

A decomposition method, such as Benders decomposition and Column and Constraint Generation (C&CG), could be used to solve the bi-level problem. Appendix 6.6 describes the C&CG algorithm. For applying C&CG, the subproblem must be bounded and its feasibility must be guaranteed. To this direction, an auxiliary non-negative variable is introduced into the described problem.

This variable is used in power balance constraint and is penalized in the calculation of the operational cost. In that way, the feasibility of the subproblem is guaranteed for any combination of generators and line failures due to extreme weather [31]. The cost of the auxiliary variable should be selected as high as needed compared to the *VoLL*, in order for the solution of the master problem to lead to the zeroing of the auxiliary variable [42].

6.3.1.3 Case Study Applications

The aforementioned resilience SCUC is applied to a modified IEEE Reliability Test System [45]. A simulation period of one day (24 hours) and a time resolution of one hour are considered. A Sequential Monte Carlo simulation is used in order to evaluate its effectiveness. During the Monte Carlo simulation, the day-ahead UC provided by the application of the resilience SCUC is used as input and the problem that minimizes operational costs is solved. Based on the forecasted wind speed, the failure probabilities of the lines at each simulation step are obtained using the fragility curves. The status of the lines at each Monte Carlo simulation is determined based on the comparison of the line failure probabilities with a random number, as described in Section 3.2.2. It is noted that π_{thres} in (6.3) is used in order to allow the operator to consider in the optimization problem only the system degrade states with probability higher than a predefined threshold. This does not imply that during Monte Carlo simulation failures with lower probability than π_{thres} do not occur, but the line failures considered are obtained only from the comparison between their failure probabilities, obtained by fragility curves, and the randomly generated numbers [42].

The selection of π_{thres} is a trade-off between the security criteria to be fulfilled and the cost for meeting these criteria and can be found through a cost-benefit analysis. A reasonable selection of π_{thres} is the higher value that leads to minimal load shedding. Lower π_{thres} values would lead to similar load shedding, but at higher cost. Considering all potential system degraded states, i.e. including the ones with very low probability, would require a very high computational time and lead to a very conservative solution without contributing to load shedding minimization. Different values of threshold π_{thres} are tested in this case study [42].

The Monte Carlo simulation is also run for a traditional constrained UC (TCUC), without considering the impact of the weather event on the transmission lines, and a conservative constrained UC (CCUC) that considers all the affected lines off-line during the whole examined period [42].

The IEEE Reliability Test System and the area affected by the extreme weather are shown in Figure 6.3 [42]. In the rest of the system, it is assumed that wind speeds are not high, and the normal failure probability of these lines is taken as 0.01%. The affected lines are B12–B23, B13–B23, B14–B16, B15–B16, and the double circuit B15–B21. It is noted that in case all the lines affected by the weather are damaged, the system is separated into the northern and southern subsystems. Similar generators at a bus are represented by their single machine equivalent. The system is composed of 14 generators, 24 buses, and 38 transmission lines. It is assumed that at t=0, all generators have satisfied their minimum up and down times constraints, so they can be readily switched on or off. Lines and towers across the system are considered identical and therefore the same fragility curves, shown in Figure 3.5 [42] for the base case scenario, are used for all the components. The length of the lines has been taken directly from [45]. It is assumed that for each line, there is one tower per 300m connected in series [46]. All lines are overhead, except for those between B1–B2 and B6–B10 which are underground cables. The underground cables are not affected by weather and thus, they are not considered in the impact modeling of the weather event. The weather characteristics are assumed homogeneous within the affected area, and therefore, the system components within this area are exposed to the same weather conditions. Figure 6.4 [42] presents the hourly wind speed profile in the affected area. The maximum wind speed is 45 m/s. Even if the wind is not forecasted

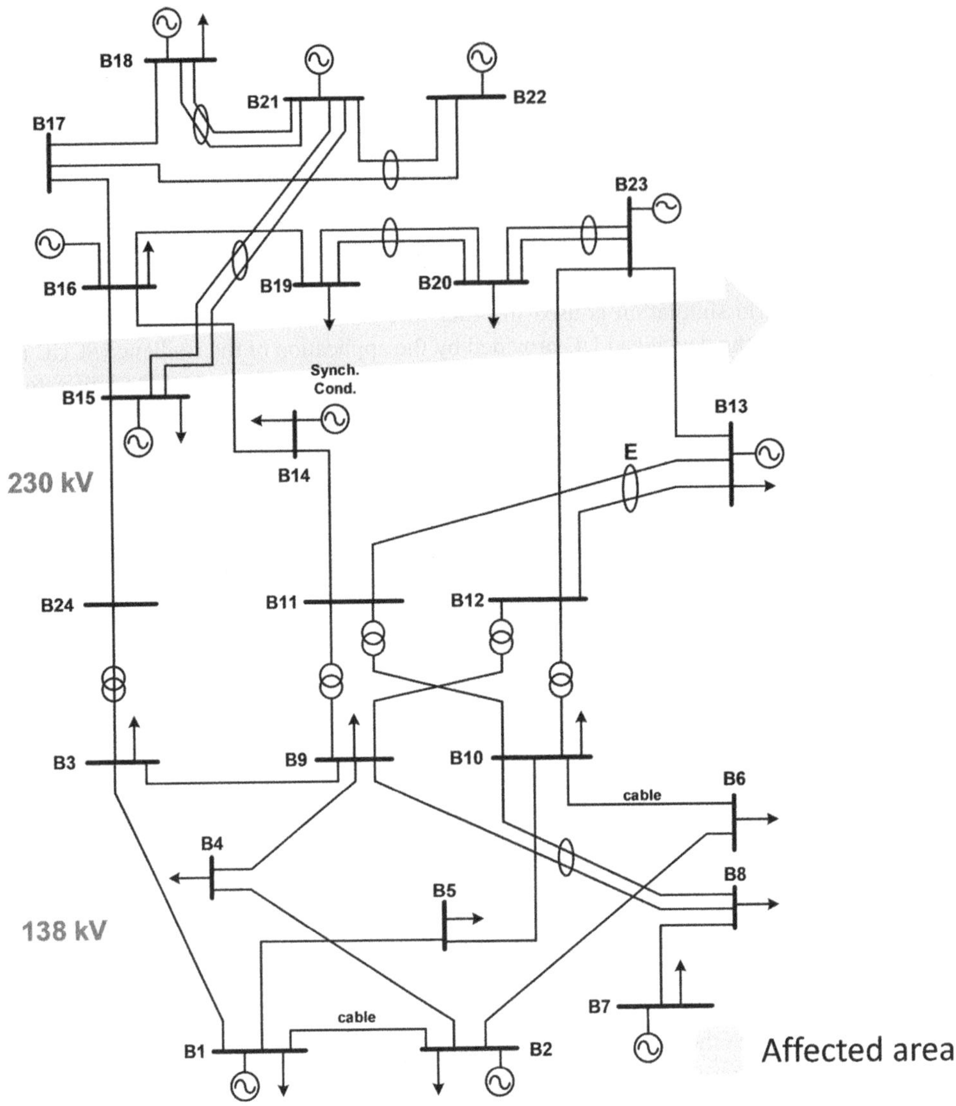

Figure 6.3 IEEE Reliability Test System and the affected area. *Source:* [42] / IEEE.

with high accuracy, in the context of robust optimization, the maximum wind forecasting error is used to consider the worst prevailing conditions. The hourly load demand is adopted from [47]. The cost of the auxiliary variable and the value of lost load, *VoLL*, are set at 7000 $/MWh and 5000 $/MWh, respectively [42].

In order to verify the effectiveness of the resilience SCUC, it is applied for the wind profile shown in Figure 6.4 and the selection of the predefined threshold π_{thres} is investigated. The results of the Monte Carlo simulation are presented in Table 6.1 for π_{thres} equal to 0.1%, 0.5%, and from 1% to 7% with a step of 2%. Results for π_{thres} equal to 7.5% are also presented, since this was found to be the maximum value that results in a feasible optimization problem (using a step of 0.1%). The penultimate row of Table 6.1 shows the results of CCUC, while the last row presents the results of

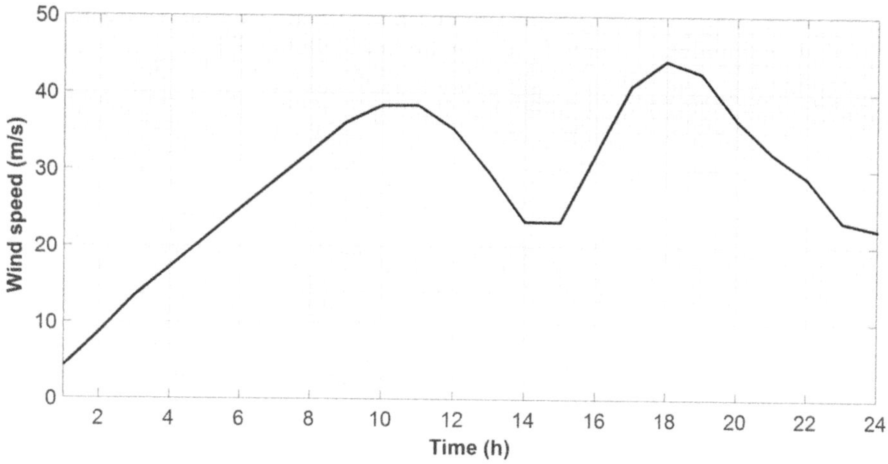

Figure 6.4 Wind speed profile in the affected area. *Source:* [42] / IEEE.

Table 6.1 Monte Carlo simulation results.

π_{thres} (%)	Load shedding (MW)	MC iterations with load shedding (%)	UC cost ($\$\times10^3$)	Total cost ($\$\times10^5$)
0.1	0	0.00	6.10	3.521
0.5	0	0.00	6.10	3.479
1	0	0.00	6.10	3.477
3	0.45	1.00	6.10	3.331
5	1.02	4.32	6.10	3.337
7	155.34	32.23	6.10	10.920
7.5	207.70	41.86	6.10	13.525
CCUC	0	0	6.10	4.315
TCUC	5, 150.30	100	0.60	259.611

Source: [42] / IEEE.

TCUC. The total cost shown consists of UC, generation, and load shedding costs. The third column shows the percentage of Monte Carlo iterations, where load shedding is detected.

The contribution of the resilience SCUC to the power system is revealed by comparing the amount of the load that is shed to keep the balance between generation and demand, due to the line failures. It is seen that the solution of TCUC leads to significantly higher load shedding during Monte Carlo simulation (at least 24 times), while the application of the resilience UC achieves much lower load shedding and therefore significant resilience enhancement. The high difference in load shedding results from the separation of the system due to line failures in the affected area. It is noted that dynamic frequency response (FR) constraints in case of system separation are not considered [42]. For more details on this, interested readers can go to Section 6.5.

Figure 6.5 [42] shows the optimal schedule of generators for the TCUC. It also presents the specific subsystem where the generators are located, in case of system separation. The dots and the circles denote the on and off status of the generators, respectively. It is observed that the power

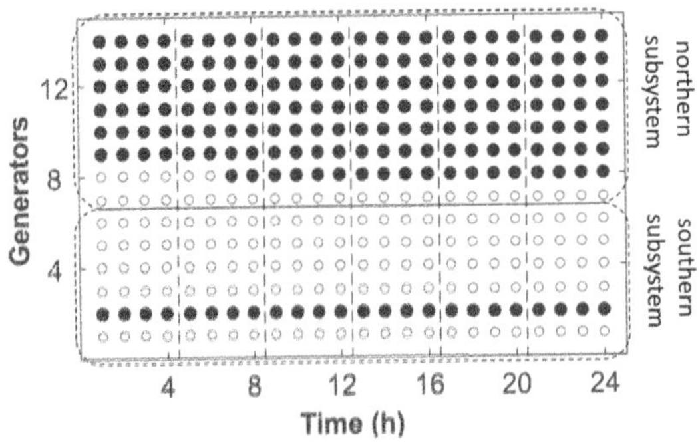

Figure 6.5 Optimal scheduling for traditional constrained UC (TCUC) [42].

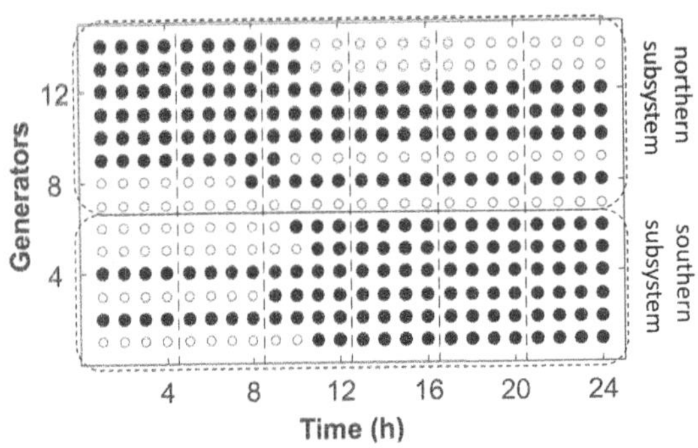

Figure 6.6 Optimal scheduling for π_{thres} equal to 1%. *Source:* [42] / IEEE.

system demand is satisfied mainly by the generators in the northern part of the system. The generation cost of these generators is lower, and their capacity is high enough to meet the demand. The line failures during the extreme weather event lead to the separation of the system, and the demand of the southern subsystem cannot be satisfied by the committed generators of the northern subsystem. As a result, a high amount of load is shed to keep the balance in the southern subsystem, and therefore, the total cost increases significantly [42].

Furthermore, it is observed that for π_{thres} values higher than 1%, as the value of the predefined threshold π_{thres} increases, load shedding increases too. This occurs because as π_{thres} increases, a lower number of line failures is taken into account in the optimization problem. As can be seen from Table 6.1, the UC cost is similar for any of the examined values of π_{thres}. The reason is because the same generators are committed for all the examined values of π_{thres}, but as π_{thres} decreases, the generators start up earlier. This is shown in Figures 6.6 and 6.7 [42] which present the optimal schedule of generators for π_{thres} equal to 1% and 7.5%, respectively. For π_{thres} equal to 1% and 7.5%, the worst-damage scenario based on π_{thres} value causes separation of the system at 11 and 18 hours, respectively. As a result, for π_{thres} equal to 1%, the model commits all the generators located at the southern part of the system before 11 hours, in order to meet the demand of the southern subsystem after its separation. For π_{thres} equal to 7.5%, most of the generators of the southern part start up at 18 hours. Due to this difference, during the Monte Carlo simulation, the generator schedule, provided for π_{thres} equal to 1%, lowers more effectively the impact of the weather event and as a result, load shedding and the total cost of the system decrease. Furthermore, comparing Figure 6.5 with Figures 6.6 and 6.7 [42], it is observed that when the impact of the weather event is considered, a much higher number of generators are committed in the southern part of the system in order to satisfy its demand after system separation.

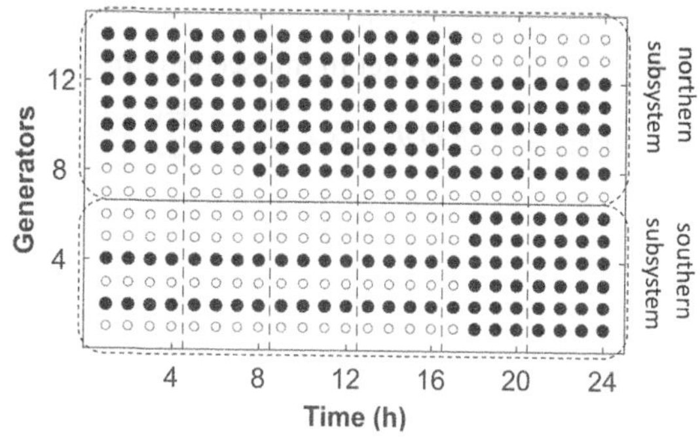

Figure 6.7 Optimal scheduling for π_{thres} equal to 7.5%. *Source:* [42] / IEEE.

By comparing the results of Table 6.1, it is observed that for π_{thres} equal to 3%, the total cost has the lowest value. Load shedding is also low, which means that during the Monte Carlo simulation, in a small percentage of iterations, the proposed generator schedule is not able to satisfy the demand. Since resilience is directly associated with minimum load shedding [43], [46], the optimal value of the π_{thres} can be selected equal to 1%, 0.5%, or 0.1%. As mentioned before, a reasonable π_{thres} selection would be 1%, since this is the higher π_{thres} value that leads to the minimum load shedding and therefore lower total cost compared to lower π_{thres} values. Nevertheless, a lower π_{thres} value equal to 0.1% or 0.5% increases slightly the total cost, providing a higher level of security and could be equally selected [42].

In addition, it is observed that load shedding is zero for the CCUC. However, the total cost is 19.4% higher compared to the total cost for π_{thres} equal to 1%, where the load shedding is also zero. This difference results from the later commitment of the expensive generators located in the southern part of the system during the weather event, obtained by the examined method.

6.3.2 Machine Learning Techniques to Deal with Multiple Line Failures Under Extreme Events

This section presents how ML techniques can be used to solve a resilience SSCUC that aims to minimize the expected value of the objective function, usually the operational cost, subject to physical and security constraints. As explained before, the primary goal of the SCUC during emergencies is to minimize load shedding due to potential damages provoked by external threats. However, the generation and UC costs could be added to the objective function.

6.3.2.1 Mathematical Formulation

The objective function of the SSCUC can be defined as:

$$\min \sum_{s \in S} pr_s \left(C_s^{UC} + C_s^{O}\right) \tag{6.8}$$

Compared to (6.1), the objective function of the SSCUC takes into account the probability, pr_s, of each scenario under consideration, $s \in S$. Additionally, since a subset of scenarios with their corresponding probabilities is considered, the goal is to find the optimal solution that minimizes the expected cost and not the optimal solution that minimizes the cost under the worst-damage scenarios as in (6.1).

Several constraints, as the ones mentioned for the first and third levels of the robust optimization presented before, could be included in the SSCUC. Considering DC power flow, the set of constraints that can be used to calculate the power flow for the monitored lines of the set $\mathcal{M}_{(s)}$ (set $\mathcal{M}_{(s)}$ is discussed later in this section) are the following [16]:

$$pf_{(s,m,t)} = \left(\boldsymbol{PTDF}_{(m)} \times \boldsymbol{P}_{(s,t)}\right) + \sum_{o \in \mathcal{O}_{(s,t)}} \left[\left(PTDF_{(m,frm_{(o)})} - PTDF_{(m,to_{(o)})}\right) FC_{(s,t,o)}\right],$$

$$\forall s \in \mathcal{S}, t \in \mathcal{T}, m \in \mathcal{M}_{(s)} \tag{6.9}$$

$$\left(\boldsymbol{PTDF}_{(o)} \times \boldsymbol{P}_{(s,t)}\right) - FC_{(s,t,o)} + \sum_{o' \in \mathcal{O}_{(s,t)}} \left[\left(PTDF_{(o,frm_{(o')})} - PTDF_{(o,to_{(o')})}\right) FC_{(s,t,o')}\right]$$

$$= 0, \forall s \in \mathcal{S}, t \in \mathcal{T}, o \in \mathcal{O}_{(s,t)} \tag{6.10}$$

For the set of line outages, noted by $\mathcal{O}$, (6.9) and (6.10), calculate the power flow for monitored lines of the set $\mathcal{M}_{(s)}$, taking into account the effects of line outages. In (6.9), the first term on the right-side of the equation calculates line power flow concerning the normal network (without line outages) by using the power transfer distribution factor (PTDF) concept [48], while the second term, which is based on flow-canceling transaction concepts [27], considers the effect of outages of lines in $\mathcal{O}(s, t)$ on the power flow of each line in $\mathcal{M}_{(s)}$. $FC(s, t, o)$ in (6.9) is an imaginary power flow variable in flow-canceling transaction method and its calculation depends on the operation point of the system [16]. Note that, at the same time, the solution of the SSCUC problem depends on the effect of line outages through variables FC, while for the calculation of each FC the system operation point should be known. Hence, it is not possible to calculate the $FC(s, t, o)$ outside the SSCUC problem. To overcome this challenge, a system of equations and unknowns is added to the SSCUC as a constraint, which is presented by (6.10). In (6.10), the variables FC depend on each other through arrays of $\boldsymbol{PTDF}$ matrix. It is noted that this approach is valid as long as no islanding results from the outages. Readers are referred to [26] for the complete formulation and algorithm description [16].

Note that, while $\mathcal{M}_{(s)}$ can include all lines, it can be any subset of lines that are selected to be monitored. To be sure that the results stay accurate, the constraints of the removed lines from $\mathcal{M}_{(s)}$ must be inactive or redundant (e.g., the constraints of the lines that are not expected to reach their capacity limits) [16]. Removing inactive and redundant constraints can also lead to the removal of quasi-active constraints. The quasi-active constraints refer to those not-binding constraints when they are included in the problem, but removing them will affect the results as their corresponding variables violate their limits. For instance, the capacity of a line may not be fully used when the relevant constraints of this line are included in the problem, but removing them would lead to congestion of that line. Therefore, quasi-active constraints should be included in the problem [16].

When the full set of equations is used for large-scale networks with multiple outages, the number of variables and constraints becomes extremely large. Notably, (6.9) and (6.10) are the main equations responsible for growing constraints in numbers and complexity, as they combine standard power flow calculations considering the effects of outages. In some studies, such as in [40], the authors suggest combining (6.9) and (6.10) into one simpler equation based on detecting parallel lines and lines with power flows within their acceptable limits. However, these techniques consider only one outage and are not applicable for multiple line outages [16].

A promising fact in solving the SSCUC problem is that, while the original problem includes many variables and constraints, not all variables must be calculated within the optimization process, nor all the constraints reach their limits. For example, constraints (6.9) and (6.10), as well as the relevant variables, of the lines whose capacity is not fully utilized, could be removed from the problem, speeding up the computational time without affecting the final accuracy. This suggests that if it is possible to distinguish between the essential variables and constraints to be included in the optimization problem and those that are not, the problem can be solved faster. It is worth mentioning that variables and constraints excluded from the optimization problem can be calculated outside the optimizer to verify the complete set of results (post-processing step) [16]. In that post-processing step, the quasi-active constraints can be detected as described in Section 6.3.2.2.

The distinction between the essential variables and constraints to be included in the optimization problem and those that are not, with the aim of reducing the calculation time and/or hardware requirements, could be done using ML models. The ML models could be used as an assistant to SSCUC to facilitate the solution process and not to predict the optimal solution of SSCUC as it is unlikely to train an ML model to perfectly predict the optimal solution of SSCUC. In that case, the objective is to use a trained ML to predict the constraints that are not binding and can be removed from the model without impacting the final solution to solve the case accurately at increased speed [16].

6.3.2.2 Machine Learning Assisted Stochastic Unit Commitment

The candidate inputs and outputs of such an ML model should be selected judiciously. In such a model, data expressing uncertainties are the main input features used to train ML models. In SUC, uncertainties are represented by scenarios [16]. Therefore, the input features should be scenarios or data derived from them. Note that the part of the data that is the same for all scenarios carries no useful information for predictions.

The main uncertainty in case of an extreme event is the status of the vulnerable lines. Therefore, the status of these lines could be considered as input to the ML model. For example, if 100 lines out of 597 lines in a network are considered vulnerable to extreme weather, some of these 100 lines will have a nonzero failure probability in different conditions. If any of these conditions are modeled in the SUC problem, the only difference between scenarios should be related to those 100 lines and not all the 597 lines.

Regarding the outputs of the trained ML, in order to predict the non-binding constraints that can be removed from the SUC problem, it is possible to define a set of fictional binary variables that determine whether each constraint (such as line flow limits) is binding (related binary variable is equal to 1) or not (related binary variable is equal to 0). These binary variables can be used as outputs for the trained ML model(s) [16]. It is worth mentioning that when working with binary variables, the supervised ML classification seems to be a good fit, as binary variables can be translated into two classes, 0 and 1.

As mentioned before, if the trained ML can predict constraints that are not binding, those constraints can be removed from the model without impacting the final solution. Equations (6.9) and (6.10) enforce constraints to the problem that are more complex than others. Hence, a trained ML that removes unnecessary constraints in these equations can reduce calculation time of the SSCUC. Relevant constraints of the lines that are not monitored, such as the one about the capacity of the lines, should also be removed [16]. Note that, as mentioned in [40], the final solution must be checked for quasi-active constraints or binding constraints removed from the problem.

In the following, it is assumed that a trained ML model can predict which line violates its thermal limit, represented by $\mathcal{M}$ in the formulation. The feasibility study is done in the following steps [16]:

1) SSCUC is solved with only suspected lines (lines that violate their thermal limit according to the trained ML model) included in (6.9) and (6.10). The effect of the accuracy of these suspected lines is discussed later.
2) Using results from step 1, power flows are calculated, and all line flows are compared with their corresponding limits to find if any violation exists. Note that violations also may include quasi-active constraints that may have been removed from the subset of lines.

3) In case of violation, lines with flows exceeding their thermal limits are added to $\mathcal{M}$ (constraints (6.9) to (6.10) related to these lines, as well as other relevant constraints such as the one about line capacity limits, are added to the optimization problem).
4) Steps 1 to 3 are executed iteratively until there are no lines with flows exceeding their thermal limits.

6.3.2.3 Case Study Applications

A synthetic grid on the footprint of South Carolina with 500 buses, 597 lines, and 90 generation units is used as the test case to examine the impact of the accuracy of the ML output on the SUC solution. The complete system information can be found in [47], [48]. A daily load profile, as in [49], is used.

It is first evaluated whether a trained ML model that is able to predict the suspected lines can assist in solving the SSCUC faster and how the accuracy of this model affects the computational time and the accuracy of the solution. It is assumed that the set of suspected lines can be perfectly predicted. To consider different accuracy levels, random lines are added to or removed from this perfectly predicted set. Table 6.2 [16] presents the average times over ten simulations for each error in the prediction accuracy of the suspected lines. FN is the abbreviation for False-Negative (FN) and means that lines that should have been predicted as suspected to violate their limits have not been predicted. The column "without FN" shows the results when only FP (False-Positive) errors are considered, which means that the lines that are not suspected have been predicted as suspected and the relevant constraints have been included in the SSCUC. The column "with FN" shows the results when both, FN and FP errors are considered. In cases with FN, the SUC model should be solved at least one additional time, as at least one line violates its limitation in the first iteration [16].

According to Table 6.2 [16], predicting suspected lines with good accuracy can save significant computational time, while the quality of the final solution is the same as the original solution. Thus, predicting lines suspected to violate their limits seems to be an appropriate candidate output of the trained ML. In general, larger errors in the "without FN" column lead to higher solution times since a higher number of constraints is considered, making the problem more difficult to solve and larger errors in the "with FN" column means that a larger number of suspected lines is not included in the set of monitored lines and the SUC should be solved at least twice. Comparing the two columns of Table 6.2, it is observed that the ML training process should be tuned to prevent

Table 6.2 Solution time of SUC considering different levels of error in the prediction of suspected lines.

Case	Solution time (s)	
	without FN	with FN
Original SUC	169	
Assisted, suspected lines with 0.0% error	8	
Assisted, suspected lines with 10% error	25	48
Assisted, suspected lines with 25% error	51	98
Assisted, suspected lines with 50% error	87	187
Assisted, suspected lines with 60% error	104	197
Assisted, suspected lines with 75% error	134	256
Assisted, suspected lines with 100% error	308	

Source: [16] / IEEE.

FN predictions even at the cost of more FP errors. As an example, in Table 6.2, 25% of error with FN is worse than 50% error without FN [16].

Having verified that a trained ML model, which is able to predict the suspected lines, can assist in solving the SSCUC faster, the proper training of this model is discussed. To train any ML model, a set of observations should be created. In this study, each observation represents inputs and corresponding outputs related to a possible hurricane that affects the network, for which temporal line failure probabilities can be calculated. Each observation can be defined as a multi-scenario SSCUC problem. However, it seems impractical and inefficient to train ML models based on multi-scenario SUC. This is because the number of engaged variables increases with the number of scenarios, translating into a larger number of required observations. Moreover, solving each observation takes a long time. Also, if an ML model is trained for a specific number of scenarios, it might not be effective for SSCUC with any other number of scenarios in its original form. Therefore, the method to train the ML model should be practical for SSCUC with any number of scenarios [16].

When multiple scenarios are created representing a single hurricane, scenarios represent different possible damage levels provoked by that hurricane. One scenario represents the worst-possible case (worst-case scenario). The level of damage modeled within all the other scenarios is between no damage to maximum possible damage, represented by the worst-case scenario. While scenarios represent different levels of damage by a single hurricane, it is possible to assume that each scenario represents maximum damage by a different hurricane. For example, the worst-case scenario of one hurricane can be the second-worst or n^{th}-worst possible scenario of another hurricane of higher intensity [16]. With this assumption, it is possible to train ML models based on single-scenario problems, representing the worst-case of each observation, and later use the trained models to predict suspected lines for each scenario of the SSCUC to be solved individually. This is also consistent with the provided formulation, as each scenario has its own set of suspected lines that violate their capacities.

To create a set of observations, the following steps could be implemented [16]:

1) *Determining worst-case scenario:* For each hurricane (observation), create the worst-case scenario, i.e. the one with the highest possible number of failures. The solution of a deterministic SCUC for this worst-case scenario will identify the lines that reach their capacity.
2) *Determining Input:* For the selected scenario in step 1, any failed element at any time is marked. The input corresponding to the observation is a vector with entries set as 1 if the element is marked, and 0 if not. The size of the input vector is equal to the number of vulnerable elements.
3) *Determining the corresponding output:* Solve the SCUC problem with the worst-case scenario from step 1, and mark those lines that reach their capacity at any time of the examined horizon. The output corresponding to the observation is a vector with a size equal to the total number of lines. In this vector, any line marked as congested will be assigned a value of 1, and 0 if not marked. While this procedure removes the dependency of output on time (i.e. load profile) as a capacity violation of any line at any time triggers the corresponding variable, it will automatically consider the worst possible condition (in terms of lines reaching their capacity) and reduce the chance of FN error.
4) Repeat steps 1 to 3 for all observations. Each observation includes input and its corresponding output vectors.
5) *Size-reduction:* As mentioned before, if any of the variables do not change over different observations, it does not carry any useful information in ML training. Hence, if a line never reaches its capacity over all the observations, it should be removed from the outputs. Similarly, if an element never fails, it should be removed from the input set. After the aforementioned steps, the

input matrix with rows as observations and columns as features, and output matrix with rows as observations and each column as output are created.

The previously mentioned ML model falls in the category of supervised classification. These models receive multiple inputs and predict only one output for each observation. Hence, it is necessary to train one ML model for each output (line included in the outputs after applying size-reduction described in step 5), which means the number of trained models is equal to the number of columns in the reduced form of the output matrix [16].

To predict the desired outputs for an unseen event/hurricane, the SSCUC problem should be formed with its scenario set. Next, for each scenario, the input vector should be created, as explained in step 2. Then, this vector of features should be deployed into the trained ML models. The predicted output of each model is either 1 (corresponding line should be monitored) or 0 (corresponding line does not need to be monitored). After repeating the same process for all the scenarios, $\boldsymbol{M}_{(s)}$ is created [16]. Finally, the SSCUC problem is solved by following the steps presented in the feasibility study described in Section 6.3.2.2.

To demonstrate the efficiency of the ML-assisted SUC algorithm, the synthetic grid on the footprint of South Carolina mentioned before is used to simulate 24 hours. 40 out of 597 transmission lines are considered vulnerable to failure due to extreme weather events. Details of model training can be found in [16]. Once the training process is complete, the trained models are ready to assist in solving any unseen SSCUC problem. To evaluate the effectiveness of the method, different SSCUC problems are created and solved with both methods: Original SSCUC and ML-Assisted SSCUC.

The SSCUC problems are created with different numbers of line outages (i.e. different hurricane intensities). They cover zero to 30 possible outages, where each outage may happen at any random time of the examined horizon. The scenario set regarding each SSCUC problem consists of 10 scenarios. The optimization problem is formulated as described briefly in Section 6.3.2.1 and explained in detail in [26]. Figure 6.8 [16] illustrates the solution time for both methods and different SSCUC problems (for every possible number of line-outages, 10 randomly generated problems are solved).

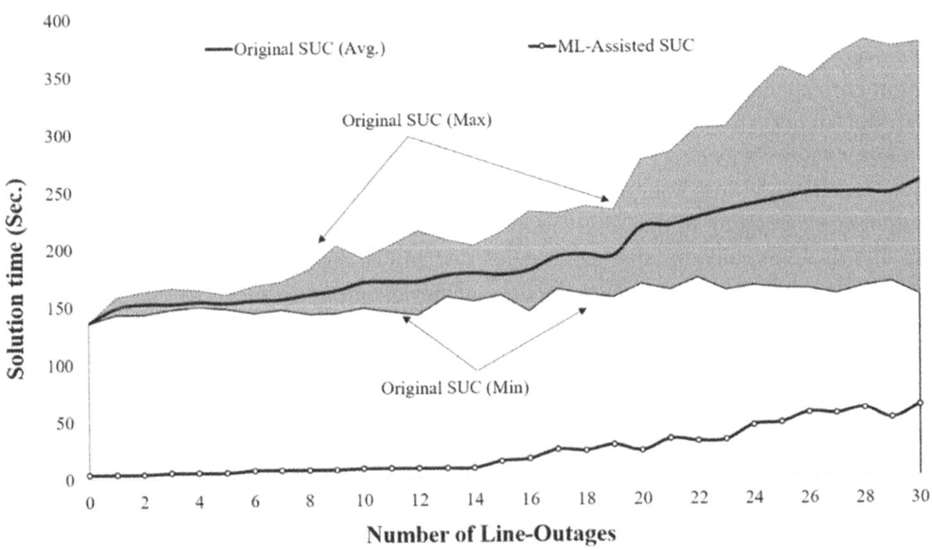

Figure 6.8 Solution time for the original SUC and the ML-assisted SUC. *Source:* [16] / IEEE.

Table 6.3 Comparison of original SUC and ML-assisted SUC: number of implemented constraints corresponding to transmission lines.

	Number of constraints regarding transmission lines	
Number of line outages	**Original SUC**	**ML-assisted SUC**
0	143,280	720
5	143,574	1,014
10	144,084	1,524
20	144,608	2,888
30	145,416	3,936

Source: [16] / IEEE.

As shown, the solution time with the ML-Assisted method is significantly reduced compared to the original SSCUC solution.

More specifically, the ML-assisted model reduces the calculation time by 90% on average (within 75% minimum and 97% maximum time reductions) for the different SSCUC problems. Simultaneously, the value of the objective function, commitment status, and line flows are similar in both methods within a defined mixed-integer problem (MIP) gap of 1% in CPLEX [16].

Among different SSCUC problems, the trained ML model perfectly predicted the suspected lines for all scenarios. While no FN was observed within the simulations, even if FN was observed and post-processing revealed a line violates its capacity, the total calculation time would still be much less than the original SSCUC. Moreover, to show the effect of predicting suspected lines on the complexity of SSCUC, the number of implemented constraints is shown in Table 6.3 [16] for the original and ML-Assisted models. This clearly shows the reason for the significant reduction in computation time when the ML-Assisted model is used.

6.4 Defensive Islanding to Enhance Power System Resilience Under Extreme Weather Events

This section discusses the contribution of defensive islanding to the mitigation of the impact of extreme events on the power system. The defensive islanding splits the system into stable and self-adequate islands in order to isolate vulnerable transmission lines, whose failure would trigger cascading events. The controlled splitting of the network into stable islands, can be carried out with limited customer interruptions compared to uncontrolled cascaded events and system splitting [46]. This preventive operational action is adaptive to the prevailing system and weather conditions by utilizing advanced monitoring capabilities and using real-time data. The application and contribution of islanding schemes have been discussed in several papers [3, 50–53].

6.4.1 Defensive Islanding Algorithm

Defensive islanding algorithm addresses the problem of finding which lines should be disconnected, in order to isolate the vulnerable lines and avoid cascading outages. The scheme comprises dynamic preventive control, which should be adaptive to the weather conditions and to the various phases of the post-event degraded state, while taking into account the current topology and loading conditions of the system [46].

The topology of network and loading conditions could be determined using data reported by phasor measurement units (PMUs) [54]. It is assumed that PMUs, distributed throughout the network, check the connectivity of the network lines and measure the lines' flow values, which are transferred to a central controller through advanced communication technologies. The PMUs' measurements can be also used for the reconnection of the islands. Frequency, voltage, and phase angle conditions must be satisfied for the reconnection, and thus monitoring is critical [46]. The PMUs' real-time measurements are used to adjust the frequency, voltage, and phase angle of islands to meet the requirements of synchronization [55].

Defensive islanding can be defined as a graph-cut problem [56]. The power system can be described by an undirected graph $\boldsymbol{G}(\boldsymbol{V}, \boldsymbol{E}, \boldsymbol{W})$ with vertex set $\boldsymbol{V}$ (buses) and edge set $\boldsymbol{E}$ (lines). The matrix $\boldsymbol{W}$ is the weighted adjacency matrix and must satisfy the following properties: $w_{ij} = w_{ji}$ (symmetric matrix), $w_{ii} = 0$ (diagonal elements of the matrix are zero), and $w_{ij} = 0$, if vertex i is not adjacent to vertex j in $\boldsymbol{G}$. The aim is to divide the system into two sub-graphs $\boldsymbol{G}_1(\boldsymbol{V}_1, \boldsymbol{E}_1, \boldsymbol{W}_1)$ and $\boldsymbol{G}_2(\boldsymbol{V}_2, \boldsymbol{E}_2, \boldsymbol{W}_2)$ by disconnecting the edges that connect these two sub-graphs. $\boldsymbol{V}_1$ and $\boldsymbol{V}_2$ are disjoint subsets and therefore the following conditions should be satisfied: $\boldsymbol{V}_1 \cap \boldsymbol{V}_2 = \emptyset$ and $\boldsymbol{V}_1 \cup \boldsymbol{V}_2 = \boldsymbol{V}$ [57]. The vulnerable transmission lines should be included in the same sub-graph or in the set of disconnected lines to avoid cascading events in the entire network due to their failure.

The set of edges removed to create the two sub-graphs is called the cutset and the sum of the weights of the edges included in this set is called the cut. In order to create two self-sufficient islands, the cut should be minimized [46].

Minimal power imbalance and minimal power-flow disruption, expressed in (6.11) and (6.12), respectively, can be used as objective function for the partition of the system into islands $\boldsymbol{V}_1$ and $\boldsymbol{V}_2$ [57].

$$\min_{V_1, V_2 \subset V} \left| \sum_{i \in V_1, j \in V_2} w_{ij} \right| \tag{6.11}$$

$$\min_{V_1, V_2 \subset V} \sum_{i \in V_1, j \in V_2} |w_{ij}| \tag{6.12}$$

where, w_{ij} can be the active or apparent power of the line that connects buses i and j.

Minimal power imbalance creates islands with a good power balance, minimizing the load shedding that should be carried out after the separation, while minimal power-flow creates islands with the minimum change from the pre-disturbance power-flow pattern. This property of the minimal power-flow disruption has also minimal impact on the transient stability of the islands, reduces the possibility of overloading the transmission lines within the islands, and facilitates the islands' eventual reintegration with the rest of the system [46].

In order to determine the cutset, several methods can be used. Ordered Binary Decision Diagram-based methods have been used in [58–60], while Koopman mode analysis has been used in [61]. Optimization-based methods have been developed in [62] and [63] to determine the islanding solution. Heuristic methods, such as particle swarm optimization and genetic algorithms, have been used in [64–66] and k-way partitioning has been used in [67]. In this section, as in [57], *constrained spectral clustering* is used to provide an islanding solution that isolates vulnerable components. More details on the application of spectral clustering to power systems can be found in [46–48].

Using constrained spectral clustering, the isolation of vulnerable lines can be achieved by isolating the "vulnerable buses" to which the vulnerable lines are connected. The formed islands and their number depend on the geographical location of the vulnerable buses, as well as the current topology of the system at the time interval of the examined horizon. These islands operate as independent systems and load shedding is carried out, if necessary, in order to form islands that satisfy the required constraints. The aim is to isolate the vulnerable buses in one island, since the isolation of vulnerable buses in more than one island may lead to increase in load shedding due to congestion [46].

To create stable islands, the islanding solution must satisfy steady state and dynamic constraints, such as load generation balance, thermal limits, and transient and voltage stability. The islanding method discussed in this section is focused on creating islands that satisfy steady-state constraints only. By satisfying these constraints and considering the benefits of minimal power flow disruption, it is assumed that the islands are also stable; thus transient and voltage stability of the islands after a contingency are not examined [46]. Hence, minimal power flow disruption is used in this section.

As previously described, defensive islanding can be defined as a graph-cut problem and constrained spectral clustering is used to provide a solution. However, any other method that can determine the split of the power system into self-sufficient islands can be used. More details on the theoretical background of constrained spectral clustering can be found in [46].

It is worth mentioning that constrained spectral clustering provides a solution that minimizes power flow disruption and divides the power system into two subsets $\boldsymbol{V}_1$ and $\boldsymbol{V}_2$. In the case study presented in Section 6.4.3, assuming that vulnerable buses should belong to $\boldsymbol{V}_1$, the quality of the islanding solution is defined as the ratio of the number of vulnerable buses to the total number of buses in $\boldsymbol{V}_1$. If the ratio is smaller than a desired value, the constrained spectral clustering is reapplied in $\boldsymbol{V}_1$. The same procedure is repeated until the quality of the islanding solution is considered satisfactory, and the lines that should be disconnected to form the islands are those obtained from the last application of constrained spectral clustering. The quality of an islanding solution takes values in the range [0, 1]. Large values limit the possible formations of islands, while low values mean that the formed island with vulnerable buses could include a higher number of non-vulnerable buses than the vulnerable ones. Regarding the case study presented in Section 6.4.3, the minimum ratio of the number of vulnerable buses to the total number of buses in $\boldsymbol{V}_1$ is selected equal to 67% (2/3=0.67). This ratio has been selected for this specific application [46]. A different ratio could be selected for another application, the selection of which requires offline simulations. In addition, the network may be divided into islands due to line failures caused by extreme weather. In this case, each island is treated as an independent system and the network may be divided into more than two islands [46].

6.4.2 Approach for Determining when to Apply Defensive Islanding

The need for splitting the network in a controllable way could be determined following a risk-based approach, showing when its application could be beneficial for enhancing system resilience. The risk-based approach can support the system operators' decision-making as to when to apply the islanding solution for creating stable and sustainable islands and preventing uncontrollable outage propagation [46].

The *Severity Risk Index* (*SRI*) proposed in [17] or any other index could be used for the risk-based approach. *SRI* considers the probability and impact of possible scenarios based on line failure probabilities. A threshold of *SRI* based on the evolving power system conditions can be effectively defined, for which the resilience of the power grid to the extreme weather event benefits from the application of defensive islanding [46]. To calculate *SRI*, the impact of extreme events on power system components should be assessed initially, and then the power system risk should be assessed.

At every time interval of the examined horizon, the failure probabilities are used to determine the vulnerable lines. All the lines having failure probabilities higher than a specified failure probability threshold are considered as vulnerable lines and are used to generate all the possible failure scenarios. More specifically, if the sequence of the contingencies is neglected and the number of vulnerable lines is N, then the number of possible failure scenarios is 2^N. From the generated failure scenarios, only those that have higher probability than a pre-specified threshold are selected (i.e. K scenarios) to reduce the number of scenarios. The aforementioned thresholds depend on the system operator and indicate the security criteria to be fulfilled. When determining the threshold, a trade-off between computational complexity and safety level has to be realized, as the computational complexity increases with the security level due to the higher number of scenarios included in the risk assessment procedure [46]. A scenario generation and reduction method as the one proposed in [17], could be used to produce a manageable number of representative scenarios and therefore reduce computational time of *SRI*.

Using the impact and probability of the selected failure scenarios, *SRI* is calculated as [46]:

$$SRI = \sum_{s \in S} pr_s Im_s \tag{6.13}$$

where, pr_s is the probability of scenario s, Im_s is the impact of scenario s and S the set of selected failure scenarios.

Defensive islanding is applied to prevent or decrease the spread of cascading outages. The *SRI* will be used to help the system operator decide when the defensive islanding should be applied. Depending on the cascading outages taken into account, the outaged lines can be determined by applying dynamic [71, 72], or static [73] simulations.

The impact of each scenario is considered equal to the amount of load shedding that is carried out for stabilizing the system. Independent islands may be formed during the risk assessment procedure due to lines failing due to an external threat or cascading. If the estimated *SRI* is larger than a pre-specified threshold SRI_{thres}, then the system operator will be notified for activating defensive islanding [46]. The selection of SRI_{thres} needs investigation and is discussed in the next section.

6.4.3 Case Study Applications

This section presents the results by applying defensive islanding to a simplified 29-bus version of the Great Britain (GB) transmission network presented in Figure 6.9 a) [46]. This is the same network used in Chapter 3. In order to account for the spatial impact of windstorms across the transmission network, the test network has been arbitrarily divided into 6 weather regions, as shown in Figure 6.9 b) [46]. A simulation period of one winter week is used, where the peak demand and high wind speeds are usually observed in GB. Further, an hourly resolution is used, which is considered adequate for modelling weather events. An adequate number of scenarios has been considered in the simulations to evaluate the effectiveness of defensive islanding.

A Sequential Monte Carlo simulation is applied to deal with the uncertainties regarding line failures. During the simulations, the weather regions 1 and 2 of Figure 6.9 b) are subject to windstorms, while in the rest of the system, it is assumed that there are no high wind speeds, and the normal failure probability is considered for the lines that are not exposed to windstorms. Each line with a failure probability higher than 1% is considered vulnerable and is used to generate the examined failure scenarios. The failure scenarios considered in the risk assessment procedure have a probability of occurring higher than 1% [46].

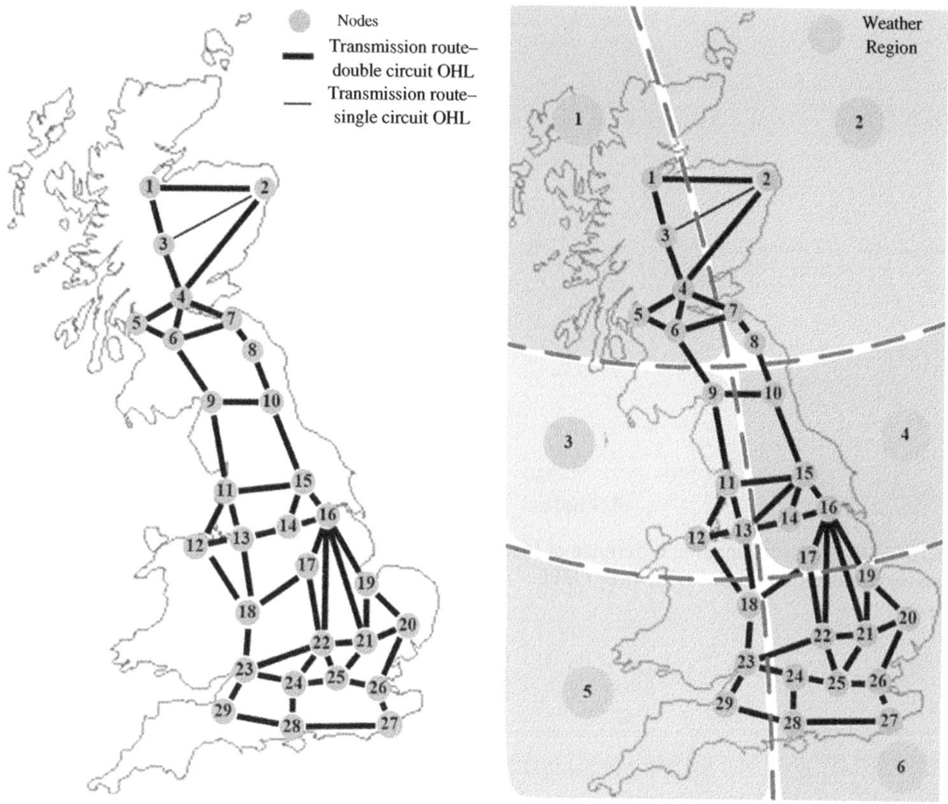

(a) Simplified GB network (b) Weather regions

Figure 6.9 The simplified 29-bus GB transmission network. *Source:* [46] / IEEE.

After applying the appropriate islanding solution, the next step is to determine which components will actually trip due to the weather event. To do this, the line failures and the restoration time of the failed lines are determined based on the methodology discussed in Section 3.2.2. Following this approach, the power system performance is evaluated at every simulation step. This helps determine if the load has to be shed for balancing the network. This procedure is repeated until the end of the simulation is reached. The system resilience is then assessed, which can be expressed by the number of customers disconnected during the weather event. This approach is conducted with and without considering defensive islanding to compare the results [46].

Figures 6.10–6.15 [46] show the *SRI* (dashed line) and the corresponding difference of load shedding that is carried out without and with defensive islanding for a range of maximum wind speeds that may occur in the affected areas. Every simulation step where $SRI > 0$ (i.e. probability and impact higher than zero) is treated as a scenario under investigation.

Those scenarios are presented in ascending order according to their *SRI* values. When the difference of load shedding is positive, the load that needs to be shed when applying defensive islanding is lower compared to the load that is shed when defensive islanding is not applied. Consequently, it is beneficial to apply defensive islanding as a preventive control method. Conversely, when the difference of load shedding is negative, load shedding increases when defensive islanding is applied [46]. It can be observed that higher wind speeds increase the number of scenarios under investigation. Additionally, the percentage of scenarios with a positive load shedding difference increases

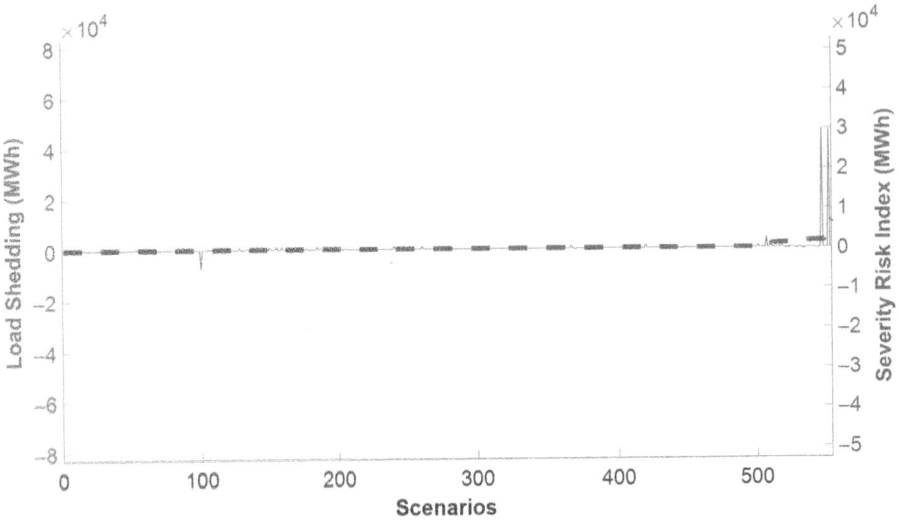

Figure 6.10 *SRI* and corresponding difference of load shedding without and with defensive islanding for maximum wind speed of 36 m/s. *Source:* [46] / IEEE.

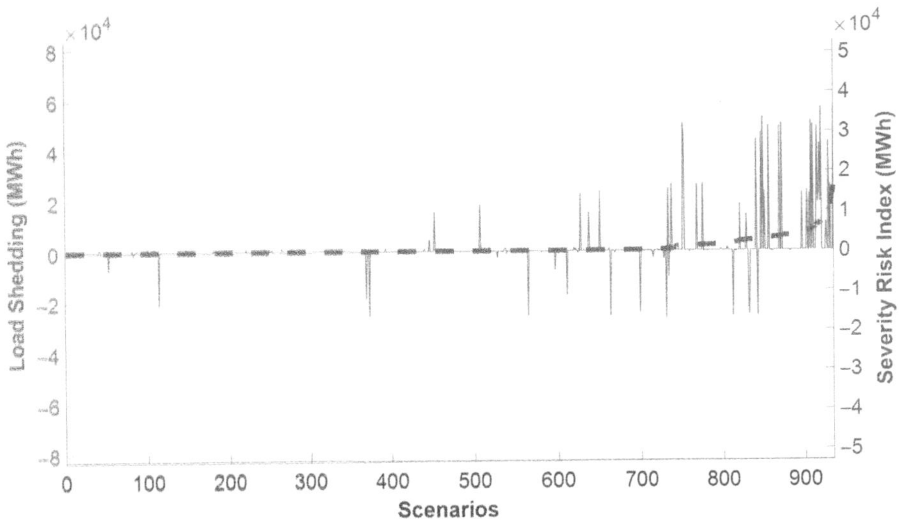

Figure 6.11 *SRI* and corresponding difference of load shedding without and with defensive islanding for maximum wind speed of 38 m/s [46].

with the increase in *SRI*. This shows that for higher *SRI* values, the network resilience benefits from defensive islanding.

By comparing Figures 6.10–6.15 [46], it can be observed that for wind speeds up to 36 m/s (Figure 6.10), the system can effectively withstand the stress imposed by the windstorm. Therefore, defensive islanding has limited value. For maximum wind speeds of 38 m/s (Figure 6.11), defensive islanding could potentially be beneficial, but proper conclusions cannot be drawn. For wind speeds higher than 38 m/s, the resilience of the test system is significantly compromised, as shown by the large number of scenarios where load shedding takes place.

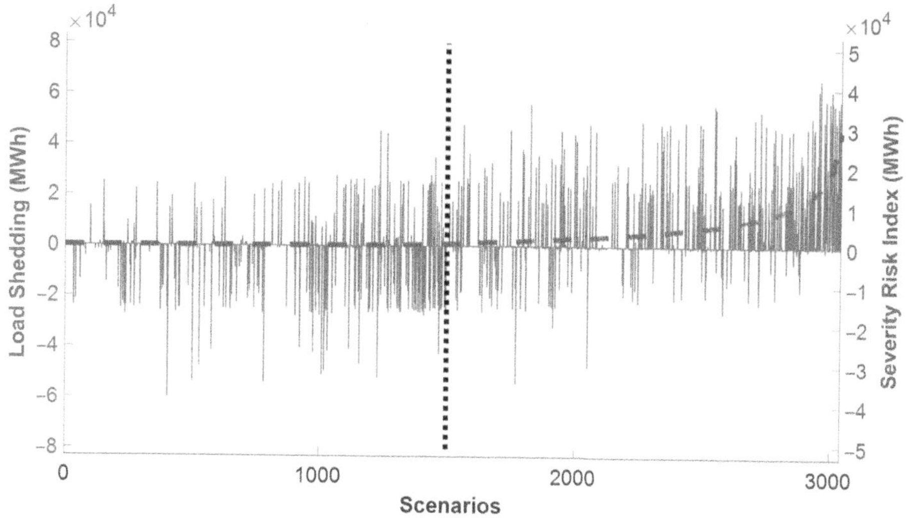

Figure 6.12 *SRI* and corresponding difference of load shedding without and with defensive islanding for maximum wind speed of 40 m/s. *Source:* [46] / IEEE.

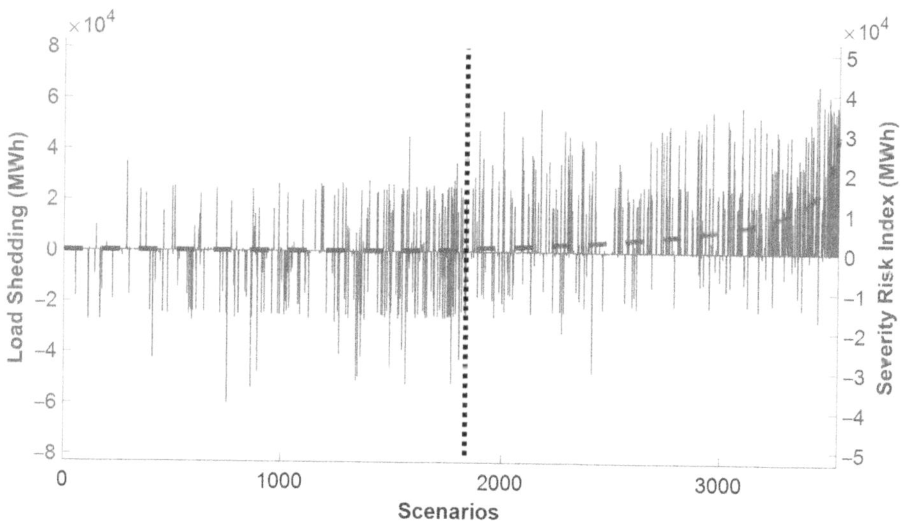

Figure 6.13 *SRI* and corresponding difference of load shedding without and with defensive islanding for maximum wind speed of 42 m/s. *Source:* [46] / IEEE.

The proper selection of $\mathrm{SRI}_{\mathrm{thres}}$ for online applications require extensive offline studies. Based on simulation results, it is possible to define the SRI_{thres}, for which defensive islanding is applied, as the value that maximizes the function F [46]:

$$F = \sum_{s \in \boldsymbol{S}^*} \left(LS_s - LS_s^{DI}\right) \tag{6.14}$$

where $\boldsymbol{S}^*$ is the set of all the scenarios with *SRI* values higher than the selected SRI_{thres}. The *LS* and LS^{DI} are the MWh amount of load shedding which is carried out without and with defensive islanding, respectively.

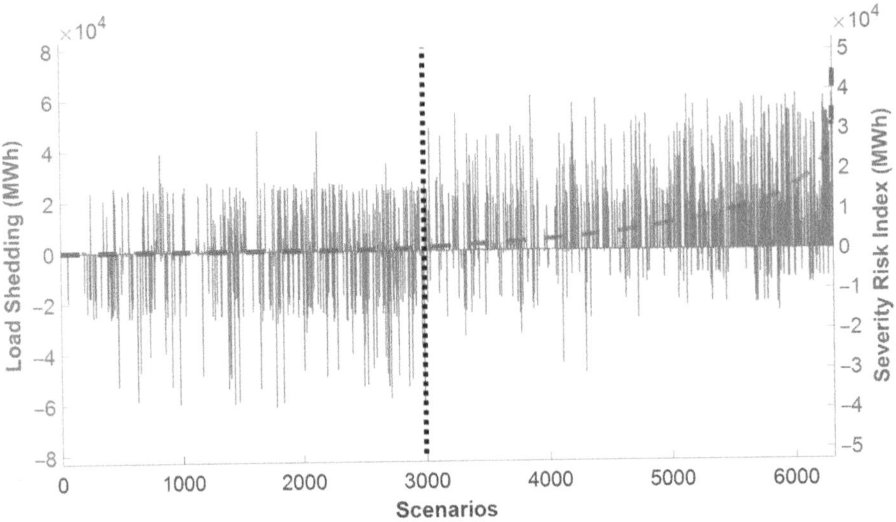

Figure 6.14 *SRI* and corresponding difference of load shedding without and with defensive islanding for maximum wind speed of 44 m/s. *Source:* [46] / IEEE.

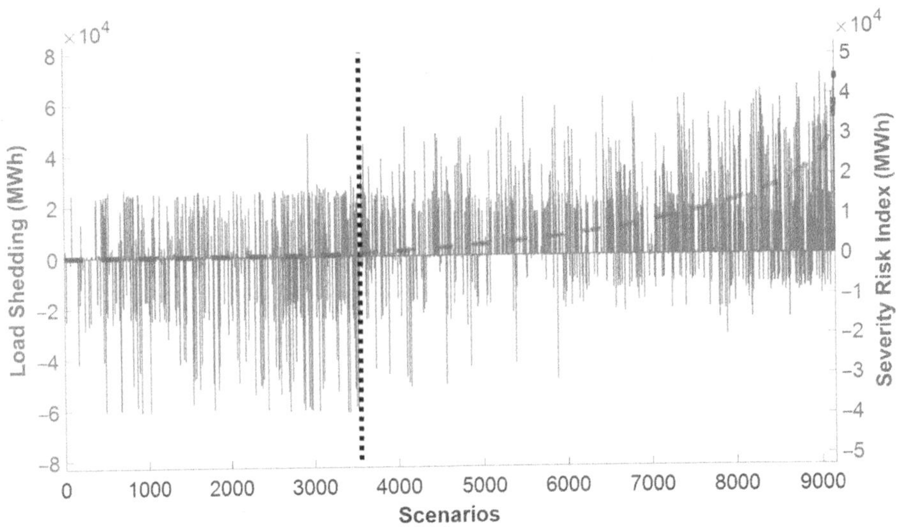

Figure 6.15 *SRI* and corresponding difference of load shedding without and with defensive islanding for maximum wind speed of 46 m/s. *Source:* [46] / IEEE.

The SRI value, marked with the vertical dotted line in Figures 6.12–6.15 [46], represents the calculated SRI_{thres} based on (6.14). It is noted that when defensive islanding is applied, the load that is shed in order to form stable and self-adequate islands is included in the overall load shedding LS^{DI}.

Table 6.4 [46] shows the SRI_{thres} and the maximum SRI values for each windstorm intensity investigated in this study. It can be observed that the maximum SRI is getting larger values with the increase in wind speed. Furthermore, SRI_{thres} takes similar values in all the entire range of maximum wind speeds, given that SRI takes values in the range $[0, 4.5{\cdot}10^4]$ MWh. For a real-life application, a single characteristic threshold within the range of those SRI_{thres} could be defined. Table 6.5 [46] presents the value of the function F considering all the combinations of maximum

Table 6.4 Defined SRI_{thres} and SRI_{max} for a range of maximum wind speeds.

Maximum wind speed (m/s)	SRI_{thres} (MWh)	SRI_{max} (MWh×10^4)
36	-	0.68
38	-	1.61
40	597.13	2.94
42	583.50	2.97
44	866.76	4.47
46	673.71	4.50

Source: [46] / IEEE.

Table 6.5 Function F for all the combinations of wind speeds and SRI_{thres}.

	Maximum wind speed (m/s)			
	40	42	44	46
SRI_{thres} (MWh)	F (MWh·10^6)			
597.13	5.589	6.985	11.912	14.206
583.50	5.588	7.004	11.937	14.223
866.76	5.565	6.933	12.102	14.118
673.71	5.573	6.981	12.070	14.310
Mean Value (680.28)	5.580	6.959	12.015	14.164

Source: [46] / IEEE.

wind speeds and SRI_{thres}. The mean value of the SRI_{thres} values of the different wind speeds is also used. The results in Table 6.5 demonstrate that the selection of any of these values as the characteristic threshold of the test system leads to similar results. In addition, it is observed that as the wind speed increases, defensive islanding leads to higher values of function F.

In Figures 6.12–6.15 [46], it can be seen that for $SRI > SRI_{thres}$, defensive islanding benefits the network resilience (i.e. load shedding>0) in the majority of the scenarios. Moreover, defensive islanding seems beneficial in a number of scenarios for SRI lower than SRI_{thres}. This observation is summarized in Table 6.6, which shows the statistical analysis of the simulation results (in percentage of the total scenarios considered in the analysis). The first and second rows of each maximum wind speed in Table 6.6 [46] refer to SRI values lower and higher than the SRI_{thres}, respectively. The second and third columns is the percentage of the number of scenarios where LS is lower and higher than LS^{DI}, respectively. The fourth column is the percentage of scenarios where load shedding is not needed with or without defensive islanding. The fifth column is the percentage of scenarios where the load shedding with and without defensive islanding is equal to or higher than zero. In these scenarios, the load shedding that is carried out in both cases is due to bus isolation after the uncontrolled outage of lines.

As can be seen from Table 6.6, for $SRI > SRI_{thres}$ the percentage of the scenarios where the defensive islanding is beneficial, i.e., $LS > LS^{DI}$, increases with the wind speed from 46.34% for 40 m/s to 64.27% for 46 m/s. Further, it is more than double (i.e. approximately 4.5 times larger for maximum

Table 6.6 Statistical analysis of the simulation results.

Max wind speed (m/s)	$LS < LS_{DI}$ (%)	$LS > LS_{DI}$ (%)	$LS = LS_{DI} = 0$ (%)	$LS = LS_{DI} > 0$ (%)
40	25.95	20.54	43.95	9.56
	18.96	46.34	18.71	15.99
42	23.91	24.57	40.03	11.49
	18.17	49.15	16.49	16.19
44	22.79	27.38	35.27	14.56
	15.19	55.77	10.30	18.74
46	21.01	32.34	28.51	18.14
	13.93	64.27	5.61	16.19

Source: [46] / IEEE.

wind speed of 46 m/s) than the percentage of cases where the defensive islanding results in higher load shedding (i.e. $LS < LS^{DI}$) for all the wind speeds. In contrast, for $SRI < SRI_{thres}$, this difference is much smaller [46]. Therefore, this analysis shows that defensive islanding is beneficial when the risk of losing customers is high, which might be the case during severe windstorms.

6.5 Resilience Enhancement in Low-Carbon, Low-Inertia Power Systems

The previous sections have not considered how deep penetration of renewable energy sources (RES) might affect the resilience of the system. However, as it will also be further discussed in Chapter 8, RES have been rapidly installed worldwide with the aim of reducing greenhouse gas emissions and their penetration is expected to increase dramatically in the coming years. This means that fossil fuel-fired synchronous generators are rapidly being withdrawn from the system, leading to the reduction of available inertia and FR capabilities, which poses new challenges to transmission system operators (TSOs), particularly to TSOs in electrically isolated synchronous islands such as Australia, GB, Ireland, Texas, etc. [74]. Under extreme weather events, when the damage of several system components is likely, the challenges in operating such low-inertia systems become even greater [75].

In August 2018, a storm in northeastern Australia caused the complete disconnection of the Queensland-New South Wales interconnector (see Figure 6.16). This loss resulted in an 828 MW power imbalance, prompting demand shedding across various regions of Australia and leading to the tripping of the South Australia–Victoria interconnector, which had been transferring 245 MW at the time [75]. As a result, the Australian east coast's interconnected power system—one of the longest in the world at approximately 4,000 km—was split into three separate islands, which continued operating independently for several minutes. Adapted from [76].

Losing an interconnector is a rare event that is normally referred to as a "non-credible" contingency. However, during extreme weather events, it may be desirable to appropriately schedule FR resources to *preventively* enhance system resilience by assuming that an interconnector's loss might happen. In fact, under such conditions, generally "non-credible" contingencies might become possible, and they can have a tremendous impact on system frequency, especially in low-inertia conditions.

Recent advancements in the co-optimization of inertia and FR resources within UC and economic dispatch frameworks [77–81] have established a consistent approach for evaluating FR needs and capabilities to mitigate the effects of generation contingencies in low-inertia power systems [75]. Typically, these response resources are allocated based on quasi-steady-state frequency (QSSF) criteria [82, 83]. Some models also incorporate inertia allocation to ensure compliance with rate of change of frequency (ROCOF) requirements [84–87]. Furthermore, certain formulations simultaneously optimize inertia and primary frequency response (PFR) to help constrain the frequency nadir [80, 88]. During extreme events, it may also become important to pay attention to transmission contingencies that could lead to system split, as in the case of Australia mentioned above, which can have devastating effects. Similar system splits are indeed also feared and have been experienced in Europe [89].

This section describes the development of means to *operationally* enhance system resilience when, in real-time, e.g., due to intense storms, lightning strikes, etc., system separation becomes a possible event. The process involves jointly optimizing pre-contingency control variables and post-contingency response measures to reduce the operational costs of a system that faces risk of splitting. A separation-constrained optimal power flow (SECOPF) model is described in the next subsections that aims to cost-effectively allocate inertia and PFR resources both within individual areas and across the entire system [75]. The model addresses both low- and high-frequency events resulting from load variations, generation outages, or interconnector failures. In the case of a low-frequency event (such as a generation loss), upward or "raise" PFR resources are deployed, while during high-frequency events (such as load disconnection), downward or "lower" PFR resources are activated to stabilize the frequency. If an interconnector failure causes the system to split into two regions, the exporting area will experience over-frequency, while the importing area will face under-frequency conditions. The model also incorporates secondary frequency response (SFR) resources to handle high and low frequency disturbances.

6.5.1 Dynamic Network Model

The network model should capture the spatial frequency dynamics during transients and be capable of deriving ROCOF requirements for both the overall system and local areas under load or generation contingencies and in the event of system separation. However, the emphasis here is on determining requirements to mitigate the effects of separation events [75].

Regarding the post-separation contingency network dynamics, it is supposed at time $t = 0^-$, line/interconnector $\tilde{l}$ (connecting areas β and γ, that we assume for simplicity modeled as two buses) is disconnected, and such disconnection partitions the system into two subsystems. The post-contingency frequency dynamics for $t \geq 0$ can be derived, for area β, starting from classical two-area frequency dynamic models with DC power flow approximation [44]:

$$2H_\beta \dot{f}_\beta(t) + D_\beta P^L_\beta f_\beta(t) = p^{inj}_\beta(t) + pf^0_{\tilde{l}} \tag{6.15}$$

where H is the total kinetic energy (MEs/MVA), f is the angular frequency (rad/s), D is the composite load-damping factor of the area, P^L is the load in area, p^{inj} is the active power injection from the area to the rest of the network and $pf^0_{\tilde{l}}$ is active power flow of the interconnector $\tilde{l}$ at $t = 0^-$. The same applies for area γ, but with $pf^0_{\tilde{l}}$ of opposite sign.

Concerning ROCOF, by definition of the *perturbation* variables, $f_\beta(0) = g_\beta(0) = fl_l(0) = 0$ at $t = 0$. Substituting this into (6.15) and applying it to both areas [75]:

$$\dot{f}_\beta(0) = pf_{\tilde{l}}^0 / 2H_\beta \tag{6.16}$$

$$\dot{f}_\gamma(0) = -pf_{\tilde{l}}^0 / 2H_\gamma \tag{6.17}$$

In practice, this means that when interconnector $\tilde{l}$ is disconnected, the local ROCOFs at buses β and γ are proportional to the line flow $pf_{\tilde{l}}^0$ on the interconnector right before the disconnection with opposite signs, and inversely proportional to the after-separation local inertia. That is why it becomes important to make sure that the interconnector flow is limited and/or the local inertia in the relevant areas is high enough in order to prevent excessive ROCOF values in either or both areas.

For ROCOF security, one can require that the ROCOF values $\dot{f}_\beta(0)$ and $\dot{f}_\gamma(0)$ stay within a predefined range, for example, to prevent impact on ROCOF-sensitive protection relays of distributed generators. This constrains the total inertia in each subsystem after the separation contingency, as well as the scheduled line flow $\left|pf_{\tilde{l}}^0\right|$ on the interconnector $\tilde{l}$.

Regarding QSSF, focusing again on area β, from the post-contingency dynamics and under droop control, if we call ρ_β the aggregated droop value of all area's generators, the aggregate generation FR $g_{qss,\beta}$ in the quasi-steady state (qss) is [75]:

$$p_{qss,\beta}^{inj} = -\rho_\beta f_{qss,\beta} \tag{6.18}$$

Generator response forms other than droop control can also be modeled, with the relevant FR amount replacing (6.18) by an appropriate quasi-steady state behavior. Following the analysis described in [75], the QSSF in the area is therefore [75]:

$$f_{qss,\beta} = pf_{\tilde{l}}^0 / \left(D_\beta P_\beta^L + \rho_\beta\right) \tag{6.19}$$

Hence, if $pf_{\tilde{l}}^0 > 0$, i.e., power flows from bus β to bus γ before the separation contingency, then the QSSF (deviation) $f_{qss,\beta} > 0$ and hence the frequency in the subsystem β will increase above the pre-contingency frequency f^0, while $f_{qss,\gamma} < 0$ and hence the frequency in the area γ will decrease.

With regards to the after-separation frequency extrema, for the low-frequency area, it is of interest to capture the minimum frequency level, that is, the frequency *nadir*, while for the high-frequency area, it is of interest to capture the maximum frequency level, that is, the *zenith*. The nadir/zenith expressions, however, cannot generally be calculated in closed form and depend in a complex, highly non-linear way on all the parameters in (6.15), including, in particular, the interconnector flow and the inertia levels in each of the two areas. The interested reader is invited to look at [75] for details.

6.5.2 Mathematical Formulation

The resilience-oriented OPF model can be formulated with a "classical" objective function, as the one described in (6.2), aimed to minimize the total cost of operation, which consists of the generation cost and load shedding cost. Some of the "typical" constraints to be included in the optimization problem are [75]:

- power balance in each area and across the interconnector;
- DC power flow interconnector limit;
- curtailment limits of RES;
- limits of generators (including ESS).

In addition to these constraints, additional frequency security constraints must be incorporated to limit the ROCOF, frequency nadir/zenith, and QSSF within acceptable requirements in the event of system split. The objective is not only to avoid frequency excursions beyond technical thresholds following load or generation contingencies, but also to allocate response resources and control interconnector flows to prevent frequency violations during separation events. The model assumes that inertia is provided exclusively by thermal units (though it can be extended to include "virtual inertia" from sources such as batteries or hydrogen electrolyzers [90]) and PFR is delivered both by thermal units and ESS.

In general, the resulting optimization problem is a mixed-integer nonlinear programming (MINLP), where the integer variables are the commitment status of the different synchronous units (essential to schedule area-level inertia) and the main nonlinearity, even with a DC power flow approximation, emerges due to the complex nadir/zenith representation. The model can, however, be transformed into MILP by appropriately linearising offline the nadir/zenith equations via hyperplanes approximation [79]. Given the frequency resilience-oriented aim of this analysis, the main decision variables of interest are the aggregate level of inertia in each area, depending on the number and type of scheduled synchronous units, and the flow across the interconnector, which again also depends on the energy output of the scheduled units, besides the aggregate among of FR allocated in each area. Again, the interested reader is invited to read [75] for details.

6.5.3 Case Study Applications

The case study application [75] examines a future scenario for the Australian National Electricity Market (NEM), aiming to evaluate resource allocation and operational costs under extreme operating conditions, particularly when certain interconnectors are at high risk of tripping. The NEM covers the states of Victoria (VIC), New South Wales (NSW), Queensland (QLD), Tasmania (TAS) and South Australia (SA) (Figure 6.16 [75]). As reported in [91], the system has an installed capacity of 50.5 GW, with an additional 2.6 GW of utility-scale solar PV and 3.1 GW of wind power under development, along with 4.8 GW of rooftop PV systems [92]. The different states are interconnected via a radial network of High Voltage Alternating Current (HVAC) and High Voltage Direct Current (HVDC) interconnectors, whose transfer limits are specified in Table 6.7 [75].

The case study examines system resilience under the loss of specific interconnectors in Figure 6.16 [75], that is, Heywood and QNI (HVAC), and Basslink (HVDC). For each operating scenario, the base case assumes these interconnectors are not considered credible contingencies. Three additional cases are then analyzed, each treating one of the interconnectors as a credible event, alongside the largest generation and load contingencies. Resources are scheduled using the discussed optimization model. To simplify the analysis, the generation contingency is fixed

Table 6.7 Studied interconnectors. Current [93] and {future} [94] ratings [75].

Interconnector	State A	State B	A to B Rating (MW)	B to A Rating (MW)
Heywood	VIC	SA	600 {650}	500 {650}
Basslink	TAS	VIC	594 {630}	478 {480}
QNI	NSW	QLD	300-600 {1055}	1078 {1658}

Source: [75] / with permission of Elsevier.

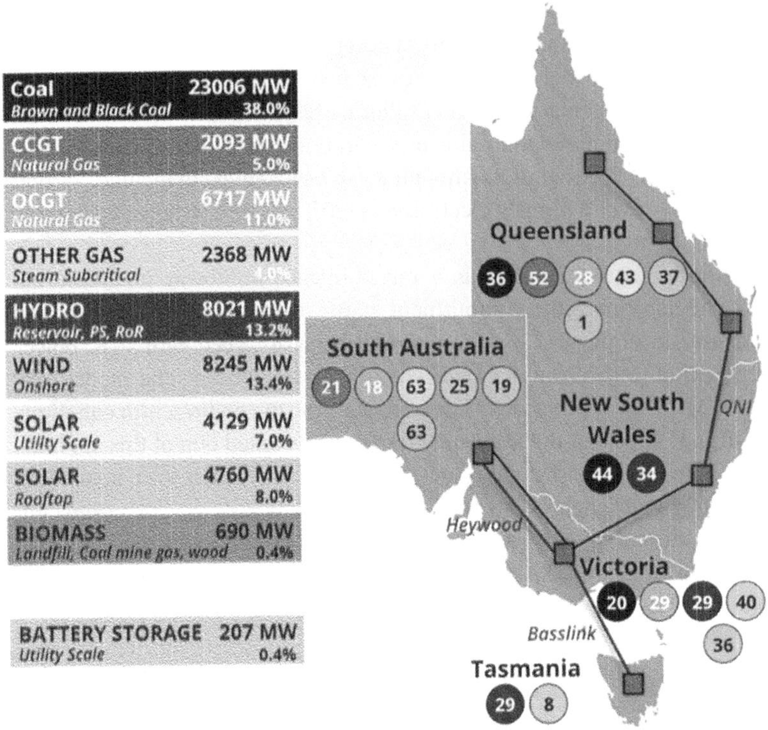

Figure 6.16 NEM multi-area interconnected system. Circles indicate the percentage of total installed and committed capacity by technology in each state by October 2018 (shown only when the capacity exceeds 10% of the state's total installed capacity). The interconnectors examined in this study are Heywood, Basslink and QNI. The utility battery ESS capacity is provided for each state. *Source:* [75, 91, 92] / with permission of Elsevier.

rather than optimized, corresponding to the loss of the Kogan Creek coal-fired power plant in QLD, which has an output just under 700 MW. The largest load contingency is set at 350 MW, representing the potential loss of a section of the Portland aluminum smelter in VIC. Flows are defined as positive when they are leaving node A in Table 6.7 [75]. These scenarios reflect expected future operating conditions based on data from [94, 95].

6.5.3.1 Base Case Results

Table 6.8 [75] presents the base case results under both low and high demand conditions (17.2 GW and 34 GW, respectively), including raise and lower PFR services (denoted as "up" and "dw") and the total aggregated inertia in each system area for the respective demand levels. Table 6.9 [75] presents the corresponding interconnector flows under the demand scenarios. To support the analysis, the frequency deviations are constrained by the following limits: a ROCOF threshold of −1 Hz/s, a QSSF threshold of 49.5 Hz and a frequency nadir limit of 49 Hz for low-frequency events, as well as a ROCOF threshold of 1 Hz/s, a QSSF threshold of 50.5 Hz and a zenith limit of 51 Hz for high-frequency events [75].

In the base case, PFR resources and inertia are allocated to recover from the largest generation and load contingencies under the assumption that the system remains fully interconnected. However, those dispatch conditions are insufficient to withstand interconnector failures. For example, under low demand conditions, SA lacks enough inertia to maintain ROCOF within acceptable limits for the loss of imports of 77.2 MW through the Heywood interconnector. Similarly, TAS would

Table 6.8 Base case results for PFR (in MW) and inertia (in GWs).

Dem	Resource	SA	VIC	TAS	NSW	QLD
	Inertia	9.6	34.3	8.5	61.2	52.2
High	PFR-dw	0.0	13.3	0.0	0.0	0.0
	PFR-up	117.3	90.9	35.9	41.5	73.6
	Inertia	0.1	29.4	11.9	10.6	27.1
Low	PFR-dw	0.0	181.3	0.0	0.0	0.0
	PFR-up	2.3	61.9	48.5	129.6	284.8

Source: [75] / with permission of Elsevier.

Table 6.9 Base case results for interconnector flows (in MW). Negative flows represent reversed flows (from node B to node A).

Dem	Heywood (VIC to SA)	Basslink (TAS to VIC)	QNI (NSW to QLD)
High	142.5	−423.1	−727.7
Low	77.2	515.9	−857.4

Source: [75] / with permission of Elsevier.

experience excessive ROCOF values in both high (−1.24 Hz/s) and low (−1.07 Hz/s) demand scenarios, given its import/export levels and allocated inertia. If any of the three interconnectors fails, QSSF requirements would be violated in the affected areas [75].

Frequency extrema are not identified for QLD and TAS, due to severe frequency excursions and limited local response capacity. SA meets the frequency nadir requirement under high demand (49.01 Hz) but not under low demand (47.18 Hz). The mainland (the remaining interconnected regions) generally remains compliant following any interconnector loss, with two exceptions: QSSF violations occur when QNI trips, leading to frequency stabilization at 49.1 Hz (high demand) and 47.58 Hz (low demand). Also, the mainland frequency stabilizes at 50.63 Hz after Basslink fails under high demand conditions [75].

6.5.3.2 SECOPF Results

The analysis is repeated under the same demand and renewable generation conditions as the base case, but with each interconnector treated individually as a separate credible contingency. For the SA-separation constrained operation, i.e., in the event of a Heywood interconnector trip, the resulting PFR resources and inertia are shown in Table 6.10[1] [75].

As shown in Table 6.10 [75], applying the SECOPF model to account for a potential loss of the Heywood interconnector leads to reduced power flows through the interconnector and increased inertia and upward PFR resources in SA under low demand conditions. These adjustments ensure that SA meets both ROCOF and QSSF requirements across both demand levels. In this scenario, the interconnector loss effectively resembles a generation contingency in SA. Under high demand, the resulting frequency nadir is 49.37 Hz, while no nadir is observed in the low demand case. No FR constraint is binding in SA due to the minimum stable generation limit of the gas units in the area. Notably, although SA's inertia levels are very low, the SECOPF model successfully identifies feasible and secure operating conditions by actively controlling flows through Heywood [75].

1 The HVDC link between SA and VIC is assumed not responsive [75].

Table 6.10 SECOPF PFR (in MW) and inertia (in GWs) for loss of Heywood.

Dem	Flow	Resource	SA	VIC	TAS	NSW	QLD
		Inertia	**9.6**	34.4	8.8	61.2	51.1
High	102.8	PFR-dw	**0.0**	0.0	13.3	0.0	0.0
		PFR-up	**112.7**	113.2	68.2	41.5	23.6
		Inertia	**0.7**	28.9	11.9	10.6	26.3
Low	11.9	PFR-dw	**0.0**	181.3	0.0	0.0	0.0
		PFR-up	**11.4**	85.1	16.2	132.2	282.3

Source: [75] / with permission of Elsevier.

Table 6.11 SECOPF PFR (in MW) and inertia (in GWs) for loss of Basslink.

Dem	Flow	Resource	SA	VIC	TAS	NSW	QLD
		Inertia	8.3	34.3	**13.8**	61.2	50.8
High	−280.5	PFR-dw	13.3	0.0	**0.0**	0.0	0.0
		PFR-up	0.0	0.0	**416.0**	0.0	0.0
		Inertia	0.1	28.1	**11.9**	14.4	24.0
Low	102.8	PFR-dw	0.0	89.7	**91.5**	0.0	0.0
		PFR-up	2.3	0.0	**0.0**	275.4	249.4

Source: [75] / with permission of Elsevier.

Similarly, when the loss of Basslink is a credible contingency (Table 6.11), the most cost-effective strategy for maintaining frequency stability in TAS involves reducing interconnector flows—while preserving the original flow directions for each case. With the adjusted transfers and reallocated FR resources, the QSSF constraint becomes binding in TAS under low demand conditions, with the frequency zenith reaching 50.71 Hz. In the high demand scenario, the frequency nadir constraint becomes binding due to the loss of 281 MW being transferred through Heywood with an inertia level of 13.8 GWs and 416 MW of local upward PFR [75].

Table 6.12 [75] shows the results for the interconnector between QLD and NSW. In this case, the OPF again reduces interconnector flows. However, due to the significant transfers in the base

Table 6.12 SECOPF PFR (in MW) and inertia (in GWs) for loss of QNI.

Dem	Flow	Resource	SA	VIC	TAS	NSW	QLD
		Inertia	9.6	35.3	8.5	62.7	**50.8**
High	−551.0	PFR-dw	0.0	0.0	0.0	0.0	**655.5**
		PFR-up	98.8	105.9	38.0	66.5	**50.0**
		Inertia	0.1	29.4	11.9	14.4	**22.5**
Low	−371.3	PFR-dw	0.0	0.0	0.0	0.0	**684.1**
		PFR-up	2.3	0.0	16.2	225.3	**283.4**

Source: [75] / with permission of Elsevier.

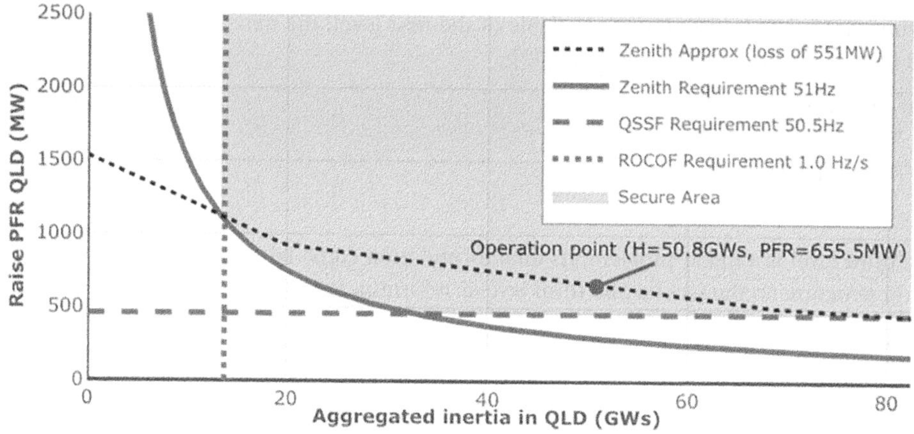

Figure 6.17 QLD frequency response (FR) security map [96] for the loss of 551MW transferred through QNI, with 9224MW of demand in QLD. *Source:* [75] / with permission of Elsevier.

case and the availability of low-cost generation in QLD, reducing the flow to near-zero levels is less cost-effective compared to other scenarios. The binding constraints are the frequency zenith limit in QLD following the tripping of QNI, for both high and low demand conditions. Additionally, the QSSF constraint becomes binding in the mainland under low demand [75].

It is important to note that the linear hyperplanes used to approximate frequency extrema are calculated to be robust (following [79]). This is illustrated in Figure 6.17 [75], where for the high demand scenario in QLD, the zenith approximation curve serves as the binding constraint, remaining consistently above the QSSF requirement curve, and is also conservative, staying above the 51 Hz zenith limit. In essence, the approximated security map [96] is always smaller than the formally calculated one. The degree of conservativeness depends on the range of contingencies considered (100 to 800 MW in this case, which is relatively broad) and the number of hyperplanes used (limited to 4 in this case). By a better knowledge of the operation point of the SECOPF (e.g. via iterative calculation of the operation point) and/or by adding more linear approximations, this result would be improved. In this example, the resulting frequency zeniths are 50.63 Hz and 50.67 Hz for the high and low demand cases, respectively [75].

6.6 Appendix

6.6.1 Reformulation and Solution Algorithm of Tri-level Problems

A general formulation of the tri-level problem described in Section 6.3.1.1 is the following [42]:

$$\min_{\boldsymbol{h}} \left\{ \mathbf{c}_1^{\mathrm{T}} \boldsymbol{h} + \max_{z \in \mathrm{W}} \min_{\boldsymbol{o} \in \mathrm{O}(z,\hat{\boldsymbol{h}})} \mathbf{c}_2^{\mathrm{T}} \boldsymbol{o} \right\} \tag{6.20}$$

$$s.t.\ \boldsymbol{A}\boldsymbol{o} + \boldsymbol{B}\boldsymbol{h} + \boldsymbol{D}\boldsymbol{z} \geq \boldsymbol{E} \tag{6.21}$$

$$\boldsymbol{G}\boldsymbol{z} \geq \boldsymbol{K} \tag{6.22}$$

where, $\mathbf{c}_1^{\mathrm{T}}$ and $\mathbf{c}_2^{\mathrm{T}}$ are matrices describing the costs, $\boldsymbol{h}$, $\boldsymbol{z}$, and $\boldsymbol{o}$, the decision variables of the first, second and third level, respectively and $\boldsymbol{A}$, $\boldsymbol{B}$, $\boldsymbol{E}$, $\boldsymbol{F}$, $\boldsymbol{G}$, and $\boldsymbol{K}$ are matrices used to describe constraints. $\boldsymbol{z}$ describes the status of the lines (binary variables).

Considering that $\hat{h}$ is a solution of the variable of the first level, the max–min problem of the second and third level is expressed as follows [42]:

$$\max_{z \in \mathbb{W}} \min_{o \in \mathbb{O}(z,\hat{h})} \mathbf{c}_2^{\mathrm{T}} \boldsymbol{o} \tag{6.23}$$

$$s.t.\ \boldsymbol{Ao} + \boldsymbol{B}\hat{\boldsymbol{h}} + \boldsymbol{Dz} \geq \boldsymbol{E} \tag{6.24}$$

$$\boldsymbol{Gz} \geq \boldsymbol{K} \tag{6.25}$$

Since the third level is a linear problem, its dual formulation could be considered. In that way, the max–min structure on the second and third level is reformulated as follows [42]:

$$\max_{z \in \mathbb{W}} \max_{\lambda \in \mathbb{F}(z,\hat{h})} (\boldsymbol{E} - \boldsymbol{B}\hat{\boldsymbol{h}} - \boldsymbol{Dz})^{\mathrm{T}} \lambda \tag{6.26}$$

$$s.t.\ \boldsymbol{A}^{\mathrm{T}} \lambda \leq \mathbf{c} \tag{6.27}$$

$$\boldsymbol{Gz} \geq \boldsymbol{K} \tag{6.28}$$

where λ is the matrix describing the dual variables of the constraints of the third level and $\mathbb{F}(z, \hat{h})$ is the set of feasible solutions of the dual problem of the third level. In this way, the tri-level problem can be reformulated as a bi-level problem. It is observed that the objective function of the dual problem has bi-linear terms due to the multiplication of $\lambda_{ij,t}$ with $z_{ij,t}$. These terms can be linearized using techniques that linearize the multiplication of binary ($z_{ij,t}$) with a real variable ($\lambda_{ij,t}$).

6.6.2 Solution Algorithm

The bi-level problem can be decoupled into a master problem and a subproblem. Decomposition methods, such as Benders decomposition and C&CG, can be used to solve the problem. Here, the C&CG method is used. C&CG is a decomposition strategy used for solving two-stage robust optimization problems that is proven to converge to optimal solution [97]. The main idea of C&CG is iteratively adding worst-damage scenarios from subproblem and the corresponding variables to the master problem until an optimal solution is found. Hence, the master problem considers a set of worst-damage scenarios $\hat{\mathbb{W}} = \{\hat{z}^s, s = 1, \dots, m\}$ and is expressed as [42]:

$$\min_{h \in \mathbb{U}, o^s \in \mathbb{O}(\hat{z}^s, h)} \mathbf{c}_1{}^{\mathrm{T}} \boldsymbol{h} + \varphi \tag{6.29}$$

$$s.t.\ \varphi \geq \mathbf{c}_2{}^{\mathrm{T}} \boldsymbol{o}^s, \forall\, s \in 1, \dots, m \tag{6.30}$$

$$\boldsymbol{Ao}^s + \boldsymbol{Bh} + \boldsymbol{D}\hat{\boldsymbol{z}}^s \geq \boldsymbol{E} \tag{6.31}$$

where φ is a scalar variable introduced to guarantee that the optimal solution of the master problem dominates the included worst-damage scenarios. For applying C&CG, the subproblem must be bounded and its feasibility must be guaranteed.

References

1 M. Panteli and P. Mancarella, "The grid: stronger, bigger, smarter?: presenting a conceptual framework of power system resilience," *IEEE Power and Energy Mag.*, vol. 13, no. 3, pp. 58–66, 2015, doi: 10.1109/MPE.2015.2397334.

2 M. Mahzarnia, M. P. Moghaddam, P. T. Baboli, and P. Siano, "A review of the measures to enhance power systems resilience," *IEEE Systems Journal*, vol. 14, no. 3, pp. 4059–4070, 2020, doi: 10.1109/JSYST.2020.2965993.

3 Y. Wang, C. Chen, J. Wang, and R. Baldick, "Research on resilience of power systems under natural disasters—a review," *IEEE Transactions on Power Apparatus and Systems*, vol. 31, no. 2, pp. 1604–1613, 2016, doi: 10.1109/TPWRS.2015.2429656.

4 M. Panteli and D. S. Kirschen, "Situation awareness in power systems: theory, challenges and applications," *Electric Power Systems Research*, vol. 122, pp. 140–151, 2015, doi: 10.1016/j.epsr.2015.01.008.

5 A. Arab, A. Khodaei, S. K. Khator, K. Ding, V. A. Emesih, and Z. Han, "Stochastic pre-hurricane restoration planning for electric power systems infrastructure," *IEEE Transactions on Smart Grid*, vol. 6, no. 2, pp. 1046–1054, 2015, doi: 10.1109/TSG.2015.2388736.

6 R. Moreno *et al.*, "Microgrids against wildfires: distributed energy resources enhance system resilience," *IEEE Power and Energy Mag.*, vol. 20, no. 1, pp. 78–89, 2022, doi: 10.1109/MPE.2021.3122772.

7 M. Panteli, D. N. Trakas, P. Mancarella, and N. D. Hatziargyriou, "Power systems resilience assessment: hardening and smart operational enhancement strategies," *Proceedings of the IEEE*, vol. 105, no. 7, pp. 1202–1213, 2017, doi: 10.1109/JPROC.2017.2691357.

8 G. Strbac, D. Kirschen, and R. Moreno, "Reliability standards for the operation and planning of future electricity networks," *FNT in Electric Energy Systems*, vol. 1, no. 1, pp. 143–219, 2016, doi: 10.1561/3100000001.

9 Y. Pipelzadeh, R. Moreno, B. Chaudhuri, G. Strbac, and T. C. Green, "Corrective control with transient assistive measures: value assessment for great Britain transmission system," *IEEE Transactions on Power Apparatus and Systems*, pp. 1–1, 2016, doi: 10.1109/TPWRS.2016.2598815.

10 Y. Chen, R. Moreno, G. Strbac, and D. Alvarado, "Coordination strategies for securing AC/DC flexible transmission networks with renewables," *IEEE Transactions on Power Apparatus and Systems*, vol. 33, no. 6, pp. 6309–6320, 2018, doi: 10.1109/TPWRS.2018.2851214.

11 D. P. Nedic, I. Dobson, D. S. Kirschen, B. A. Carreras, and V. E. Lynch, "Criticality in a cascading failure blackout model," *International Journal of Electrical Power & Energy Systems*, vol. 28, no. 9, pp. 627–633, 2006, doi: 10.1016/j.ijepes.2006.03.006.

12 D. Xianzhong and S. Sheng, "Self-organized criticality in time series of power systems fault, its mechanism, and potential application," *IEEE Transactions on Power Apparatus and Systems*, vol. 25, no. 4, pp. 1857–1864, 2010, doi: 10.1109/TPWRS.2010.2058932.

13 M. Panteli, P. A. Crossley, D. S. Kirschen, and D. J. Sobajic, "Assessing the impact of insufficient situation awareness on power system operation," *IEEE Transactions on Power Apparatus and Systems*, vol. 28, no. 3, pp. 2967–2977, 2013, doi: 10.1109/TPWRS.2013.2240705.

14 R. Moreira, R. Moreno, and G. Strbac, "Value of corrective network security for distributed energy storage applications," *IET Generation, Transmission & Distribution*, vol. 10, no. 7, pp. 1758–1767, 2016, doi: 10.1049/iet-gtd.2015.0457.

15 Y. Zhou, M. Panteli, R. Moreno, and P. Mancarella, "System-level assessment of reliability and resilience provision from microgrids," *Applied Energy*, vol. 230, pp. 374–392, 2018, doi: 10.1016/j.apenergy.2018.08.054.

16 F. Mohammadi, M. Sahraei-Ardakani, D. Trakas, and N. Hatziargyriou, "Machine learning assisted stochastic unit commitment during hurricanes with predictable line outages," *IEEE Transactions on Power Apparatus and Systems*, vol. 36, no. 6, pp. 5131–5142, 2021, doi: 10.1109/TPWRS.2021.3069443.

17 D. N. Trakas, M. Panteli, N. D. Hatziargyriou, and P. Mancarella, "Spatial risk analysis of power systems resilience during extreme events: spatial risk analysis of power systems resilience," *Risk Analysis*, vol. 39, no. 1, pp. 195–211, 2019, doi: 10.1111/risa.13220.

18 F. Mohammadi and M. Sahraei-Ardakani, "Multidimensional scenario selection for power systems with stochastic failures," *IEEE Transactions on Power Apparatus and Systems*, vol. 35, no. 6, pp. 4528–4538, 2020, doi: 10.1109/TPWRS.2020.2990877.

19 E. Du, N. Zhang, C. Kang, and Q. Xia, "Scenario map based stochastic unit commitment," *IEEE Transactions on Power Apparatus and Systems*, vol. 33, no. 5, pp. 4694–4705, 2018, doi: 10.1109/TPWRS.2018.2799954.

20 A. Papavasiliou, S. S. Oren, and B. Rountree, "Applying high performance computing to transmission-constrained stochastic unit commitment for renewable energy integration," *IEEE Transactions on Power Apparatus and Systems*, vol. 30, no. 3, pp. 1109–1120, 2015, doi: 10.1109/TPWRS.2014.2341354.

21 A. Papavasiliou, S. S. Oren, and B. Rountree, "Applying high performance computing to transmission-constrained stochastic unit commitment for renewable energy integration," *IEEE Transactions on Power Apparatus and Systems*, vol. 30, no. 3, pp. 1109–1120, 2015, doi: 10.1109/TPWRS.2014.2341354.

22 J. Xue, F. Mohammadi, X. Li, M. Sahraei-Ardakani, G. Ou, and Z. Pu, "Impact of transmission tower-line interaction to the bulk power system during hurricane," *Reliability Engineering & System Safety*, vol. 203, 107079, 2020, doi: 10.1016/j.ress.2020.107079.

23 S. Park, Q. Xu, and B. F. Hobbs, "Comparing scenario reduction methods for stochastic transmission planning," *IET Generation Transmission and Distribution*, vol. 13, no. 7, pp. 1005–1013, 2019, doi: 10.1049/iet-gtd.2018.6362.

24 Y. Wang, Y. Liu, and D. S. Kirschen, "Scenario reduction with submodular optimization," *IEEE Transactions on Power Apparatus and Systems*, vol. 32, no. 3, pp. 2479–2480, 2017, doi: 10.1109/TPWRS.2016.2603448.

25 J. Hu and H. Li, "A new clustering approach for scenario reduction in multi-stochastic variable programming," *IEEE Transactions on Power Apparatus and Systems*, vol. 34, no. 5, pp. 3813–3825, 2019, doi: 10.1109/TPWRS.2019.2901545.

26 F. Mohammadi and M. Sahraei-Ardakani, "Tractable stochastic unit commitment for large systems during predictable hazards," *IEEE Access*, vol. 8, pp. 115078–115088, 2020, doi: 10.1109/ACCESS.2020.3004391.

27 P. A. Ruiz, E. A. Goldis, A. Rudkevich, M. C. Caramanis, C. R. Philbrick, and J. M. Foster, "Security-constrained transmission topology control MILP formulation using sensitivity factors," *IEEE Transactions on Power Apparatus and Systems*, pp. 1–1, 2016, doi: 10.1109/TPWRS.2016.2577689.

28 A. Street, A. Moreira, and J. M. Arroyo, "Energy and reserve scheduling under a joint generation and transmission security criterion: an adjustable robust optimization approach," *IEEE Transactions on Power Apparatus and Systems*, vol. 29, no. 1, pp. 3–14, 2014, doi: 10.1109/TPWRS.2013.2278700.

29 A. Moreira, A. Street, and J. M. Arroyo, "An adjustable robust optimization approach for contingency-constrained transmission expansion planning," *IEEE Transactions on Power Apparatus and Systems*, vol. 30, no. 4, pp. 2013–2022, 2015, doi: 10.1109/TPWRS.2014.2349031.

30 Q. Wang, J.-P. Watson, and Y. Guan, "Two-stage robust optimization for N-k contingency-constrained unit commitment," *IEEE Transactions on Power Apparatus and Systems*, vol. 28, no. 3, pp. 2366–2375, 2013, doi: 10.1109/TPWRS.2013.2244619.

31 C. Zhao and R. Jiang, "Distributionally robust contingency-constrained unit commitment," *IEEE Transactions on Power Apparatus and Systems*, vol. 33, no. 1, pp. 94–102, 2018, doi: 10.1109/TPWRS.2017.2699121.

32 P. Xiong and P. Jirutitijaroen "An adjustable robust optimization approach for unit commitment under outage contingencies," in *in 2012 IEEE Power and Energy Society General Meeting*. San Diego, CA: IEEE, 2012, pp. 1–8, doi: 10.1109/PESGM.2012.6345297.

33 R. Fernandez-Blanco, Y. Dvorkin, and M. A. Ortega-Vazquez, "Probabilistic security-constrained unit commitment with generation and transmission contingencies," *IEEE Transactions on Power Apparatus and Systems*, vol. 32, no. 1, pp. 228–239, 2017, doi: 10.1109/TPWRS.2016.2550585.

34 D. Pozo and J. Contreras, "A chance-constrained unit commitment with an $n-K$ security criterion and significant wind generation," *IEEE Transactions on Power Apparatus and Systems*, vol. 28, no. 3, pp. 2842–2851, 2013, doi: 10.1109/TPWRS.2012.2227841.

35 K. A. Papadogiannis and N. D. Hatziargyriou, "Optimal allocation of primary reserve services in energy markets," *IEEE Transactions on Power Apparatus and Systems*, vol. 19, no. 1, pp. 652–659, 2004, doi: 10.1109/TPWRS.2003.820702.

36 E. S. Karapidakis and N. D. Hatziargyriou, "Online preventive dynamic security of isolated power systems using decision trees," *IEEE Transactions on Power Apparatus and Systems*, vol. 17, no. 2, pp. 297–304, 2002, doi: 10.1109/TPWRS.2002.1007896.

37 E. M. Voumvoulakis and N. D. Hatziargyriou, "Decision trees-aided self-organized maps for corrective dynamic security," *IEEE Transactions on Power Apparatus and Systems*, vol. 23, no. 2, pp. 622–630, 2008, doi: 10.1109/TPWRS.2008.920194.

38 G. Dalal, E. Gilboa, S. Mannor, and L. Wehenkel "Unit commitment using nearest neighbor as a short-term proxy," in *in 2018 Power Systems Computation Conference (PSCC)*. Dublin, Ireland: IEEE, June 2018, pp. 1–7, doi: 10.23919/PSCC.2018.8442516.

39 G. Dalal, E. Gilboa, S. Mannor, and L. Wehenkel, "Chance-constrained outage scheduling using a machine learning proxy," *IEEE Transactions on Power Apparatus and Systems*, vol. 34, no. 4, pp. 2528–2540, 2019, doi: 10.1109/TPWRS.2018.2889237.

40 S. Pineda, J. M. Morales, and A. Jimenez-Cordero, "Data-driven screening of network constraints for unit commitment," *IEEE Transactions on Power Apparatus and Systems*, vol. 35, no. 5, pp. 3695–3705, 2020, doi: 10.1109/TPWRS.2020.2980212.

41 G. Dalal and S. Mannor "Reinforcement learning for the unit commitment problem," in *in 2015 IEEE Eindhoven PowerTech*. Eindhoven, Netherlands: IEEE, 2015, pp. 1–6, doi: 10.1109/PTC.2015.7232646.

42 D. N. Trakas and N. D. Hatziargyriou, "Resilience constrained day-ahead unit commitment under extreme weather events," *IEEE Transactions on Power Apparatus and Systems*, vol. 35, no. 2, pp. 1242–1253, 2020, doi: 10.1109/TPWRS.2019.2945107.

43 M. Panteli, P. Mancarella, D. N. Trakas, E. Kyriakides, and N. D. Hatziargyriou, "Metrics and quantification of operational and infrastructure resilience in power systems," *IEEE Transactions on Power Apparatus and Systems*, vol. 32, no. 6, pp. 4732–4742, 2017, doi: 10.1109/TPWRS.2017.2664141.

44 A. J. Wood, B. F. Wollenberg, and G. B. Sheblé, *Power generation, operation, and control*, Third ed. Hoboken, New Jersey: Wiley-IEEE, 2013.

45 C. Grigg *et al.*, "The IEEE reliability test system-1996. a report prepared by the reliability test system task force of the application of probability methods subcommittee," *IEEE Transactions on Power Apparatus and Systems*, vol. 14, no. 3, pp. 1010–1020, 1999, doi: 10.1109/59.780914.

46 M. Panteli, D. N. Trakas, P. Mancarella, and N. D. Hatziargyriou, "Boosting the power grid resilience to extreme weather events using defensive islanding," *IEEE Transactions on Smart Grid*, vol. 7, no. 6, pp. 2913–2922, 2016, doi: 10.1109/TSG.2016.2535228.

47 A. Kargarian and Y. Fu, "System of systems based security-constrained unit commitment incorporating active distribution grids," *IEEE Transactions on Power Apparatus and Systems*, vol. 29, no. 5, pp. 2489–2498, 2014, doi: 10.1109/TPWRS.2014.2307863.

48 H. Ronellenfitsch, M. Timme, and D. Witthaut, "A dual method for computing power transfer distribution factors," *IEEE Transactions on Power Apparatus and Systems*, pp. 1–1, 2016, doi: 10.1109/TPWRS.2016.2589464.

49 F. Mohammadi, M. Sahraei-Ardakani, Y. M. Al-Abdullah, and G. T. Heydt, "Coordinated scheduling of power generation and water desalination units," *IEEE Transactions on Power Apparatus and Systems*, vol. 34, no. 5, pp. 3657–3666, 2019, doi: 10.1109/TPWRS.2019.2901807.

50 Z. Wang and J. Wang, "Self-healing resilient distribution systems based on sectionalization Into microgrids," *IEEE Transactions on Power Apparatus and Systems*, vol. 30, no. 6, pp. 3139–3149, 2015, doi: 10.1109/TPWRS.2015.2389753.

51 Vaiman *et al.*, "Risk assessment of cascading outages: methodologies and challenges," *IEEE Transactions on Power Apparatus and Systems*, vol. 27, no. 2, pp. 631–641, 2012, doi: 10.1109/TPWRS.2011.2177868.

52 M. Amin "Challenges in reliability, security, efficiency, and resilience of energy infrastructure: toward smart self-healing electric power grid," in *in 2008 IEEE Power and Energy Society General Meeting - Conversion and Delivery of Electrical Energy in the 21st Century*. Pittsburgh, PA, USA: IEEE, 2008, pp. 1–5, doi: 10.1109/PES.2008.4596791.

53 L. R. Phillips *et al.* "Agents and islands: managing a power system before, during, and after transition to the islanded state," in *in 2006 IEEE/SMC International Conference on System of Systems Engineering*. Los Angeles, California, USA: IEEE, 2006, pp. 161–166, doi: 10.1109/SYSOSE.2006.1652291.

54 A. Ashrafi and S. M. Shahrtash, "Dynamic wide area voltage control strategy based on organized multi-agent system," *IEEE Transactions on Power Apparatus and Systems*, vol. 29, no. 6, pp. 2590–2601, 2014, doi: 10.1109/TPWRS.2014.2313607.

55 Z. Qin, S. Liu, and Y. Hou "Virtual synchroscope: a novel application of PMU for system restoration," in *in 2011 International Conference on Advanced Power System Automation and Protection*. Beijing, China: IEEE, 2011, pp. 193–197, doi: 10.1109/APAP.2011.6180407.

56 R. Moreno and A. Torres "Security of the power system based on the separation into islands," in *2011 IEEE Pes Conference On Innovative Smart Grid Technologies Latin America (ISGT LA)*. Medellin, Colombia: IEEE, 2011, pp. 1–5, doi: 10.1109/ISGT-LA.2011.6083210.

57 L. Ding, F. M. Gonzalez-Longatt, P. Wall, and V. Terzija, "Two-step spectral clustering controlled islanding algorithm," *IEEE Transactions on Power Apparatus and Systems*, vol. 28, no. 1, pp. 75–84, 2013, doi: 10.1109/TPWRS.2012.2197640.

58 Q. Zhao, K. Sun, D.-Z. Zheng, J. Ma, and L. Qiang, "A study of system splitting strategies for island operation of power system: a two-phase method based on OBDDs," *IEEE Transactions on Power Apparatus and Systems*, vol. 18, no. 4, pp. 1556–1565, 2003, doi: 10.1109/TPWRS.2003.818747.

59 K. Sun, D.-Z. Zheng, and Q. Lu, "A simulation study of OBDD-based proper splitting strategies for power systems under consideration of transient stability," *IEEE Transactions on Power Apparatus and Systems*, vol. 20, no. 1, pp. 389–399, 2005, doi: 10.1109/TPWRS.2004.841239.

60 K. Sun, D.-Z. Zheng, and L. Qiang, "Splitting strategies for islanding operation of large-scale power systems using OBDD-based methods," *IEEE Transactions on Power Apparatus and Systems*, vol. 18, no. 2, pp. 912–923, 2003, doi: 10.1109/TPWRS.2003.810995.

61 F. Raak, Y. Susuki, and T. Hikihara, "Data-driven partitioning of power networks via koopman mode analysis," *IEEE Transactions on Power Apparatus and Systems*, vol. 31, no. 4, pp. 2799–2808, 2016, doi: 10.1109/TPWRS.2015.2464779.

62 T. Ding, K. Sun, C. Huang, Z. Bie, and F. Li, "Mixed-integer linear programming-based splitting strategies for power system islanding operation considering network connectivity," *IEEE Systems Journal*, vol. 12, no. 1, pp. 350–359, 2018, doi: 10.1109/JSYST.2015.2493880.

63 P. A. Trodden, W. A. Bukhsh, A. Grothey, and K. I. M. McKinnon, "Optimization-based islanding of power networks using piecewise linear AC power flow," *IEEE Transactions on Power Apparatus and Systems*, vol. 29, no. 3, pp. 1212–1220, 2014, doi: 10.1109/TPWRS.2013.2291660.

64 P. Paradell *et al.* "Increasing resilience of power systems using intentional islanding; a comparison of Binary genetic algorithm and deep learning based method," in *2021 IEEE International Conference on Cyber Security and Resilience (CSR)*. Rhodes, Greece: IEEE, 2021, pp. 491–496, doi: 10.1109/CSR51186.2021.9527918.

65 D. N. Trakas, E. M. Voumvoulakis, and N. D. Hatziargyriou, "Controlled islanding of power networks using machine learning algorithm," in *MedPower 2014*, Athens, Greece: Institution of Engineering and Technology, 2014, p. 50 (6 pp.)-50 (6 pp.). doi: 10.1049/cp.2014.1683.

66 W. Liu, L. Liu, and D. A. Cartes "Slow coherency and angle modulated particle swarm optimization based islanding of large scale power systems," in *in 2007 International Joint Conference on Neural Networks*. Orlando, FL, USA: IEEE, Aug. 2007, pp. 2087–2092, doi: 10.1109/IJCNN.2007.4371280.

67 G. Xu, V. Vittal, A. Meklin, and J. E. Thalman, "Controlled islanding demonstrations on the WECC system," *IEEE Transactions on Power Apparatus and Systems*, vol. 26, no. 1, pp. 334–343, 2011, doi: 10.1109/TPWRS.2010.2047413.

68 J. Quirós-Tortós, R. Sánchez-García, J. Brodzki, J. Bialek, and V. Terzija, "Constrained spectral clustering-based methodology for intentional controlled islanding of large-scale power systems," *IET Generation, Transmission & Distribution*, vol. 9, no. 1, pp. 31–42, 2015, doi: 10.1049/iet-gtd.2014.0228.

69 D. N. Trakas, E. M. Voumvoulakis, and N. D. Hatziargyriou "Decentralized control of power system zones based on probabilistic constrained load flow," in *2014 International Conference on Probabilistic Methods Applied to Power Systems (PMAPS)*. Durham, United Kingdom: IEEE, 2014, pp. 1–6, doi: 10.1109/PMAPS.2014.6960652.

70 H. Mehrjerdi, S. Lefebvre, M. Saad, and D. Asber, "A decentralized control of partitioned power networks for voltage regulation and prevention against disturbance propagation," *IEEE Transactions on Power Apparatus and Systems*, vol. 28, no. 2, pp. 1461–1469, 2013, doi: 10.1109/TPWRS.2012.2225154.

71 Y. Dai, R. Preece, and M. Panteli, "Risk assessment of cascading failures in power systems with increasing wind penetration," *Electric Power Systems Research*, vol. 211, 108392, 2022, doi: 10.1016/j.epsr.2022.108392.

72 Y. Dai, M. Noebels, M. Panteli, and R. Preece "Evaluating the effect of dynamic and static modelling on cascading failure analysis in power systems," in *in 2021 IEEE Madrid PowerTech*. Madrid, Spain: IEEE, 2021, pp. 1–6, doi: 10.1109/PowerTech46648.2021.9495001.

73 J. D. Ming Ni, V. V. McCalley, and T. Tayyib, "Online risk-based security assessment," *IEEE Transactions on Power Apparatus and Systems*, vol. 18, no. 1, pp. 258–265, 2003, doi: 10.1109/TPWRS.2002.807091.

74 L. Meegahapola, P. Mancarella, D. Flynn, and R. Moreno, "Power system stability in the transition to a low carbon grid: A techno-economic perspective on challenges and opportunities," *WIREs Energy and Environment*, vol. 10, no. 5, 2021, doi: 10.1002/wene.399.

75 S. Püschel-Løvengreen, M. Ghazavi Dozein, S. Low, and P. Mancarella, "Separation event-constrained optimal power flow to enhance resilience in low-inertia power systems," *Electric Power Systems Research*, vol. 189, 106678, 2020, doi: 10.1016/j.epsr.2020.106678.

76 Aemo, "Final report – Queensland and South Australia system separation on 25 August 2018," Australian Energy Market Operator, 2019. [Online]. Available: https://www.aemo.com.au/-/media/Files/Electricity/NEM/Market_Notices_and_Events/Power_System_Incident_Reports/2018/Qld---SA-Separation-25-August-2018-Incident-Report.pdf.

77 F. Teng, V. Trovato, and G. Strbac, "Stochastic scheduling with inertia-dependent fast frequency response requirements," *IEEE Transactions on Power System*, vol. 31, no. 2, pp. 1557–1566, 2016.

78 L. Zhang, Y. Zhou, D. Flynn, J. Mutale, and P. Mancarella, "System-level operational and adequacy impact assessment of PV and distributed energy storage, with consideration of inertial constraints, dynamic reserve and interconnection flexibility," *Energies*, vol. 10, no. 989, pp. 1–40, 2017, doi: 10.3390/en10070989.

79 S. Püschel-Løvengreen and P. Mancarella, "Frequency response constrained economic dispatch with consideration of generation contingency size," in 20th Power Systems Computation Conf., 2018.

80 F. Teng, Y. Mu, H. Jia, J. Wu, P. Zeng, and G. Strbac, "Challenges on primary frequency control and potential solution from EVs in the future GB electricity system," *Applied Energy*, vol. 194, pp. 353–362, 2017, doi: 10.1016/j.apenergy.2016.05.123.

81 D. Ortiz-Villalba, J. Llanos, Y. Muñoz-Jadan, R. Moreno, C. Rahmann, and B. C. Pal, "Optimizing system operation with nadir considerations via simulations of detailed system dynamic responses," *Electric Power Systems Research*, vol. 212, 108533, 2022, doi: 10.1016/j.epsr.2022.108533.

82 National Grid, "*Future Requirements for Balancing Services*," 2016.

83 AEMO, "*Frequency control ancillary services*," 2012.

84 P. Daly, D. Flynn, and N. Cunniffe, "Inertia considerations within unit commitment and economic dispatch for systems with high non-synchronous penetrations," *IEEE PowerTech*, 2015, doi: 10.1109/PTC.2015.7232567.

85 AEMO, "Inertia Requirements Methodology," no. June, 2018.

86 National Grid, "*System Needs and Product Strategy*," 2017.

87 H. Gu, R. Yan, and T. K. Saha, "Minimum Synchronous Inertia Requirement of Renewable Power Systems," *IEEE Transactions on Power Systems*, vol. 8950, 2017, doi: 10.1109/TPWRS.2017.2720621.

88 A. S. Ahmadyar, S. Riaz, G. Verbic, J. Riesz, and A. Chapman, "Assessment of minimum inertia requirement for system frequency stability," *2016 IEEE International Conference on Power System Technology, POWERCON* 2016, pp. 1–6, 2016, doi: 10.1109/POWERCON.2016.7753912.

89 ENTSO-E Technical Experts' Team, "Continental Europe Synchronous Area Separation on 24 July 2021," ENTSO-E, Mar. 2022. [Online]. Available: https://eepublicdownloads.azureedge.net/clean-documents/Publications/2022/entso-e_CESysSep_210724_02_Final_Report_220325.pdf.

90 M. G. Dozein, A. M. De Corato, and P. Mancarella, "Virtual inertia response and frequency control ancillary services from hydrogen electrolyzers," *IEEE Transactions on Power Apparatus and Systems*, vol. 38, no. 3, pp. 2447–2459, 2023, doi: 10.1109/TPWRS.2022.3181853.

91 AEMO, "Generation Information Page." 2018. [Online]. Available: https://www.aemo.com.au/Electricity/National-Electricity-Market-NEM/Planning-and-forecasting/Generation-information

92 Australian PV Institue, "Mapping Australian Photovoltaic installations." 2018. [Online]. Available: https://d284f79vx7w9nf.cloudfront.net/attachments/analysis/7/states_e395.csv.

93 AEMO, "*Interconnector Capabilities for the National Electricity Market*," 2017.

94 Jacobs Group, "*Modelling illustrative electricity sector emissions reduction policies Climate Change Authority*," 2016.

95 AEMO, "*Integrated System Plan*," 2018.

96 Melbourne Energy Institute of University of Melbourne, "Power System Security Assessment of the future National Electricity Market," June 2017. [Online]. Available: https://www.energy.gov.au/sites/default/files/independent-review-future-nem-power-system-security-assessment.pdf.

97 B. Zeng and L. Zhao, "Solving two-stage robust optimization problems using a column-and-constraint generation method," *Operations Research Letters*, vol. 41, no. 5, pp. 457–461, 2013, doi: 10.1016/j.orl.2013.05.003.

7

Resilience by Distributed Energy Resources and Microgrids

7.1 Local and Bulk System Resilience

The Supergrid, the construction and expansion of densely interconnected high-voltage (HV) transmission system, enables the exploitation of counter-cyclicality among primary renewable energy sources (RES) and the system to benefit from diversity in demand and supply across wider geographical areas, while allowing sharing of short and long-term reserves across neighboring countries (local level volatility reduces across dispersed areas). Moreover, increasing the share of RES requires extended HV transmission networks, in order to exploit the renewable potential often found in remote locations, such as offshore. Extended Transmission infrastructures, however, are also more vulnerable to catastrophic events due to climate change. Damages to the transmission system, naturally, affect the power supply of larger areas and wider populations; the cost of strengthening the relevant infrastructure to face such high-impact, low-probability (HILP) events is also high. It is therefore sometimes more efficient to support resilience also from resources connected to the distribution network [1]. Distribution networks can not only support the resilience of the transmission system, but also of the neighboring distribution networks.

As discussed in Chapter 2, Section 2.6, the system resilience and security that is traditionally delivered through asset redundancy, can also be provided through control of distributed energy resources (DER), supported by advanced communication and information technologies. These include:

- Demand-side response (DSR), through utilizing the inherent demand-side flexibility, particularly demand associated with heat and transport. DSR can be used for real-time system management, while ensuring that the intended service quality is not adversely affected.
- Electric energy storage (EES) technologies that can be used to support demand–supply balancing or control of network flows and hence increase utilization of electricity infrastructure assets [2].
- Distributed generators (DGs), that mainly include photovoltaics (PVs), wind turbines (WTs) and microturbines (MTs), that can be used to facilitate more secure and cost-effective real-time demand–supply balance and control of network flows.
- Network technologies, such as advanced measurement and network sensors, advanced power electronics, and various novel control and protection schemes, that enhance the utilization and resilience of the network assets through a more sophisticated real-time control of the system.

DER technologies, mostly connected at the distribution networks, allow the local generation, storage and control of energy for local power supply in HILP events, instead of relying on long transmission lines, thus decreasing the vulnerability of the network and improving its resilient response.

Fundamentals of Power System Resilience: Disruptions by Natural Causes, First Edition.
Mathaios Panteli, Rodrigo Moreno, Dimitris Trakas, Magnus Jamieson, Pierluigi Mancarella, Goran Strbac, and Nikos Hatziargyriou.

7.1.1 Overview of Microgrids

Microgrids (MGs) are considered as the most effective way to coordinate DER. In the IEEE standard 2030.7 [3], a MG is defined as "a group of interconnected loads and DER with clearly defined electrical boundaries that acts as a single controllable entity with respect to the grid and can connect and disconnect from the grid to enable it to operate in both grid-connected or island modes." Similarly, CIGRE defines MGs as "sections of electricity distribution systems containing loads and DER, (such as DGs, storage devices, or controllable loads) that can be operated in a controlled, coordinated way, either while connected to the main power network and/or while islanded" [4]. In both definitions, the two distinct features that distinguish MGs from any distribution lines with DER, are their capabilities to (i) operate islanded, thus achieving increased reliability and resilience, and (ii) appear to the upstream network as controlled, coordinated units [2].

Thanks to these characteristics, MGs can disconnect from the upstream grid, operate autonomously, help mitigate grid disturbances and serve as a grid resource for a more resilient response and faster restoration. In this context, MGs, with appropriate enabling technologies, facilitate the paradigm shift in delivering resilience and security of supply from redundancy in assets and preventive control to more intelligent operation through corrective control actions supported by a range of enabling technologies and ICT, as shown in Figure 7.1.

It should be noted that multi-energy MGs consisting of DERs combined with cooling, heat, and power (CCHP) plants to simultaneously provide electricity and thermal energy supply can substantially enhance the ability of MGs to enhance the resilience of the system [5]. This is an area that needs to be further researched and has not been particularly considered in this chapter.

The basic architecture of an MG is shown in Figure 7.2. It comprises a hierarchically controlled distributed structure composed of a network of local controllers connected to each MG element, and a higher control layer termed the microgrid central controller (MGCC), installed at the medium voltage (MV)/low voltage (LV) substation, that is interfaced with the overall distribution management system (DMS). Given the different characteristics of the DER, the local level can be distinguished in load controllers (LC) for the DSR units and microsource controllers (MC) for the DGs and EES.

Three control architectures have been developed for power systems: the centralized, the decentralized and the distributed ones [6]. The main feature of the decentralized architecture is that the control system is composed of several individual controllers, which do not share information, independently of whether or not the selection of the controlled variables takes into account the interactions within the system. Recent interest triggered by advances in technology and telecommunications has focused research attention on distributed control structures. In this architecture,

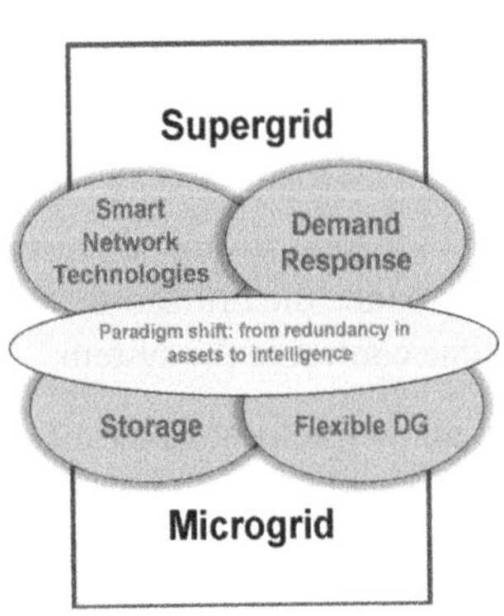

Figure 7.1 Integration of super and microgrids (MG) to facilitate cost-effective and resilient evolution to low-carbon future. *Source:* Adapted from [1].

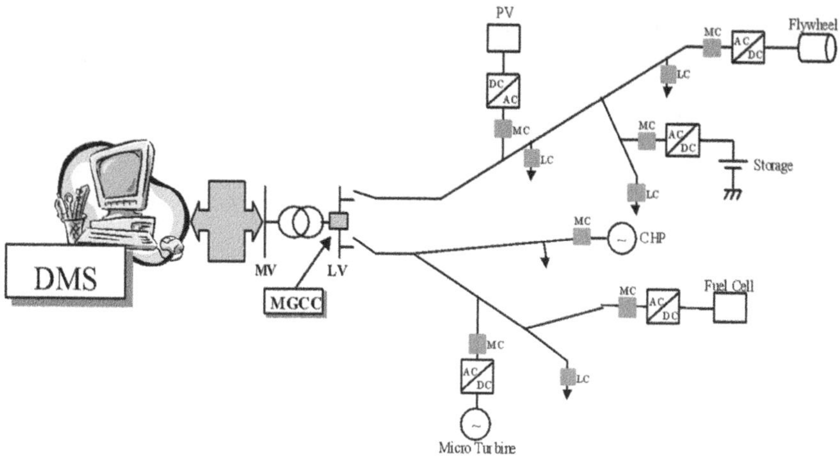

Figure 7.2 Overview of a MG architecture. *Source:* [2] / John Wiley & Sons. Photo: Aytan / Adobe Stock Photos.

it is assumed that some information is shared among controllers, so that each of them has some knowledge about the behavior of the others, thus raising the overall performance. However, there is always a trade-off between communication burden and performance. Centralized architecture consists of a single controller that manages and communicates with all the other components control decisions based on knowledge of all control inputs, optimized in a single optimization problem. In the context of MGs, all three approaches have been investigated. The practical implementation depends on the level of responsibilities assigned to the MGCC and the local DER controllers, and the telecommunication requirements.

The basic challenges that need to be faced in the operation of the MGs are voltage control and frequency/load-generation balance. Voltage control is the main challenge, when the MG operates interconnected to the grid. This can be achieved through combined control of MV/LV on load tap changing and control of the active and reactive power of the DGs. Local control is effected through the use of a P/V droop control solution. Frequency control is the main challenge in islanded operation.

In Section 7.2, selected real-world cases are reviewed, where DER and MGs have been employed during catastrophic events. As highlighted in previous chapters, MGs are able to effectively restore power supply in major events of load disruptions (see e.g. [1]). Furthermore, increasing penetration of DERs, including electrification of transport and heat sectors, and transition to digitalized energy paradigm, can significantly enhance resilience of supply very cost-effectively. In this regard, as also discussed in Chapter 8, resilience should be considered more broadly beyond the power sector and taking into account interdependencies among multiple energy sectors. By doing so, future energy system decarbonization strategies that rely on electrification of transport and heat sectors and resilience objectives can also be effectively linked to facilitate the development of cost-effective, zero-carbon technology and system evolution paths.

7.1.2 Resilient Decarbonization of Multi-Energy Microgrids

Multi-energy microgrids (MEMGs), integrating local energy systems that include electricity, gas/hydrogen, cooling, heating, transport, and other energy forms [5], could address the two main challenges facing the future energy industry: the need to decarbonize the energy sector

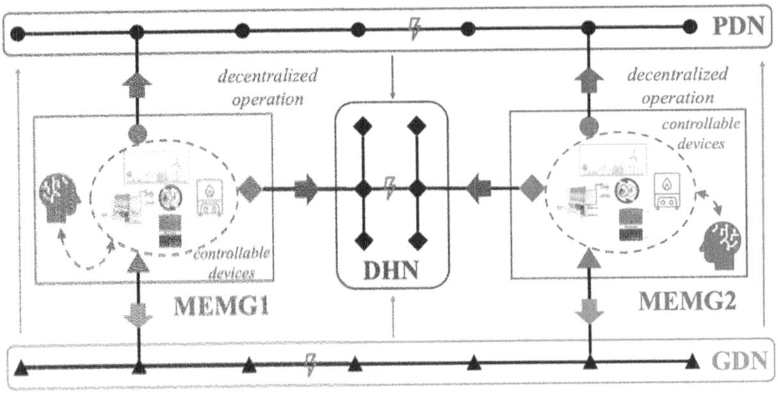

Figure 7.3 Illustration of the coordination scheme of multi-MEMGs in an integrated energy system toward load restoration after extreme events.

and enhance system resilience. Furthermore, coordinated operation of multiple local networked MEMGs can further enhance resilience [7].

One core aspect is related to decarbonization of heat sector, which has been based on the use of natural gas for water and space heating. In a number of countries, one of the most cost-effective paradigms for heat sector decarbonization is based on full electrification, which would involve application of heat pumps. On the other side, if some part of the heat sector is decarbonized through district heating system supplied via hydrogen-based combined heat and power (CHP) plant, the resilience can be significantly increased. Figure 7.3 presents a case of two MEMGs that include power distribution networks (PDNs), gas/hydrogen distribution networks (GDNs), and district heating networks (DHNs) that are controlled via integrated energy systems (IES) based concepts [8–10].

In case when the district area is fully out of electricity supply due to a major fault in electricity transmission network, local CHP plant that is used to provide heat to DHNs, would produce electricity that can be used to provide electricity supply to critical loads. Initial analysis demonstrated that in the application of district heating, as a part of heat supply in urban areas, would only increase the cost of decarbonization marginally (for about 10%), while radically enhancing resilience through significantly reducing the amount of critical demand curtailment (for about 80%). As again discussed in Chapter 8, this demonstrates that decarbonization and resilience objectives need to be considered simultaneously, while, currently, energy decarbonization pathways and system resilience are considered in isolation.

While there is a clear need for comprehensive demonstration of the operation and control of alternative MEMG concepts, it is evident that the integration of future MEMGs introduces a series of challenges in enhancing resilience and supporting energy system decarbonization. This is primarily due to the ongoing shift toward a decentralized operational paradigm, so addressing the network dynamics and uncertainties is very complex. In this context, the application of data-driven artificial intelligence (AI) approaches emerges as a promising solution to developing efficient decentralized real-time control framework aimed at effectively coordinating the activities of MEMGs in the context of resilient and decarbonized energy networks. For instance, with advanced AI-driven smart control, these essential loads, such as hospitals, police stations, data centers, and traffic lights, can be prioritized. As an AI-driven approach, reinforcement learning (RL) has been successfully applied to many decision-making problems in power and energy systems [11]. In contrast to constructing an optimization problem, RL first formulates the decision-making problem as a Markov decision process (MDP) and then solves this MDP using advanced AI-based RL algorithms. Furthermore, as an online learning method, RL learns the system's uncertainties and dynamics directly from the dataset itself and can adapt to various state conditions in milliseconds without any optimizations. Thus, RL can be assumed to be a valid solution for the real-time control-related decision-making process, which is crucial for the decentralized smart control for enhancing resilience in interconnected MEMGs [12].

7.1.3 Electrified Transport Sector and Mobile Sources Enhancing Resilience

Linked to energy sector decarbonization and digitalization, a new category of DERs known as mobile power sources (MPSs) [13] has also emerged as a promising solution to a resilient and decarbonized power systems. Primarily, MPSs, such as private electrical vehicles (EVs), public electric buses (EBs), emergency mobile generators (EMGs), mobile energy storage systems (MESSs), are characterized by their mobility and flexibility, making them very valuable assets for radically enhancing resilience of supply, after major infrastructure outages at the national level [14, 15]. These mobile sources can be strategically positioned in areas with the greatest need and provide supply to essential demand via V2G (e.g. medical facilities, community centers, emergency services, etc.) and hence enhance significantly resilience of supply. For example, the plan is to electrify all buses and taxis in London by 2030, which would bring around 40 GW of batteries in the transport sector and could be used to radically enhance resilience of supply in case of a major outage of transmission network that would leave London disconnected from electricity supply infrastructure. Furthermore, electrified transport sector, including EVs and EBs, could provide major flexibility services for balancing of renewable generation-dominated electricity system and hence support cost-effective energy system decarbonization. Furthermore, EVs can be routed to the locations experiencing demand peaks and network congestion to provide energy supplies without causing network overloads [16]. This optimization minimizes the need to activate high-cost and high-pollution generators to meet locational demand peaks, effectively avoiding the requirement for network reinforcement and providing cost-effective solutions [17].

7.2 Resilience Support by Real-World Microgrids

This section reviews several major natural disasters, in which MGs have played a significant role in enhancing their resilience [18]. This has been achieved by satisfying locally critical power demand, relieving the system stress, thus avoiding large-scale blackouts, and facilitating contingency handling and service restoration. It also shows cases where individual resilient MGs have withstood severe disruptions and served local power demand efficiently in extreme events.

7.2.1 Princeton Microgrid Against Hurricanes – United States

The Princeton's microgrid, shown in Figure 7.4, uses multiple fuel sources, multiple power-generating assets, energy storage, CHP production, and modern digital controls. It is supplied by a gas turbine with a nominal power of 15 MW. On sunny days, this power is supplemented by a 4.5-MW solar field. Princeton's MG normally operates connected with the local utility. When the price of utility power is lower than Princeton's cost to generate, the MG draws from the utility grid, while when Princeton's MG can produce power less expensively than the utility, it supplies as much as possible of the electricity needs of the university and exports surplus power to earn

Figure 7.4 The Princeton University MG.

revenues. During Hurricane Sandy, the Princeton University MG produced the heating, cooling, and electricity for the campus, allowing most of the university to operate for almost two days, while more than 8 million customers were without electricity [19].

7.2.2 The Sendai Microgrid During and After the Tsunami and Large-Scale Earthquake – Japan

On March 11, 2011, a tsunami and large-scale earthquake struck the Tohoku area and caused severe damage to many cities and towns in Japan. The Sendai MG, depicted in Figure 7.5, is designed as an ideal power supply system that can simultaneously provide services with multiple power quality levels. The MG was developed by NTT Facilities and was installed on the campus of Tohoku Fukushi University in Sendai City. It is an integrated 1 MW power system consisting of power electronics equipment, EES (batteries), and DGs (gas engine generators, solar cells, and fuel cells) [20]. It can easily interconnect and disconnect with existing utility power grids. In normal operation, the Sendai microgrid is connected to the utility's grid and improves the level of power quality for the hospital, welfare care facilities, and university buildings on the campus. It can disconnect from the utility grid after a power outage and continue supplying power to essential loads or facilities without interruption.

After the 3/11 disaster, services continued to be supplied with high-quality power by using the energy from solar cells and storage batteries. In addition, since the gas supply network in the city of Sendai was intact, the gas engine generators could perform blackout restart after power failure at the utility grid and function as the main power supply of the MG. Moreover, the gas engine generator sets, and fuel cells of the Sendai MG worked as CHP to produce both electric power and heat for hot water and space heating. Thus, the Sendai microgrid ensured that many patients in the hospital and in the medical and welfare buildings were able to survive and maintain their health.

7.2.3 Blue Lake Rancheria Microgrid Against Wildfires – United States

A series of major wildfires has repeatedly hit the state of California, especially during the last decade. Due to the disastrous results of these wildfires, distribution companies started to deploy

Figure 7.5 The Sendai MG.

measures to efficiently harden the grid and to find ways to preventively enhance its resilience. Blue Lake Rancheria (BLR) region of Humboldt County in California has a demonstration MG that includes 420 kW of PVs, a 500-kW/950-kWh battery bank, and a 1-MW backup generator, all connected to the Pacific Gas and Electric (PG&E) distribution grid at 12.5 kV, through a computer-controlled circuit breaker, as depicted in Figure 7.6 [21]. A second MG is also formed by a fueling station and an attached convenience store, including PV, batteries and some building system controls. Although the main MG was not initially introduced as a preventive measure against wildfires, it was eventually pushed to that direction, because of the utility's objective of enhancing resilience against outages and potential cost savings, estimated at up to 200,000$/year [22]. So, despite the application of public safety power shut-offs during extreme conditions, customers connected to MG retained access to power. In that way, the BLR MG strengthened resilience against outages and contingencies of the bulk transmission system, supporting actions of wildfire risk mitigation. Also, the Blue Lake event center and hotel can provide shelter to evacuees in an emergency situation [22]. The hotel MG was first activated when islanded against a small bush fire in 2017 and acted as a medical center for patients, a command center for emergency crews and a shelter for evacuees. During the wildfire of October 2019, BLR's MG is estimated to have saved 4 human lives and served about 10,000 people, equal to 10% of the county's population. Californians had recognized the importance of deteriorating fire risk, so the California Public Utilities Commission (CPUC) established exemplar MGs, such as BLR, that can reinforce the grid resilience before the summer of 2020 [22].

7.2.4 Texas Microgrids Against Winter Storms – United States

In February 2021, a very strong winter storm hit Texas, causing rolling blackouts that led more than 4,300,000 citizens to lose access to electricity for hours or even days [23]. The unusual and severe cold resulted in a major power demand rise up to a new winter peak of 69,222 MW. As large power

Figure 7.6 Overview of the BLR community MG. Photo: Schatz Energy Research Center, Humboldt State University.

plants began to trip off-line due to the severe rise of demand and extreme cold and many wind farms lost their productivity, as their turbines succumbed to icing, the electric demand exceeded the available supply, raising the fear of wide-scale blackouts throughout the power system. Utilities were forced to begin controlled blackouts and to shut down circuits to many customers, to reduce demand in order to ensure hospitals and other emergency services receive power, while power plant operators struggled to bring their units back online. Despite the small number of MGs in Texas, at this time of major disruption, they managed to alleviate some amount of power demand, combining heat and power, relieving stress on the main grid, and extending its capacity. In addition, MGs vital aid managed to provide power to hospitals, stores and shelters during the crisis, mitigating its catastrophic humanitarian and economic impacts and anticipating a total power system blackout [23].

7.2.5 Louisiana Microgrids Against Hurricanes – United States

The state of Louisiana has experienced two major destructive hurricanes (both category 4 hurricanes) during the last years, Hurricane Laura, in August 2020 and Hurricane Ida, in August 2021. MGs played a significant role in keeping the electricity flowing for their customers during Ida, providing power to stores, distribution centers, data centers and water municipalities, while the main grid was knocked out by the disaster, inducing 1,200,000 people to lose access to power [24]. Indicatively, after three days into the storm, the local utility managed only to restore electricity to 167,000 customers [25]. PowerSecure Inc. had 15 permanent MG installations as well as 23 mobile MGs in the storm's path, most of them with a capacity of 1,250 kW. A year earlier, PowerSecure Inc. MGs withstood a frontal attack from Hurricane Laura even when the backside of its eyewall went over them. MGs managed to operate during the power outage, feeding crucial community structures that procured shelter, food, water, medicines, and other essentials to citizens.

Figure 7.7 Overview of the Kythnos MG.

7.2.6 Microgrids Against Earthquakes – Haiti

A devastating 7.2 earthquake hit Haiti in August 2021, killing 2200 people and destroying up to 55,000 homes [26]. After the event, the 100 kW community MG in Les Anglais [27] was temporarily de-energized to check for downed wires. Given the magnitude of the destruction, the fear of not being able to rapidly repair the damages and re-energize the MG was derived, while waiting for potential aftershocks. Fortunately, the MG consisting of PVs, a diesel generator and storage units, did re-energize and was able to provide power to a cell tower, allowing residents to reclaim access to electricity and to hold funerals for those who died in the earthquake. A 95 kW MG in Tiburon, a small fishing town in the southern peninsula of Haiti, also operated during the disaster, providing power to 1200 people and survived the earthquake with no visible damages [27]. Due to the critical role of MGs during the disaster, interest has been raised into installing more community MGs in different regions of Haiti and expanding the ones of Les Anglais and Tiburon, by adding more DER to extend their capacity in order to serve more people.

7.2.7 The Kythnos Microgrid During Natural Disasters – Greece

The Kythnos Microgrid was installed in 2001 in the Gaidouromantra valley (Figure 7.7), forming probably the first MG in Europe supplied by 100% renewable generation most of the year [28]. The system was built in the framework of European-funded research projects, for the electrification of 12 vacation houses in a remote settlement. The generation system consists of seven distributed PV arrays with 11 kWp total installed power and two lead-acid battery banks: one of a nominal capacity of 1000 Ah/48 V (main) and a second one of 480 Ah/60 V (secondary). The main system is managed through three single-phase battery inverters, which form the three-phase power supply. The secondary is managed through one single-phase inverter for the power supply of communication and control systems. A three-phase diesel generator of 9 KVA was also installed as a backup supply, in case battery charging is very low. A system house of 30 m^2 was built in the middle of the settlement to house the battery inverters, the battery banks, the diesel genset and its tank, the computer equipment for monitoring, and the communication hardware.

Throughout the years of its operation, the installation sustained and overcame one severe water flood due to very heavy rain that destroyed and covered with rocks and earth yards, plants and

Figure 7.8 Use of mobile generators for restoration (Courtesy: Public Power Corporation, Greece). Photo: Hatziargyriou / IEEE (Book Author).

rock partition walls between plots. The MG did not suffer any damage and continued its operation supplying the loads. In June 2012, a wild bush fire spread in the southern part of the island, passed through the valley and destroyed two houses, a PV array, and one wooden pole of the electrical grid [29]. The system started to operate two days after the fire, by the time faults were cleared. The PV installation was replaced, and the houses rebuilt.

7.2.8 Support from Mobile Generators – Ad Hoc Microgrids Against Wildfires – Greece

A series of catastrophic wildfires in Greece resulted in 125,000 ha of forest and cultivable land and dozens of homes burned for more than 3 weeks and extensive damage to the network. In Evia, 7 MV lines feeding 13,000 consumers were destroyed; in Attica, 9 MV lines affecting 38,000 consumers. Twelve mobile generators (Figure 7.8) with a total capacity of 2.12 MW, formed ad-hoc low-voltage MGs to provide the essential electricity needs before network restoration [30].

7.2.9 Support from Local Generation Islanded Operation of Constitución Against Wildfires – Chile

A series of wildfires in January 2017 affected the line San Javier-Constitución 66 kV (single circuit), leaving the city of Constitución operating as an island (Figure 7.9).

The island was supplied by local generation between January 25 and 31, with the Celco power plant providing the needed frequency control. Hence, residential and industrial consumers (pulp and paper production) continued to be supplied through local thermal generation using biomass and diesel [31]. Although the justification to install these distributed generating units was not their capability to provide reliability and resilience services, this may change going forward.

The above practical examples are indicative of the capabilities of MGs in enhancing power system resilience. In the following sections, systematic methods that can be used for planning and operation of MGs to support power system resilience are analyzed.

Figure 7.9 Representation of wildfires in Chile on January 26, 2017.

7.3 DER/Microgrids Role in Strengthening Resilience

7.3.1 Resilience Enhancement at All Power System Levels

MGs, while transferring from grid-connected to islanded mode, constitute an effective resiliency resource, both for transmission and distribution systems [18], as shown in Figure 7.10, which illustrates how MGs can be exploited to enhance transmission and distribution system resilience and how individual MG resilience is strengthened.

Transmission system resilience enhancement: In case of multiple line failures in transmission system and isolation of part of it, MGs connected to transmission system directly (MG1) or MGs connected to the downstream distribution system (MG2–MG4) can act as power resources to feed the demand of the isolated part (transmission and distribution system) locally and minimize load shedding. The dashed arrows show the direction of power from MGs to distribution and transmission systems and indicate that MGs act as power resources.

Distribution system resilience enhancement: In case of isolation of distribution system due to a failure, MGs connected to distribution system (MG2–MG4) can act as power sources to meet locally the demand of the distribution system. In addition, when multiple failures occur in the distribution system, it could be sectionalized into self-adequate MGs utilizing the multiple DERs connected in order to minimize load shedding.

Individual MG resilience enhancement: When a failure at distribution system leads to the isolation of the MG (MG4), it can operate in islanded mode by scheduling its resources properly to serve at least its critical load.

The following sections focus on the *Distribution system resilience enhancement*.

7.3.2 Resilience Enhancement in All Resilience Phases

Considering the time evolution of a catastrophic event, a power system might go through different states. These are depicted in the resilient trapezoid model that has been analyzed in Chapters

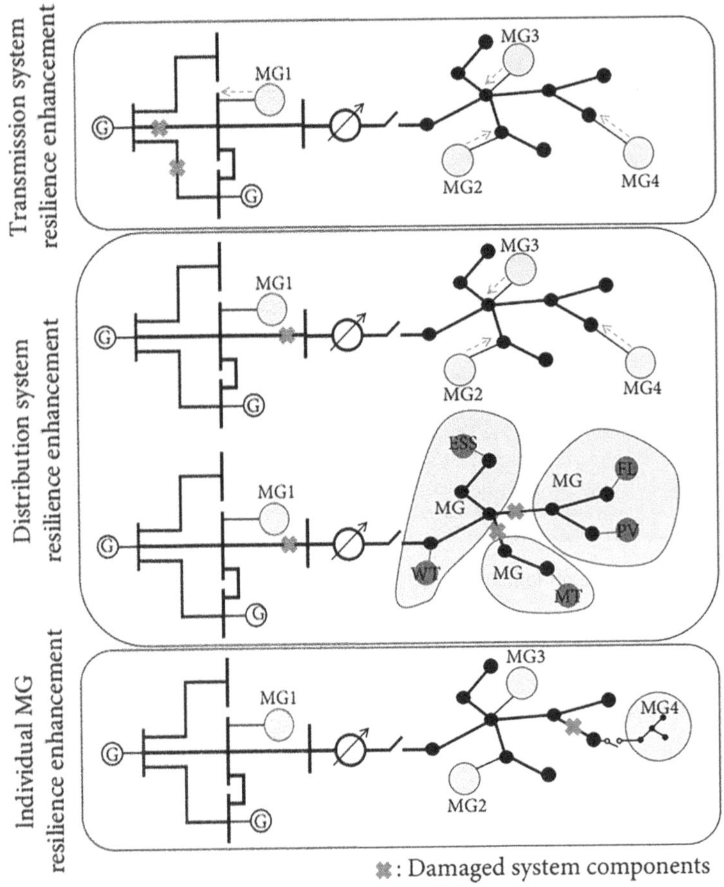

Figure 7.10 Transmission and distribution resilience enhancement using MGs and individual MG resilience enhancement.

2 and 3. The trapezoid consists of the pre-disturbance resilient state, the state during disturbance progress (when the event occurs), the post-disturbance degraded state (after the event and before the initiation of restoration), the restorative state and the post-restoration state, when the power system is fully restored and operates under normal conditions. The formation and operation of MGs to boost power system resilience requires both investment-planning and advanced operational measures. Both types of measures should take into account the uncertainties brought mainly by the stochastic nature of RES and the potential damages in the event of external threats.

MGs can play a crucial role in all three phases of the resilient trapezoid of Figure 2.6. They can be used for MG-based resilient planning (long-term measures, pre-disturbance state), preventive and emergency MG schemes (short-term operational measures, on-event and post-event, respectively) and MG restoration techniques (operational measures, post-event restorative state). Several survey papers have been published dealing with MG-oriented power system resilience, e.g. [32] addresses different MG-based resilience scenarios, objectives, metrics and control methods for distribution systems resilience enhancement. Ref. [33] reviews methodologies for operation and control of networked MGs for improving grid resilience, robustness and efficiency. Various proactive strategies, outage management, and advanced operation methods for resilience improvement are described in [34], focusing on self-sustainable, networked, and dynamic MGs formation frameworks.

When isolated from the upstream grid with the aid of tie switches, MGs can operate in a self-sufficient way or can be coupled with neighboring MGs, forming multi-MGs (MMGs) in order to supply critical loads (e.g. hospitals and data centers) by sharing power resources [34, 35]. Defensive islanding aims to the formation of efficient, self-adequate MGs or to the isolation of the vulnerable areas with the minimum load curtailment to avoid cascading events. The main goal of these strategies is to prevent total or partial blackouts and minimize or even eliminate load curtailments by optimally scheduling and dispatching DERs and by prioritizing the supply of critical loads. Section 7.4 deals with the implementation of such preventive and corrective islanding techniques, while MG-aided restoration to increase resilience is discussed in Section 7.5.

In already formed MGs, e.g. at pre-disturbance state, it is crucial that their resources and storage capacities are optimally scheduled in order to be prepared for potential events [35]. Via resilient-based preventive scheduling, MGs adapt the commitment of dispatchable DGs, DSR and EES to achieve feasible islanding during any potential disruptive event. In that way, the transition between grid-connected and islanded mode can be made with minimum disruption to consumers. Optimal DER scheduling is described in Section 7.6.

7.4 Preventive and Corrective Microgrid Formation

This section deals with the formation of MGs as a preventive or corrective measure, in case a distribution system faces an imminent disaster, days or hours before the event or after the event hits the network (Figure 7.10). As the disturbance progresses, it is probable that the load cannot be supplied by the main grids, or the damages to the distribution facilities might create isolated areas without power. With the emergence of smart grid technologies, many electric distribution utilities have implemented or are planning to implement remotely controlled, automatic switching devices on distribution feeders, as part of their grid modernization strategy to achieve "self-healing" distribution grids. These automatic switching devices make it possible to form MGs. The efficient formulation of MGs based on several DGs in the distribution system is not an easy task, especially when facilities are destroyed due to a disaster. The scope of this section focuses on the worst scenario, where the main substation of the distribution system is out of service. It should be noted that the presented methodology is also applicable to scenarios, where the main substation remains operating, in case the distribution system is preparing for an imminent disaster. This is considered as a special case of the presented analysis, by changing some specific formulations accordingly.

7.4.1 The Base Model

Considering a radial power distribution system consisting of $|N|$ nodes denoted by the set of nodes N, and $|E|$ power lines represented by the set of edges E. When there is a major fault at the main grid, as a result of a natural disaster, the distribution system can no longer be energized by the main grid through the substation and feeders. Instead, the DGs installed at certain nodes, which are denoted by the set N_D, can be utilized to continue supplying critical loads in the distribution system before the restoration of the main grid is complete. Two assumptions are made previous to modeling the MG formation problem.

1) Rather than acting as a backup generator just supplying the load at the DG node, each DG can supply multiple loads by separately forming $|N_D|$ MGs. Let's assume that the number of MGs with DGs to achieve grid resilience has been defined in advance. Under this assumption, each MG $k \in K$ consists of a set of nodes N_k satisfying $N_k \subset N$ and $N_{k_1} \cap N_{k_2} = \emptyset, \forall k_1, k_2 \in K$. Each MG is energized by a DG, so $|K| = |N_D|$. For notational simplicity, the kth MG is supplied (or mathematically defined) by the kth DG, i.e. $N_D(k)$.

2) Based on the fast development of distribution automation technology, it is assumed that the necessary remotely controlled automatic switches have been implemented on distribution feeders, so that lines can be opened/closed, and loads can be connected/disconnected to form multiple MGs. Depending on the distribution automation progress, the remotely controlled automatic switches can be fully or partly installed at each line and node, to provide different flexibilities in controlling the topology of the distribution network. Due to the occurrence of natural disasters, some switches may be at fault.

   Let $y_{ij} \in \{0, 1\}$ denote the binary decision variables indicating whether the switch associated with branch (i, j) is open ($y_{ij} = 0$) or closed ($y_{ij} = 1$). Similarly, let $x_i^L \in \{0, 1\}$ denote the binary decision variables representing whether the switch connecting the load and node i is open ($x_i^L = 0$) or closed ($x_i^L = 1$). By controlling these switches, $|K|$ self-adequate MGs are formed to maximize the total critical loads to be restored.

   The post-distribution MG formation proposed in [36] and further developed in [37] is presented below as the base model. $|K|$ MGs, each with one DG, are formed by controlling the switches and output of DGs by solving the optimization problem, to maximize the total critical loads to be restored after natural disasters, and to guarantee that each MG is a self-adequate system.

7.4.1.1 Partitioning Constraints [37]

Node Clustering Constraints If the load at node i can be energized by MGs, this node can only belong to one of the $|K|$ MGs. It should be noted that some nodes may remain islanded, without being connected to any MG due to permanent line faults. These faulted islands should be removed from the set of nodes to model the node clustering constraints. This can be achieved by finding the connected components of an undirected graph representing the distribution system topology and removing the nodes and edges of the connected component without a DG. The aforementioned node set N and branch set E are obtained after this network simplification process.

Define auxiliary binary decision variables $x_{i,k} \in \{0, 1\}$ indicating whether node i belongs to MG k, ($x_{i,k} = 1$ if node i belongs to MG k, and $x_{i,k} = 0$ otherwise), the node clustering constraints can be expressed as

$$\sum_{k \in K} x_{i,k} = 1, \quad \forall i \in N \tag{7.1}$$

The kth DG node surely belongs to MG k, which can be written as the following equality constraint:

$$x_{N_D(k),k} = 1, \quad \forall k \in K \tag{7.2}$$

where $N_D(k)$ is the node where the kth DG is installed.

A sample distribution system is provided in Figure 7.11 for a better understanding of the variables in the base model. As shown, there are, in total, 11 branches and 12 nodes. Besides, there are 3 DGs installed at nodes 3, 6, and 10, indicating that the number of MGs to be formed is 3. The sets of MGs, DG nodes, nodes, and branches are specified as follows.

$$K = \{1 \ 2 \ 3\}$$

$$N_D = \{3 \ 6 \ 10\}$$

$$N = \{1 \ 2 \ 3 \ 4 \ 5 \ 6 \ 7 \ 8 \ 9 \ 10 \ 11 \ 12\}$$

$$E = \{(1,2) \ (1,3) \ (3,4) \ (4,5) \ (4,6) \ (6,7) \ (7,8) \ (8,9) \ (8,10) \ (10,11) \ (11,12)\}$$

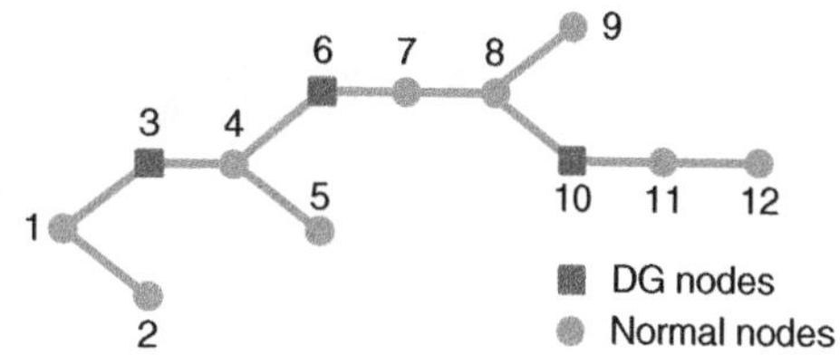

Figure 7.11 A sample distribution system.

Let **x** and **y** denote the matrices of node and branch energization statuses:

$$\mathbf{x} = \begin{bmatrix} x_{1,1} & x_{2,1} & x_{3,1} & x_{4,1} & x_{5,1} & x_{6,1} & x_{7,1} & x_{8,1} & x_{9,1} & x_{10,1} & x_{11,1} & x_{12,1} \\ x_{1,2} & x_{2,2} & x_{3,2} & x_{4,2} & x_{5,2} & x_{6,2} & x_{7,2} & x_{8,2} & x_{9,2} & x_{10,2} & x_{11,2} & x_{12,2} \\ x_{1,3} & x_{2,3} & x_{3,3} & x_{4,3} & x_{5,3} & x_{6,3} & x_{7,3} & x_{8,3} & x_{9,3} & x_{10,3} & x_{11,3} & x_{12,3} \end{bmatrix}$$

$$\mathbf{y} = \begin{bmatrix} y_{12} & y_{13} & y_{34} & y_{45} & y_{46} & y_{67} & y_{78} & y_{89} & y_{810} & y_{1011} & y_{1112} \end{bmatrix}$$

The number of rows of **x** corresponds to the MGs, and the number of columns to the nodes. The kth row of **x** indicates the membership of nodes of MG k, e.g. $x_{1,1} = 1$ indicates that node 1 belongs to MG 1. As constraint (7.1) requires that a node can only belong to one MG, the other two entries in the 1st column of **x** should be 0, $x_{1,2} = 0$ and $x_{1,3} = 0$. The entries of the **y** matrix indicate the energization status of the branches, e.g. $y_{12} = 1$ denotes that branch (1, 2) is energized. On this basis, an instance of **x** and **y** indicates a specific MG formation scheme.

Based on constraint (7.2), the kth DG must belong to MG k, requiring that in the kth row of **x**, the variable regarding the kth DG node should take the value of 1. Specifically, in the **x** matrix, the entries regarding the DG nodes, i.e. nodes 3, 6, and 10, should take the following values.

$$\mathbf{x} = \begin{bmatrix} x_{1,1} & x_{2,1} & 1 & x_{4,1} & x_{5,1} & 0 & x_{7,1} & x_{8,1} & x_{9,1} & 0 & x_{11,1} & x_{12,1} \\ x_{1,2} & x_{2,2} & 0 & x_{4,2} & x_{5,2} & 1 & x_{7,2} & x_{8,2} & x_{9,2} & 0 & x_{11,2} & x_{12,2} \\ x_{1,3} & x_{2,3} & 0 & x_{4,3} & x_{5,3} & 0 & x_{7,3} & x_{8,3} & x_{9,3} & 1 & x_{11,3} & x_{12,3} \end{bmatrix}$$

The above matrix **x** clearly defines that node 3 belongs to MG 1, node 6 belongs to MG 2, and node 10 belongs to MG 3, respectively.

An MG formation scheme and the corresponding values of **x** and **y** are shown in Figure 7.12. Since there are 3 DGs, based on the assumption that each DG supplies one MG, there are 3 MGs, i.e. MG 1, 2, and 3 supported by the DGs at nodes 3, 6, and 10, respectively. As shown in Figure 7.12, MG 1 comprises nodes 1–5, indicating that the first 5 entries in the 1st row of **x** should be 1, as required by the definition of **x** that each row defines the energized nodes of the MG. Similarly, the 6th to 9th entries in the 2nd row of **x** should be 1 because nodes 6–9 are energized by MG 2, and the 10th to 12th entries in the 3rd row should be 1 because nodes 10–12 are energized by MG 3. In addition, the binary energization statuses of branches are given by the **y** matrix. The first 4 branches constitute MG 1, so the first 4 entries in **y** are 1, and the 6th to 8th and the 10th to 11th entries should be 1 based on the topology of MGs 2 and 3. It is worth noting that the entries of **y** do not have the subscript k, as it is not necessary to define them for each MG.

MG Connectivity Constraints For a radial (tree) distribution network, each MG can be considered as a subtree network with the root node being the node where the DG is installed. Due to the connectivity feature of a tree, one node can belong to MG k only if its parent node belongs to MG k, which can be expressed by the following inequality constraints:

$$x_{i,k} \leq x_{p_k(i),k}, \quad \forall i \in N, k \in K \tag{7.3}$$

where $p_k(i)$ denotes the parent node of node i regarding MG k.

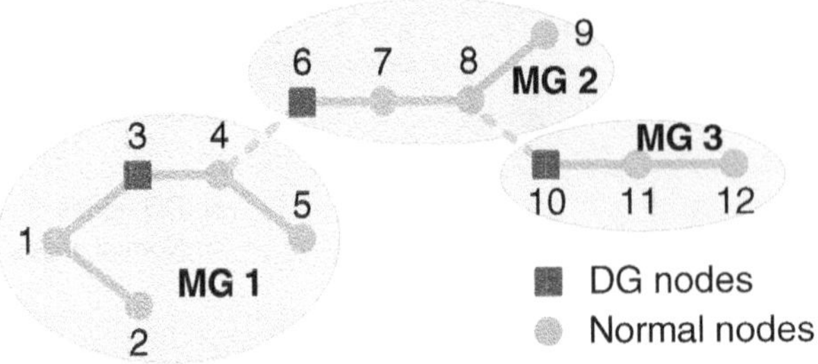

$$\mathbf{x} = \begin{bmatrix} 1 & 1 & 1 & 1 & 1 & 0 & 0 & 0 & 0 & 0 & 0 & 0 \\ 0 & 0 & 0 & 0 & 0 & 1 & 1 & 1 & 1 & 0 & 0 & 0 \\ 0 & 0 & 0 & 0 & 0 & 0 & 0 & 0 & 0 & 1 & 1 & 1 \end{bmatrix}$$

$$\mathbf{y} = \begin{bmatrix} 1 & 1 & 1 & 1 & 0 & 1 & 1 & 1 & 0 & 1 & 1 \end{bmatrix}$$

Figure 7.12 An MG formation scheme of the sample distribution system.

The value of $p_k(i)$ should be determined for every MG. For instance, in the sample distribution system shown in Figure 7.12, $p_1(i)$ is determined based on the flows on the distribution network rooted at the 1st DG node, i.e. node 3. Similarly, the value of $p_2(i)$ and $p_3(i)$ is decided by the flows on the distribution network rooted at nodes 6 and 10, respectively. The value of $p_k(i)$ is given by the matrix **p** below.

$$i = 1\ 2\ 3\ 4\ 5\ 6\ 7\ 8\ 9\ 10\ 11\ 12$$

$$\mathbf{p} = \begin{bmatrix} p_1(i) \\ p_2(i) \\ p_3(i) \end{bmatrix} = \begin{bmatrix} 3 & 1 & \emptyset & 3 & 4 & 4 & 6 & 7 & 8 & 8 & 10 & 11 \\ 3 & 1 & 4 & 6 & 4 & \emptyset & 6 & 7 & 8 & 8 & 10 & 11 \\ 3 & 1 & 4 & 6 & 4 & 7 & 8 & 10 & 8 & \emptyset & 10 & 11 \end{bmatrix}$$

The entry in the kth row and the ith column of **p** indicates the parent node of node i, based on the distribution network rooted at the kth DG node. In the 1st row of **x**, for example, the 1st entry indicates that the parent node of node 1 is node 3, and the 2nd entry indicates that the parent node of node 2 is node 1, based on the distribution network rooted at node 3 (the 1st DG node). Based on the value of **p**, constraint (7.3) requires that, for example, if node 2 belongs to MG 1, node 1 (the parent of node 2) has to belong to MG 1, and node 3 (the parent of node 1) also has to belong to MG 1, which is already imposed by constraint (7.2).

Accordingly, the entries $x_{p_k(i),k}$ are collected by the matrix **xp**

$$i = 1\ 2\ 3\ 4\ 5\ 6\ 7\ 8\ 9\ 10\ 11\ 12$$

$$\mathbf{xp} = \begin{bmatrix} x_{p_1(i),1} \\ x_{p_2(i),2} \\ x_{p_3(i),3} \end{bmatrix} = \begin{bmatrix} x_{3,1} & x_{1,1} & \emptyset & x_{3,1} & x_{4,1} & x_{4,1} & x_{6,1} & x_{7,1} & x_{8,1} & x_{8,1} & x_{10,1} & x_{11,1} \\ x_{3,2} & x_{1,2} & x_{4,2} & x_{6,2} & x_{4,2} & \emptyset & x_{6,2} & x_{7,2} & x_{8,2} & x_{8,2} & x_{10,2} & x_{11,2} \\ x_{3,3} & x_{1,3} & x_{4,3} & x_{6,3} & x_{4,3} & x_{7,3} & x_{8,3} & x_{10,3} & x_{8,3} & \emptyset & x_{10,3} & x_{11,3} \end{bmatrix}$$

MG Branch-Node Constraints The relation between nodes and branches in an MG imposes the constraint that, if both nodes i and j belong to MG k, i.e. $x_{i,k} = x_{j,k} = 1$, then the branch connecting nodes i and j should also belong to MG k. Since the parent–child relationship has been defined in (7.3) already, each branch has specifically one parent and one child regarding its two end nodes. Together with (7.3), which requires the parent node to be energized by the same MG when the children node is energized, the relation between branches and nodes can be simplified, i.e. only if the children node of (i, j) belongs to MG k, branch (i, j) belongs to MG k. If branch (i, j) belongs to any of the MGs, the switch on this line, if it exists, should be in the closed state. Thus, the branch-node constraints can be expressed as

$$y_{ij} \leq \sum_{k \in K} x_{c_k(i,j),k}, \quad \forall (i,j) \in E \tag{7.4}$$

where $c_k(i, j)$ denotes the children node of branch (i, j) regarding MG k.

Like that of $p_k(i)$, the value of $c_k(i, j)$ should also be determined for every MG. The value of $c_k(i, j)$ is shown by the matrix **c** below.

$$ij = 12 \quad 13 \quad 34 \quad 45 \quad 46 \quad 67 \quad 78 \quad 89\,810\,1011\,1112$$

$$\mathbf{c} = \begin{bmatrix} p_1(i,j) \\ p_2(i,j) \\ p_3(i,j) \end{bmatrix} = \begin{bmatrix} 2 & 1 & 4 & 5 & 6 & 7 & 8 & 9 & 10 & 11 & 12 \\ 2 & 1 & 3 & 5 & 4 & 7 & 8 & 9 & 10 & 11 & 12 \\ 2 & 1 & 3 & 5 & 4 & 6 & 7 & 9 & 8 & 11 & 12 \end{bmatrix}$$

The entry in the kth row and the ith column of **c** indicates the children node of the two end nodes of the ith branch, based on the flow on the distribution network rooted at the kth DG node. For example, the 1st entry in the 1st row indicates that the children node of the two end nodes of branch (1, 2), i.e. nodes 1 and 2, is node 2, based on the distribution network rooted at node 3 (the 1st DG node). Based on the value of **c**, constraint (7.4) for branch (1, 2) and branch (4, 6) is shown below, respectively.

$$y_{12} \leq x_{2,1} + x_{2,2} + x_{2,3} = 1 + 0 + 0 = 1$$

$$y_{46} \leq x_{6,1} + x_{4,2} + x_{4,3} = 0 + 0 + 0 = 0$$

Based on the matrices **x** and **y** for the MG formation scheme shown in Figure 7.12, branch (1, 2) is able to be energized, while branch (4, 6) has to be de-energized to separate the different MGs.

Distribution System Condition Constraints Due to the catastrophic event, some branches may be at fault, i.e. the branches are in the open state, and the repair process needs a substantial amount of time, or some switches fail and are in the open state. In this case, let's denote these branches by the set E_O. We can write the constraints as $y_{ij} = 0, \forall (i, j) \in E_O$. Branches with switches in the closed state are denoted by the set E_C and can be treated as always in the closed state, i.e. $y_{ij} = 1, \forall (i, j) \in E_C$. Similar distribution system condition constraints can be applied to the switches at nodes, where N_O denotes the set of fault switches that disconnect permanently the load from the node, and N_C denotes the set of loads whose fault switches are in the closed state and are connected to the node. The constraints regarding the binary decision variables based on the actual conditions of the distribution system can be written as follows:

$$\begin{cases} y_{ij} = 0, & \forall (i,j) \in E_O \\ y_{ij} = 1, & \forall (i,j) \in E_C \\ x_i^L = 0, & \forall i \in N_O \\ x_i^L = 1, & \forall i \in N_C \end{cases} \tag{7.5}$$

7.4.1.2 Operating Constraints [37]

The formed MGs should also satisfy operating constraints, e.g. power balance and voltage limit at each node. The linearized DistFlow equations (7.6) and (7.7) have been employed for computing the nodal voltage magnitude and power flow on branches,

$$\begin{cases} \sum\limits_{h:(h,i)\in E} p_{hi} - \sum\limits_{j:(i,j)\in E} p_{ij} = p_i^{\mathrm{G}} - p_i^{\mathrm{L}} \\ \sum\limits_{h:(h,i)\in E} q_{hi} - \sum\limits_{j:(i,j)\in E} q_{ij} = q_i^{\mathrm{G}} - q_i^{\mathrm{L}} \end{cases}, \quad \forall i \in N \tag{7.6}$$

$$\begin{cases} v_i - v_j \geq 2(R_{ij}p_{ij} + X_{ij}q_{ij}) - M(1 - y_{ij}) \\ v_i - v_j \leq 2(R_{ij}p_{ij} + X_{ij}q_{ij}) + M(1 - y_{ij}) \end{cases}, \quad \forall (i,j) \in E \tag{7.7}$$

where p_{ij} and q_{ij} are the active and reactive power flow on branch (i, j), respectively; p_i^{G} and q_i^{G} are the active and reactive power generated by the DG at node i, respectively; p_i^{L} and q_i^{L} are the active and reactive power consumed by the loads at node i; v_i is the squared voltage magnitude at node i; R_{ij} and X_{ij} are the resistance and reactance of branch (i, j), respectively.

The power flow on branches is calculated by (7.6), and the voltage magnitude difference of an energized branch is computed by (7.7).

The secure operating region, respecting the operating limits of voltage magnitude and branch flow, is

$$V_{\min}^2 \sum_{k\in K} x_{i,k} \leq v_i \leq V_{\max}^2 \sum_{k\in K} x_{i,k}, \quad \forall i \in N \tag{7.8}$$

$$-y_{ij}(S_{ij,\max})^2 \leq p_{ij}^2 + q_{ij}^2 \leq y_{ij}(S_{ij,\max})^2, \quad \forall (i,j) \in E \tag{7.9}$$

where $V_{\min}$ and $V_{\max}$ are the minimum and maximum voltage magnitude, respectively; $S_{ij,\max}$ is the maximum apparent power allowed to flow on branch (i, j).

The DG power generation and load are modeled, as follows.

$$\begin{cases} P_{i,\min}^{\mathrm{G}} \leq p_i^{\mathrm{G}} \leq P_{i,\max}^{\mathrm{G}} \\ Q_{i,\min}^{\mathrm{G}} \leq q_i^{\mathrm{G}} \leq Q_{i,\max}^{\mathrm{G}} \end{cases}, \quad \forall i \in N_{\mathrm{D}} \tag{7.10}$$

$$\begin{cases} p_i^{\mathrm{L}} = P_i^{\mathrm{L}} x_i^{\mathrm{L}} \\ q_i^{\mathrm{L}} = Q_i^{\mathrm{L}} x_i^{\mathrm{L}} \end{cases}, \quad \forall i \in N \tag{7.11}$$

where $P_{i,\min}^{\mathrm{G}}$ and $P_{i,\max}^{\mathrm{G}}$ are the minimum and maximum active power outputs of the DG at node i, respectively; $Q_{i,\min}^{\mathrm{G}}$ and $Q_{i,\max}^{\mathrm{G}}$ are the minimum and maximum reactive power outputs of the DG at node i, respectively; P_i^{L} and Q_i^{L} are the rated active and reactive power demands of the load at node i, respectively.

As shown in (7.10), the active and reactive power outputs of DGs are restricted within the operating limits. The electric loads are modeled in (7.11) using the ON/OFF binary variable, which is detailed as

$$x_i^{\mathrm{L}} \leq \sum_{k\in K} x_{i,k}, \quad \forall i \in N \tag{7.12}$$

Equation (7.12) requires that loads located at the node can be connected, only if a node is energized.

7.4.1.3 Objective Function [37]

In general, loads can have different priorities regarding their importance in sustaining critical functions. For example, hospitals or public lighting might be more crucial, especially under emergency conditions. Let w_i denote the priority weight associated with the load at node i, a larger value of w_i indicating higher priority. The total priority weighted load served can be expressed as

$$\max \sum_{i \in N} w_i p_i^L \tag{7.13}$$

The MG formation optimization aims to maximize the weighted load supplied, subject to the constraints described in Sections 7.4.1.1 and 7.4.1.2.

7.4.1.4 Numerical Results

The presented MG formation scheme is verified in two modified IEEE radial distribution system test networks, i.e. IEEE 37-and 123-node test feeders, assumed symmetrical. The system parameters are specified as follows:

1) In the IEEE 37-node test feeder, 3 DGs are installed at three nodes (702, 710, and 728), indicating 3 MGs to be formed. As a result of a catastrophic event, 4 line switches are in open status (at fault). There are five lines with uncontrollable switches that must be in closed status, depicted in Figure 7.13a. Other line switches are assumed to be controllable, which can be operated in open/closed status to reconfigure the faulted distribution system into MGs.
2) In the IEEE 123-node test feeder, 8 DGs are installed at eight nodes, i.e. 4, 27, 42, 57, 64, 70, 87, and 97, indicating 8 MGs to be formed. As a result of a catastrophic event, 7 line switches are in open status (at fault). There are 40 lines with uncontrollable switches that must be in closed status, depicted in Figure 7.13b. Other line switches are assumed to be controllable, which can be operated in open/closed status to reconfigure the faulted distribution system into MGs.

   In the two test feeders, the substation is separated by a faulted switch, and thus it is unable to supply loads. The capacity of DGs and the active and reactive power demands of loads are generated randomly, based on the post-disruption scenario that the load demand is significantly greater than the power generation. Besides, loads are assigned with different weight coefficients to simulate their real-world importance, and the weight coefficients are generated randomly from 1 to 30. By solving the base model, the following MG formation schemes are determined by operating the controllable switches with the objective of energizing as many loads as possible, considering their critical levels (weight coefficients).

As shown in Figure 7.13a, 3 MGs are formed in the faulted 37-node test feeder. The distribution lines (703, 727) and (708, 709) are switched off to separate the faulted system into 3 MGs. Considering the limited DG resources, the provided MG formation scheme energizes 14 out of 37 loads in total. The energized load demand is 575.94 out of 987.20 kW, accounting for 58.3% of the total load demand. Most of the energized loads have a higher weight coefficient (critical level) than the other de-energized loads. Except for the 10 unrecoverable nodes separated by the faulted switches, the provided MG formation scheme energizes all the remaining nodes and lines to minimize the outage area.

Similarly, in Figure 7.13b, 8 MGs are formed in the faulted 123-node test feeder. The distribution lines (8, 13), (18, 135), (25, 26), (57, 60), (67, 97), (67, 160), and (76, 86) are switched off to separate the faulted system into 8 MGs. Considering the limited DG resources, the provided MG formation scheme energizes 57 out of 123 loads in total. Among all the 123 nodes, only 14 nodes are de-energized, which are separated by the faults on neighboring distribution lines.

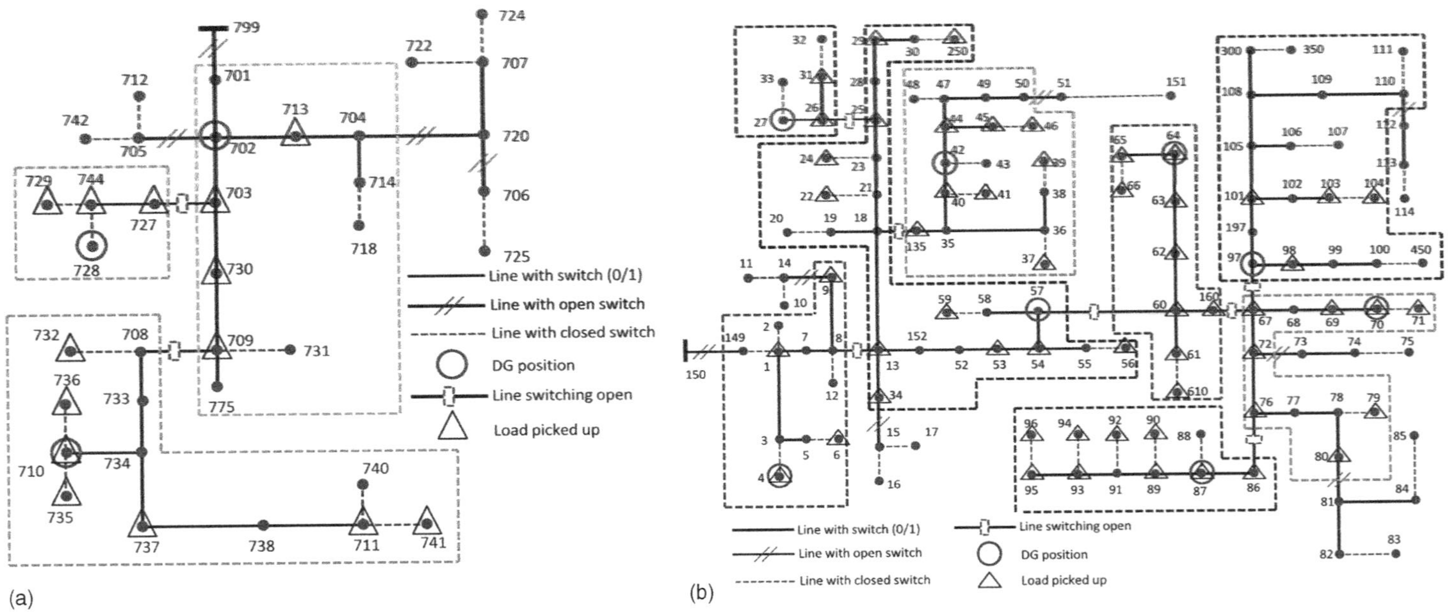

Figure 7.13 MG formation schemes of the IEEE test feeders. (a) The IEEE 37-node test feeder. (b) The IEEE 123-node test feeder.

7.4.2 Advances of the Base Model [37]

7.4.2.1 Reformulating the Partitioning Constraints

There are two major limitations that restrict the wide application of the base model:

1) The number of MGs must be the same as that of DGs, which means that each MG has to be supported by only one DG. In actual implementation, however, some DGs lack voltage and frequency control ability, and thus, they cannot support an MG as the only power source. Restricting the number of power sources in an MG to one, is also not good engineering practice, especially when the available DGs are variable and intermittent. The power outputs of DGs, like WT and PV, are stochastic, and additional sources with adjustable power outputs are needed to compensate for the stochasticity. In general, MGs supplied by multiple DGs are more reliable than those with only one DG.
2) With the exception of the nodes isolated by faults (or switching operations required to isolate faults), all the other nodes are required to be energized. This requirement aims to provide an MG formation scheme with maximum coverage of the original network. However, the power generated by DGs is limited and usually unable to meet the rated demands of all loads in the original network. Moreover, the more nodes and branches are energized, the higher the risk of a potential fault, as the catastrophic failure unfolds. An improvement to this is to relax the hard constraint on node energization, to allow flexible boundaries among MGs.

In order to develop a flexible post-disruption MG formation model, the partitioning constraints (7.1)–(7.5) in the base model are reformulated in this section. Explanations of the basic variables and sets are given below.

1) The maximum number of MGs to be formed is considered as the number of DGs with voltage and frequency control ability. The set of potential MGs is also denoted as K, like in the base model, whose element k is the index of the kth potential MG. Accordingly, each potential MG k is assigned with a potential root r_k (the controllable DG), and the set of potential roots is denoted as R.
2) The binary status of nodes, i.e. $x_{i,k}$, remains unchanged as in the base model, taking 1 if node i is selected and energized by MG k, and 0 otherwise. Note that the binary status of a potential root, i.e. $x_{rk,k}$ indicates the existence of the potential MG k. It is 0, if the kth MG does not exist and its potential root r_k (the kth DG node) could be included as a normal node by other MGs, and 1 otherwise.
3) Unlike the base model, the status of branches is reformulated for every potential MG, i.e. $y_{ij,k}$, taking 1 if branch (i, j) belongs to MG k, and 0 otherwise. As for faulted or uncontrollable switches, their statuses are fixed by adding constraints $y_{ij,k} = 0/1$.

Based on the sets, parameters, and variables defined in the base model, the following constraints are developed to allocate nodes and branches among multiple MGs. Equation (7.14) allows for flexible boundaries among MGs, wherein normal nodes are not necessarily energized, while, on the contrary, critical nodes including DGs and critical loads, need to be energized. Besides, if potential MG k does not exist (i.e. $x_{rk,k} = 0$), then no nodes could be energized in it. The binary status between branches and nodes is restricted by (7.15), meaning that if a branch is energized by MG k, its two ends should also be energized by the same MG k. Lastly, the status of faulted nodes, faulted or uncontrollable switches, is specified by (7.16).

$$\begin{cases} \sum_{k\in K} x_{i,k} = 1, & \forall i \in N_{\mathrm{CL}} \cup N_{\mathrm{D}} \\ \sum_{k\in K} x_{i,k} \leq 1, & \forall i \in N \backslash (N_{\mathrm{CL}} \cup N_{\mathrm{D}}) \\ x_{i,k} \leq x_{r_k,k}, & \forall i \in N, k \in K \end{cases} \tag{7.14}$$

$$\begin{cases} y_{ij,k} \leq x_{i,k} \\ y_{ij,k} \leq x_{j,k} \end{cases}, \quad \forall (i,j) \in E, k \in K \tag{7.15}$$

$$\begin{cases} \sum_{k\in K} x_{i,k} = 0, & \forall i \in N_{\mathrm{F}} \\ \sum_{k\in K} y_{ij,k} = 0, & \forall (i,j) \in E_{\mathrm{O}} \\ \sum_{k\in K} y_{ij,k} = 1, & \forall (i,j) \in E_{\mathrm{C}} \cup E_{\mathrm{U}} \end{cases} \tag{7.16}$$

where N_{CL} and N_{F} are the sets of nodes with critical loads and faults, respectively; E_{U} is the set of uncontrollable branches.

The base model requires that the sample distribution system has to be divided into 3 MGs, as the number of DGs is 3. In the reformulated partitioning constraints, this requirement is relaxed, and the number of MGs is optimally determined by maximizing the objective function. A two-MG formation scheme is provided for comparison between the improved partitioning constraints and that of the base model, and the value of matrix **x** is given accordingly.

As shown in Figure 7.14, there are in total 2 MGs, i.e., MG 2 and MG 3, while MG 1 does not exist. This is possible because the new partitioning constraints do not require that the kth DG node has to belong to MG k. Instead, the improved partitioning constraints allow the DG nodes to be included by other MGs. That is, $x_{3,1}$, $x_{6,2}$, and $x_{10,3}$ are allowed to take the value of 0. In the two-MG formation scheme, the value of $x_{3,1}$ is 0, indicating that MG 1 does not exist, while the values of $x_{6,2}$ and $x_{10,3}$ are both 1, indicating that MG 2 and 3 exist. As a result, no nodes can be energized by MG 1 because $x_{3,1} = 0$, while MGs 2 and 3 can energize nodes 1–6 and 10–12 because $x_{6,2} = 1$ and $x_{10,3} = 1$, respectively.

The above formulation allows for adding constraints within a specific MG, as explicitly defining the boundary of each subgraph, i.e. the value of $x_{i,k}$ and $y_{ij,k}$ regarding MG k. With the explicit

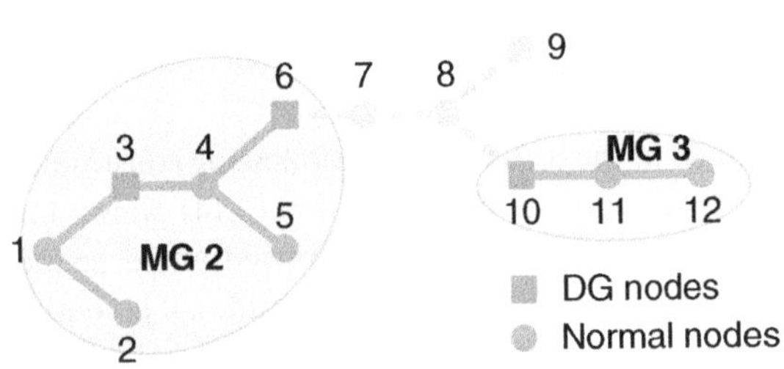

$$\mathbf{x} = \begin{bmatrix} 0 & 0 & 0 & 0 & 0 & 0 & 0 & 0 & 0 & 0 & 0 & 0 \\ 1 & 1 & 1 & 1 & 1 & 1 & 0 & 0 & 0 & 0 & 0 & 0 \\ 0 & 0 & 0 & 0 & 0 & 0 & 0 & 0 & 0 & 1 & 1 & 1 \end{bmatrix}$$

Figure 7.14 A two-MG formation scheme of the sample distribution system.

boundary given, constraints regarding MG's stability, dynamic and transient behaviors can be further integrated into (7.14)–(7.16) for future studies. This includes, for instance, constraints for MG's controllability assessment or future interconnection among MGs, i.e. networked MGs. In the presented partitioning constraints, these features can be well addressed with variables subdivided into every potential MG.

The energization status of nodes and branches at the distribution system level is denoted by x_i and y_{ij}, respectively. They are in the following form.

$$\begin{cases} x_i = \sum_{k \in K} x_{i,k}, & \forall i \in N \\ y_{ij} = \sum_{k \in K} y_{ij,k}, & \forall (i,j) \in E \end{cases} \tag{7.17}$$

In conclusion, the original partitioning constraints (7.1)–(7.5) are reformulated as (7.14)–(7.17), to overcome the limitations of the base model, namely the requirement that every MG is supplied by one DG and all healthy nodes are energized. In the improved partitioning constraints, the number of MGs is not pre-determined, and DGs are automatically allocated into multiple MGs, with the aim to supply as many critical loads as possible.

7.4.2.2 Further Advances in MGs Formulation [38]

Integrating MG Stability Assessment Previous Sections 7.4.1 and 7.4.2, only the steady-state security constraints, including nodal voltage magnitudes and power flows on branches, have been considered. Islanding of a distribution network, however, poses great challenges to the stability of these systems. The inclusion of the stability constraints is directly related to the dynamic characteristics of the system, given that the presence of DGs imposes new restrictions on the MG formation process. The degree of complexity involved and the stochastic data necessary to incorporate dynamic analysis are extremely high, and thus it is necessary to explore alternative approaches.

Advances in Modeling DGs In the base model, the power generated by DGs is modeled as a dispatch able quantity between lower and upper limits. This model is not sufficient to address the various types of DGs and their control schemes.

Operation Mode The operation of DGs depends on their role as grid-following or grid-forming units [39]. Grid-following units control the output of active and reactive power by injecting current at a given phase angle, and they cannot directly provide voltage and frequency support to the MG. Fundamentally, if the grid-following unit loses a voltage/frequency source, it must shut down. Grid-forming units are intrinsically different from the grid-following ones. A grid-forming unit is a controllable voltage source behind a coupling reactance much like grid-tied synchronous generators. Voltage source inverters with droop characteristic allow for direct control of voltage and frequency. During the process of MG formation, the droop-controlled grid-forming units will increase or decrease their output power instantaneously to balance loads and maintain local voltage and frequency. On this basis, the grid-following DGs should be modeled as constant PQ sources, while the grid-forming DGs as PV sources.

Intermittence and Stochasticity For intermittent DGs, like wind turbines and solar panels, the power generation constraint (7.10) in the base model is not sufficient to address their non-consistent behaviors. Assuming that the maximum renewable generation of intermittent

DGs, referred to as the available active power hereafter, can be predicted using state-of-the-art forecast algorithms [40], their power generation constraints can be formulated as

$$\begin{cases} 0 \le p_i^{\mathrm{G}} \le P_{i,\mathrm{av}}^{\mathrm{G}} \\ 0 \le q_i^{\mathrm{G}} \le Q_{i,\max}^{\mathrm{G}} \end{cases}, \quad \forall i \in N_{\mathrm{W}} \cup N_{\mathrm{P}} \tag{7.18}$$

where N_{W} and N_{P} are sets of nodes with wind farms and PV power plants installed, respectively; $P_{i,\mathrm{av}}^{\mathrm{G}}$ is the available active power of the wind farm /PV power plant at node i, while $Q_{i,\max}^{\mathrm{G}}$ is the maximum reactive power of the wind farm/PV power plant at node i, as restricted by its inverter capacity.

Another challenging problem is the stochastic power generation from DGs. Even with state-of-the-art forecasting algorithms, the intermittent DG generation is subject to prediction errors. A chance-constraint model can be used to address this issue.

Advances in Load Modeling In the base model, electric loads are modeled using the ON/OFF state of the load switch, i.e. the load demand is either 0, or set to the rated power, if it is supplied. Considering ongoing technologies, like demand-side management and load power forecast, load modeling can be improved as follows:

Modeling Cold Load Pickup Cold load pickup (CLPU) should be considered when restoring a system containing a high penetration of thermostatically controlled devices following an extended outage. An aggregated load under CLPU conditions will present significantly increased load demand and inrush current, which could overload transformers and lines and affect protective applications. Coping with CLPU issues involves modeling the loads as a function of time, which further necessitates the capability of handling inter-temporal models [41].

Modeling Controllable Loads One of the most efficient and practical approaches to maintain the balance between supply and demand in an MG is to apply DSR. A MILP-based demand response program to adjust loads in MG is presented in [42]. Accordingly, the load is constrained between a minimum active power representing the critical part and its rated value.

Modeling of Load Stochasticity In practice, loads fluctuate around their scheduled value. Denoting the stochastic fluctuation by ω_i^{L}, the load power is reformulated as below.

$$\begin{cases} \hat{p}_i^{\mathrm{L}} = p_i^{\mathrm{L}} + \omega_i^{\mathrm{L}} \\ \hat{q}_i^{\mathrm{L}} = q_i^{\mathrm{L}} + \tan(\cos^{-1}(\varphi_i))\omega_i^{\mathrm{L}} \end{cases} \tag{7.19}$$

where $\hat{p}_i^{\mathrm{L}}$ and $\hat{q}_i^{\mathrm{L}}$ are the real active and reactive power of the load at node i; φ_i is the power factor of the load at node i. It is assumed in equation (7.19) that a load's reactive power fluctuates in proportion to its active power based on a constant power factor per time step. Like the stochastic DG power model, a stochastic variable ω_i^{L} can be modeled by chance constraints, to derive a solvable optimization model.

7.4.2.3 Integration with DER Scheduling

The MGs DER should be properly scheduled for feasibility check based on results from fast power calculation. The following MG scheduling models [43] can be used to optimally schedule each MG k:

$$\min \; C_k = \sum_{i \in N_k} c_i^{\mathrm{L}} \left(P_i^{\mathrm{L}} - p_i^{\mathrm{L}}\right) + \sum_{i \in N_k \cap N_{\mathrm{D}}} c_i^{\mathrm{G}} p_i^{\mathrm{G}} + \sum_{i \in N_k \cap N_{\mathrm{ESS}}} c_i^{\mathrm{CH}} p_i^{\mathrm{CH}} + c_i^{\mathrm{DIS}} p_i^{\mathrm{DIS}} \tag{7.20}$$

$$\min \; \Delta V_k = \frac{1}{|N_k|} \sum_{i \in N_k} |V_i - 1| \tag{7.21}$$

$$\min \; p_k^{\text{loss}} \tag{7.22}$$

$$\min \; q_k^{\text{loss}} \tag{7.23}$$

where C_k is the total operation cost of MG k; c_i^{L}, c_i^{G}, c_i^{CH} and c_i^{DIS} are the outage cost of the load at node i, the operating cost of the DG at node i, the operating cost of charging and discharging the ESS at node i, respectively; p_i^{CH} and p_i^{DIS} are the charging and discharging power of the ESS at node i; ΔV_k are the average voltage deviation of MG k; p_k^{loss} and q_k^{loss} are the active and reactive power loss of MG k, respectively. Equations (7.20)–(7.23) minimize the total operating cost, the average voltage deviation, and active and reactive power losses, respectively. They can be used individually as the overall objective of the MG scheduling problem, or a set of weight coefficients can be assigned to them to derive a combined, single objective. Constraints include nodal power equilibria (power flow equations), limits of nodal voltages and power flows and operating ranges of DGs, ESS, and controllable loads. It should be noted that the AC power flows are nonlinear and non-convex. Therefore, techniques regarding either nonlinear optimization like the interior point algorithm, or power flow approximation, as described in Section 7.4.1.2, can be employed to solve the MG scheduling model. The problem of optimally scheduling DER in a formed MG is further analyzed in Section 7.6 for an example of resilience strengthening against an imminent disaster.

7.5 Microgrids for Resilience-Oriented Restoration

7.5.1 Microgrids Aided Distribution System Restoration

In the base model, it is assumed that MGs are formed, as a result of the disconnection of the main substation due to catastrophic events. The MG formation method presented in Section 7.4.1 can also address faults located within the distribution system. In both cases, the formed MGs should keep operating, and meanwhile, a distribution system restoration plan should be carried out to clear the faults. Efficient management of repair crews during the execution of the restoration plan plays a vital role in the restorative phase. After all faults are cleared, the MGs are interconnected to restore the distribution system to its normal post-restoration state. In this section, a distribution system restoration model [41] is presented that provides a complete solution for MG formation and fault clearance.

To simplify the distribution system restoration problem and ensure the feasibility of the solution, the following assumptions are made:

1) The distribution system restoration is formulated as a multi-time-step problem, and the time interval between two consecutive steps is assumed fixed.
2) The distribution system restoration model can generate restoration solutions that assign each discrete time step to a system configuration. In practice, the system operating condition will change continuously from one discrete step to the next discrete step, due to continuously changing variables, such as energized CLPU loads discussed in Section 7.4.2.2.3. Therefore, it is necessary to ensure that the intermediate system operating conditions always satisfy the constraints.
3) The communication network is available during the restoration process. This assumption ensures that the control commands can be performed timely at each time step. If a component is not controllable due to loss of communication, it will be classified as a failed component.
4) The energized system is maintained radial during the restoration process, and there is a single substation node or black start DG in the system. It should be noted that the operation

of meshed systems will provide extended flexibility for distribution system restoration, but presents substantial challenges for problem formulation and protection coordination. Thus, only radial topology is considered in the presented model.

7.5.1.1 Operating Constraints

Power Flow For the operating constraints the simplified DistFlow equations (7.6) and (7.7) of Section 7.4.1.2 are extended by the introduction of binary variables:

$$\begin{cases} \sum\limits_{h:(h,i)\in E} p_{hi,t} - \sum\limits_{j:(i,j)\in E} p_{ij,t} = p^{G}_{i,t} + p^{DIS}_{i,t} - p^{CH}_{i,t} - p^{L}_{i,t} \\ \sum\limits_{h:(h,i)\in E} q_{hi,t} - \sum\limits_{j:(i,j)\in E} q_{ij,t} = q^{G}_{i,t} - q^{L}_{i,t} \end{cases}, \quad \forall i \in N, t \in T \tag{7.24}$$

$$\begin{cases} v_{i,t} - v_{j,t} \geq 2(R_{ij}p_{ij,t} + X_{ij}q_{ij,t}) - M(1 - y_{ij,t}) \\ v_{i,t} - v_{j,t} \leq 2(R_{ij}p_{ij,t} + X_{ij}q_{ij,t}) + M(1 - y_{ij,t}) \end{cases}, \quad \forall (i,j) \in E, t \in T \tag{7.25}$$

$$(p_{ij,t})^2 + (q_{ij,t})^2 \leq y_{ij,t}(S_{ij,\max})^2, \qquad \forall (i,j) \in E, t \in T \tag{7.26}$$

where $p_{ij,t}$ and $q_{ij,t}$ are the active and reactive power flow on branch (i, j) at time t, respectively; $p^{G}_{i,t}$ and $q^{G}_{i,t}$ are the active and reactive power generated at node i, $p^{CH}_{i,t}$ and $p^{DIS}_{i,t}$ are the charging and discharging active power of the ESS at node i, and $p^{L}_{i,t}$ and $q^{L}_{i,t}$ are the active and reactive loads at node i at time t, respectively. $v_{i,t}$ is the squared value of the voltage magnitude of node i at time t; $S_{ij,\max}$ is the maximum thermal capacity of branch (i, j); $y_{ij,t}$ is the binary energization status of branches, extended as a multi-time-step variable, unlike the single-time-step y_{ij} in the base model.

Constraint (7.24) expresses the total apparent power equilibrium at each node. Constraint (7.25) defines the voltage magnitude difference between two end nodes of each energized line. Variable M takes large values to guarantee that the constraints are active, only when the corresponding binary variables are equal to one. Constraint (7.26) ensures that only the energized lines can carry power, and their loading does not exceed the maximum capacity.

Moreover, the voltage magnitude of each node should be maintained within acceptable limits at each time during the restoration process. This constraint can be formulated as:

$$x_{i,t}V^2_{\min} \leq v_{i,t} \leq x_{i,t}V^2_{\max}, \quad \forall i \in N, t \in T \tag{7.27}$$

The above constraint also guarantees that the nodal voltage will be zero for de-energized nodes.

DG Output For each time step, the output active and reactive power of an energized DG should not violate its rated capacity, while for some DGs, like diesel generators, minimum operating limits should be ensured. In addition, an optional constraint can be added for dispatchable DG operating at fixed power factor $\cos(\varphi_i)$. Moreover, the ramp rate of changing each DG output should be within an acceptable range.

ESS Operation A critical concern for ESS operation is to maintain the state of charge or residual energy within an acceptable range. For each ESS, the operational constraints need to take into account a multi-time-step binary energization status of node i at time t, the time span of the discrete time step and the ESS charging and discharging efficiency. It is also required that an ESS can only start to charge or discharge when the connected node is energized and that charging and discharging at the same time are mutually exclusive.

Spinning Reserve The total capacity of dispatchable generation should be greater than the total demand in order to maintain a certain level of spinning reserve. The spinning reserve is needed to account for prediction errors, unexpected loss of DG, and load. It is determined by the total capacity of DGs that are currently energized and the ESS in the discharging state.

Frequency Response Rate Constraints The frequency response rate is defined by the frequency dip when the system is subject to sudden load pickup. The frequency response is mainly determined by the reaction of generation units that convert various energy sources to mechanical energy (e.g. combustion, steam and hydro turbines), since power electronics interfaced units that directly convert energy to electricity (e.g. PVs and fuel cells) are much faster. An accurate expression of this constraint requires dynamic modeling of the MG. A simple constraint can be used for approximating the frequency response by a specific value and used in the initial state of power system restoration [44]. That is, the sudden incremental amount of restored loads at each step should be smaller than or equal to the maximum allowable pickup amount, which is a percentage of the total rated capacity of all energized DGs and ESSs in the discharging state. Note that at each step, only load demand energized by switching loads from OFF status to ON status will be counted, as loads energized in previous steps will change continuously during the succeeding steps, thus having a minor impact on transient frequency.

7.5.1.2 Connectivity and Sequence Constraints

These constraints ensure connectivity among energized components at each time step. For example, an energized network at a particular time step should satisfy $x_{i,t} = 1$ for all energized nodes, and $x_{i,t} = 0$ for all the de-energized nodes. Similar constraints apply to the binary variables for branches, loads, and DGs. The connectivity constraints among the binary variables are:

A DG can only be started, when it is connected to an energized node,

$$u_{i,t}^{\mathrm{G}} \leq x_{i,t}, \quad \forall i \in N_{\mathrm{D}}, t \in T \tag{7.28}$$

A branch can only be energized when both end nodes are energized,

$$\begin{cases} y_{ij,t} \leq x_{i,t} \\ y_{ij,t} \leq x_{j,t} \end{cases}, \quad \forall (i,j) \in E, t \in T \tag{7.29}$$

A load can only be energized when it is connected to an energized node,

$$x_{i,t}^{\mathrm{L}} \leq x_{i,t}, \quad \forall i \in N, t \in T \tag{7.30}$$

Additional constraints are applied to ensure that components remain connected after being energized, i.e.

$$\begin{cases} x_{i,t-1} \leq x_{i,t}, & \forall i \in N, t \in T \\ u_{i,t-1}^{\mathrm{G}} \leq u_{i,t}^{\mathrm{G}}, & \forall i \in N_{\mathrm{D}}, t \in T \\ x_{i,t-1}^{\mathrm{L}} \leq x_{i,t}^{\mathrm{L}}, & \forall i \in N, t \in T \\ y_{ij,t-1} \leq y_{ij,t}, & \forall (i,j) \in E, t \in T \end{cases} \tag{7.31}$$

A node can only be energized when at least one of its neighboring branches is energized. This can be described as

$$x_{i,t} \leq \sum_{h:(h,i) \in E} y_{hi,t} + \sum_{j:(i,j) \in E} y_{ij,t}, \quad \forall i \in N, t \in T \tag{7.32}$$

Sequencing constraints are introduced for branches to ensure that a feasible energization sequence can be generated. Specifically, each branch can only be energized when at least one of its ends is already energized at the previous step. This constraint can be described as:

$$y_{ij,t} \leq x_{i,t-1} + x_{j,t-1}, \quad \forall (i,j) \in E, t \in T \tag{7.33}$$

7.5.1.3 Topological Constraints

To ensure that the energized system is operated in the tree topology during the restoration process, the energized system at each time step should be a connected graph, and also satisfy:

$$\sum_{i \in N} x_{i,t} - \sum_{(i,j) \in E} y_{ij,t} = 1, \quad t \in T \tag{7.34}$$

Constraint (7.34) assumes that only one substation node or black-start DG exists in the system. When there are multiple substations or black-start DGs, constraint (7.34) should be modified, according to the actual number of subsystems.

7.5.1.4 Initial Condition Constraints

The availability of each component should be used as the initial condition. This is done by directly assigning values to the corresponding decision variables. The initial conditions are defined based on the following rules:

1) The black-start DG in the distribution system is started throughout the entire restoration process, and the voltage is regulated at a constant level V_s, shown as

$$\begin{cases} u^{G}_{i,t} = 1 \\ v_{i,t} = V_s \end{cases}, \quad \forall i = N_{BS}, \quad t \in T \tag{7.35}$$

where N_{BS} is the black-start node.

2) A disconnected or damaged branch that cannot be reclosed should not be energized

$$y_{ij,t} = 0, \quad \forall (i,j) \in E_F, \quad t \in T \tag{7.36}$$

where E_F is the set of branches that cannot be energized.

3) A damaged node that fails for $x_{i,t}$ restoration should not be energized

$$x_{i,t} = 0, \quad \forall i \in N_F, \quad t \in T \tag{7.37}$$

where N_F is the set of nodes that cannot be energized.

7.5.1.5 Optimization Formulation

When a general blackout or partial power outage occurs in a distribution system, the most critical concern is to restore the maximum amount of the affected load. The objective function aims to maximize the total restored energy for the entire time horizon:

$$\max \sum_{t \in T} \sum_{i \in N} w_i p^{L}_{i,t} \Delta T \tag{7.38}$$

The problem is formulated as a mixed-integer second-order cone programming (MISOCP) model, which can be solved by commercial solvers. It is worth noting that the presented distribution system restoration methodology can be applied to both distribution systems and MGs by manipulating corresponding constraints in the MISOCP model. For example, to restore a distribution system with substation nodes energized, constraints associated with black start DGs in the MISOCP model should be updated accordingly. If the system does not contain ESS, the EES operation constraints and the ESS terms in the power flow equations should be removed from the MISOCP model. To restore a system from a local outage state, the initial condition constraints should be configured properly to incorporate the information on the energization status and availability of each component.

7.5.2 Restoration of a Microgrid

In case of power system complete or partial blackouts, an MG might need to restore itself from collapse based on its own DER. In a centralized approach, the MG restoration is guided centrally by the MGCC. The central controller is responsible for checking the conditions and applying a set of rules, which should be identified in advance. The rules and conditions define a sequence of control actions to be carried out during the restoration procedure. The main steps to be considered include building the distribution network, connecting the DGs, controlling voltage and frequency, connecting controllable loads and synchronizing the MG with the upstream network. The MG restoration procedure will be triggered if a general or local blackout occurs or if major damage to the main substation does not provide the MG supply.

It is assumed that the MG local controllers, i.e. LCs and MCs, and the MG communication infrastructure are in operation during the restoration procedures. This is particularly important for the application of a centralized approach. Thus, small auxiliary power units are required to power the communication network elements and local controllers. It should be noted that the in-feeding distribution transformer and the DER must be disconnected after system collapse. It is also assumed that the MGCC has updated pre-disturbance information stored in its database, about the status of load/generation in the MG and about the availability of DER to restart.

7.5.2.1 Microgrid Restoration Sequence

During restoration, the following sequence of actions needs to be performed [45]:

- Sectionalizing the MG around each DG with black starting capability to allow it to feed its own loads: these actions lead to an initial creation of small islands inside the MG, which will be later interconnected.
- Building the MG network: the MG main dispatchable DG or ESS is used for energizing an initial part of the MG network, followed by the operation of the remaining switches not connected to loads or DGs.
- Small islands synchronization: DGs already in standalone operation mode should then be synchronized with the MG network. The synchronization conditions (phase sequence, frequency, and voltage differences) should be verified by the local MC, after the procedure is enabled by the MGCC, in order to avoid large transient currents that may compromise inverters operation.
- Controllable loads can be connected to the MG network, if the DG or ESS has the capacity to supply these loads. The amount of power to be connected should take into account the available capacity in order to avoid large frequency and voltage deviations during load connection. Motor load starting causes large inrush currents and requires particular attention.
- Connection of noncontrollable, intermittent DGs, such as PVs and wind generators, can be connected at this stage, since the system DGs are able to smooth the produced voltage and frequency variations. MG network paths can be created so that DGs without Blackstart (BS) capability can absorb power from the grid in order to restart.
- More loads can be connected, depending on generation capability, in order to feed as much load as possible.
- Changing the control mode of MS inverters: the MG main DG or storage inverter operates as a grid-forming inverter providing system voltage and frequency references. The DGs with BS capability inverters that have been operated as voltage source inverters (VSI) need to be changed to grid-following inverters (PQ control).
- MG synchronization with the upstream network when it becomes available: the synchronization conditions should be verified again, following the order given by the MGCC. This means the

Distribution Transformer should be previously energized from the MV side with this synchronization then being performed through LV switches.

7.5.2.2 Protection Considerations

MG's protection faces several challenges. The main challenges are due to the presence of multiple DGs that provide short circuit currents in traditionally passive distribution networks, the different fault current levels in different operational modes (islanded, grid connected) of the MG and the limited short circuit current contribution by power electronics interfaced energy resources. The coordination of Fault Ride-Through curves and protection schemes is also an issue that can even affect the overall MG stability in islanded operation. These issues are of particular concern during the MG restoration, as the BS procedure involves a step-by-step connection of DER to the grid, constantly changing the short-circuit power at the points, where protection devices are installed. In order to deal with the problem of low short circuit currents in islanded MGs, there are two practical solutions:

1) Increase short circuit current levels. This can be done by installing a source of high short-circuit current (e.g. a flywheel) to satisfy the operating settings of fuses and overcurrent relays, or by increasing the fault current capability of power electronic interfaced units. The latter can be achieved by increasing their power rating or by applying extensive cooling of their switching devices. Both options lead to higher investment costs.
2) Replace existing passive protective devices with numerical protective relays, which provide advanced protection functionalities, and deploy adaptive protection based on the MG state.
   Adaptive protection requires industrial numerical relays, mainly directional overcurrent relays with several setting groups, which can be parameterized locally or remotely by control signals, according to the current MG topology. Communication infrastructure is also required in an adaptive protection system, but low-bandwidth technologies can be used since protection is still based on local functions of the relays during the fault and communication is used only for the update of the setting group when the configuration of the MG changes. The setting groups can be updated based on calculations performed online or offline. For online calculations, the MGCC receives the required signals (e.g. circuit breaker statuses) to identify the current MG topology. Then, a local computer executes fault simulations at different locations, and according to the calculated fault currents, the proper setting group is activated, e.g. by solving an optimization problem to minimize the operation times of the protective devices. Finally, digital signals are sent to the relays via a low-bandwidth communication system. For the offline approach, short circuit calculations for all possible operating conditions of the MG are performed before the adaptive protection system deployment. Based on an optimization problem solution, an adequate set of setting groups for the protective devices for each operating state is calculated offline. During the online operation, the MGCC receives the required signals (e.g. circuit breaker statuses) to identify the current MG configuration and changes, if necessary, the active setting group of the relays via a low-bandwidth communication channel [46].

7.5.2.3 Decentralized Techniques

Most of the load restoration strategies are based on centralized optimization algorithms. Accordingly, the MGCC collects all the data required and sends control commands, thus suffering from the risk of single-point failure and from heavy computation burden. In contrast, distributed optimization algorithms, such as the alternating direction method of multipliers (ADMM) [47], rely on local data collection and information exchanges between neighboring agents. They are more resilient against communication failures and can improve computation efficiency by exploiting parallel computation. Thus, they appear promising for restoration applications. The ADMM has been successfully employed to solve convex problems; however, a number of control actions used for load restoration, such as operations of voltage regulators (VRs) and discontinuous load pick-up, can only be modeled as discrete variables. Therefore, more advanced techniques, like a projection function-based ADMM (P-ADMM) [48], are needed to deal with discrete variables for decentralized restoration approaches.

A distributed risk-limiting load restoration strategy for unbalanced DSs with networked MGs based on the conditional value-at-risk (CVaR) and P-ADMM is developed in [49]. By using the CVaR as the risk-limiting index, the optimal risk-limiting load restoration problem is formulated as a MINLP model and is transformed into an MILP model by applying linearization techniques. With the sub-additivity of the CVaR and P-ADMM, the MILP model is decoupled and solved in a distributed manner, realizing distributed risk-limiting strategy among MGs.

7.6 Resilience-Oriented Preventive and Emergency DER Scheduling

This section focuses on optimal scheduling of DER and storage capacities within already formed MGs in an imminent catastrophic event, i.e. at pre-disturbance state (preventive) or after the event hits the network and the disturbance progresses (emergency). When the extreme event is expected to affect system components, the impact should be assessed in a stochastic way, due to the inability to predict the exact time, area, and duration that a natural disaster will hit and last. In such cases, it is crucial to optimally schedule the DER, in order to be prepared for potential events and minimize load outages. Besides the time, progress and duration of the event, RES' generation, such as WT and PV generation, and the estimated damage in power system infrastructure need to be modeled using a stochastic approach. In this section, a stochastic programming approach is used to determine the operation of the MG exposed to an imminent extreme event. The impact on system components, such as direct damage or a decrease in their capacity due to line thermal rating, decreases in case of a wildfire is considered.

7.6.1 Problem Formulation

In order to capture the stochastic behavior of wind speed and direction, solar radiation and load demand typical probability distribution functions are assumed. Monte Carlo simulation (MCs) is then used to produce a large number of scenarios for the stochastic parameters based on forecasted values and the typical distribution of each parameter. The use of MCs, instead of simple mean values, is able to provide results covering a whole spectrum of the most probable scenario values including worst-case probable conditions, as will be shown in Section 7.6.2. The Backward Scenario Reduction algorithm is then used in order to reduce the number of scenarios in a tractable size [50].

The objective of the stochastic programming formulation for the optimal operation of an MG can be expressed as follows [35]:

$$\min \sum_{t=1}^{N_T}\sum_{\omega=1}^{N_\Omega}\pi_\omega\sum_{i=1}^{N_B}\mathrm{VoLL}\, p_{i,\omega,t}^{\mathrm{shed}} + \sum_{t=1}^{N_T}\sum_{\omega=1}^{N_\Omega}\pi_\omega\sum_{i=1}^{N_B}\left(c^G p_{i,\omega,t}^G\right) + \sum_{t=1}^{N_T}\sum_{\omega=1}^{N_\Omega}\pi_\omega c_t^{UP}\left(p_{\omega,t}^{UP_B} - p_{\omega,t}^{UP_S}\right)$$
$$+\sum_{t=1}^{N_T}\sum_{i=1}^{N_B}\left(su_{i,t}^G + sd_{i,t}^G\right) \tag{7.39}$$

Similarly to (6.1) of Chapter 6, the objective function (7.39) aims to minimize the expected social cost, expressed as a minimum load shedding during an emergency, while respecting operating limits of the system. Since resilience is directly associated with minimum load shedding, only the expected load shedding needs to be considered in the objective function. However, in order to provide the most economic solution, further costs are considered. The first term of the objective function represents the load shedding cost, p^{shed} is the active power shedding and VoLL is the Value of lost load [$/MWh]. The second term represents the generation cost of the controllable MT units, where p^G is the active power of each controllable DG. The third term describes the cost of power exchange with the upstream system, $p^{UP_{B/S}}$ is the active power to buy/sell from/to the upstream system. The fourth term represents start-up and shut-down costs of the controllable units su^G, sd^G. The MTs are controlled according to the proposed scheduling, and therefore, the MTs' status is the same in each scenario. Thus, the start-up and shut-down costs of MTs are not associated with the scenarios. The value of lost load should be set large enough to prioritize the demand satisfaction. ω is the index for scenarios 1 to N_Ω, where N_Ω is the number of scenarios and π_ω the probability of each scenario.

Several constraints could be added to the stochastic problem. Some constraints that could be included in a problem that considers line damages are discussed, while the mathematical expression of some critical constraints for taking into account the status of the lines is presented.

A list of constraints that could be included in the problem is the following:

1) Active and reactive power balance at each bus
2) Active and reactive output limits of controllable units
3) Bus voltage limits
4) Load shedding limits
5) Energy storage system (ESS) limits
6) Power exchange limits with upstream system

The power flow through the lines should be allowed only when the line is functional. To achieve this, the following constraint should be considered [35]:

$$-M_2 y^l_{ij,\omega,t} \le pf^P_{ij,\omega,t} \le M_2 y^l_{ij,\omega,t} \tag{7.40}$$

$$-M_2 y^l_{ij,\omega,t} \le pf^Q_{ij,\omega,t} \le M_2 y^l_{ij,\omega,t} \tag{7.41}$$

$$-M_3\left(1 - y^l_{ij,\omega,t}\right) \le pf^P_{ij,\omega,t} - g_{ij}(v_{i,\omega,t} - v_{j,\omega,t} - \zeta_{ij,\omega,t} + 1) + b_{ij}(\vartheta_{i,\omega,t} - \vartheta_{j,\omega,t}) \le M_3\left(1 - y^l_{ij,\omega,t}\right) \tag{7.42}$$

$$-M_3\left(1 - y^l_{ij,\omega,t}\right) \le pf^Q_{ij,\omega,t} + b_{ij}(v_{i,\omega,t} - v_{j,\omega,t} - \zeta_{ij,\omega,t} + 1) + g_{ij}(\vartheta_{i,\omega,t} - \vartheta_{j,\omega,t}) \le M_3\left(1 - y^l_{ij,\omega,t}\right) \tag{7.43}$$

Constraint (7.40) and (7.41) allow the active and reactive power flow through the line only when the line is not damaged. The binary variable $y^l_{ij,\omega,t}$ represents the status of the line (i, j) in scenario ω at time t. The selection of the sufficiently large constant M_2 should allow the maximum flow of the line. Equations (7.42) and (7.43) are converted to equalities when the line is functional and relate voltage magnitude and angle to power flow. The linearization of power flow equations proposed in [51] has been used in (7.42) and (7.43), where $\zeta_{ij,\omega,t}$ represents the piecewise linear approximation of $\cos(\vartheta_i - \vartheta_j)$, and M_3 is a sufficiently large constant. More details about the linearization of power flow equations and their accuracy can be found in [52]. It is noted that any linearization of power flow equations could be used.

To consider the impact of extreme events on the generation units, similar constraints could be used to allow their generation only when the units are not damaged.

7.6.2 Study Case Application [35]

7.6.2.1 Test Network and Simulation Data

This section presents the results of the proposed method applied to an MG. The modified IEEE 33-bus distribution system is used [53]. The system is assumed balanced and is represented by an equivalent single-phase circuit. It is considered that a wildfire is approaching the MG and threatens the smooth function of MG lines. To consider the impact of wildfire on MG lines the dynamic line rating (DLR) of the overhead lines is considered in order to model the impact of the wildfire on the conductor's temperature. To achieve this, the non-steady-state heat balance equation of IEEE Std 738 has been used [54]. More details on how to model the impact of a wildfire on power system lines can be found in [30]. The single-line diagram of the modified network and the progression of the wildfire are presented in Figure 7.15. The peak active demand is equal to 15 MW, constant power factor loads are considered and a typical load profile for the summer season is assumed.

The MG extends to a small geographical region; therefore, its components are exposed to similar weather conditions. The location and capacity of the system components are shown in Table 7.1. The parameters of ESSs are shown in Table 7.2. For WTs, the cut-in, cut-out and rated wind speeds are 4, 20, and 12 m/s, respectively. The rated illumination intensity of the PVs is 1000 W/m^2. Weibull and von Mises distributions are used for modelling wind speed and direction. Solar radiation is modeled using Beta distribution and Normal distribution is used for load demand. These parameters are used to assess the progression of the wildfire and estimate its impact on MG lines, as well as to estimate the generation of RES. The standard deviation is considered equal to 15% of the mean value for wind speed and solar radiation and 5% of the mean value for the loads. A k-factor of 2 is assumed for the von Mises distribution. The weather parameters and wildfire data are taken from [55]. A time step of 30 min is selected, which according to [56] is a tolerable time step for DLR modeling. Typical wind and solar data are assumed [35].

Table 7.1 Location and capacity of mg sources.

Type	Bus	Capacity (MW)
WTs	14/16/31	0.8/0.8/0.8
PV	11	0.5
MTs	8/13/16/25	3/2/2/3
ESSs	19/26	0.5/0.5

Source: [35] / IEEE.

Table 7.2 Parameters of energy storage systems.

Energy storage system	E^{ST} (MWh)	$p^{ST+/-}$ (MW)	q^{ST} (MVAr)	q^{ST} (MVAr)
ESS (bus 19)	1.5	0.5	0.3	−0.3
ESS (bus 26)	1.5	0.5	0.3	−0.3

Source: [35] / IEEE.

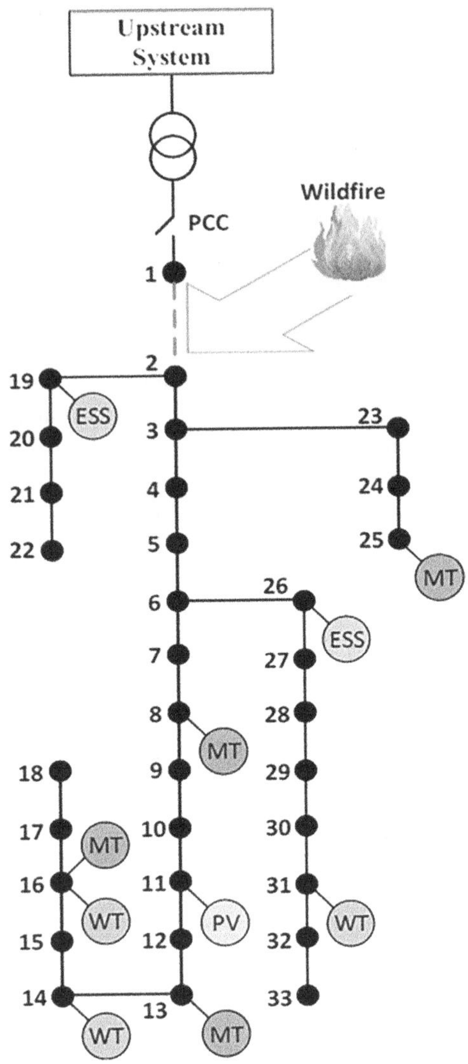

Figure 7.15 Single line diagram of the simulated MG and wildfire progression. *Source:* [35] / IEEE.

A summer day with high wind speed and ambient temperature that facilitates the development and spread of a wildfire is considered. It is assumed that the fire breaks out at $t = 1$, approaches and affects only the line between buses 1 and 2 (dashed line in Figure 7.15). The proposed method can be applied at any time during the day at which the threat of the wildfire is perceived. Therefore, in the light of a wildfire threat, the scheduled operation changes to enhance MG resilience. The initial distance from the line is $r^f_{t=1} = 1200$ m. The rest of the lines are considered at a safe distance from the fire. The selected line is the one that connects the MG with the upstream network, and therefore, the power exchange between the two systems is affected by the fire. The proposed approach is general, however, and it can be applied to the study of any line or lines approached by the fire. Assuming that the spatial data of the MG are known, e.g. by a geographic information system (GIS), the distance between the fire and the conductors can be estimated, during the progression of the wildfire. Based on the estimated distance, the impact of the wildfire on the temperature of any line can be computed [35].

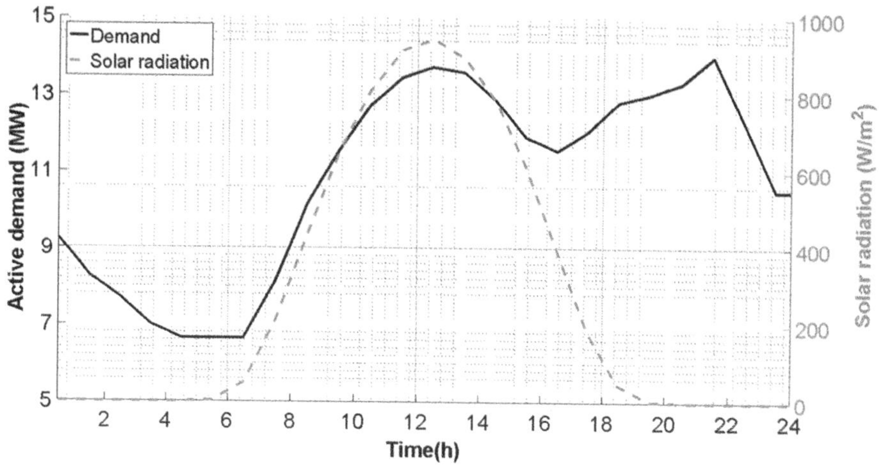

Figure 7.16 Mean value of active demand and solar radiation. *Source:* [35] / IEEE.

It is considered that at the end of the time horizon, the wildfire is extinguished, but the affected line remains out of service until it is repaired. Thus, the SOC of the ESSs is required to remain greater than 30% of the maximum capacity to contribute to demand satisfaction for the next hours after the study.

An ASCR-type conductor is considered. The diameter and the maximum permissible temperature of the line are assumed equal to 21 mm and 353 K, respectively.

The appropriate number of representative scenarios has been selected, taking into account the range of optimization results (their impact on the objective function) and the required computation time. In the presented application, 10 representative scenarios have been selected, since a higher number has an insignificant effect on the results and the computation time becomes unacceptably high for online application of the method. To produce representative scenarios, a scenario generation and reduction algorithm has been implemented [50].

The mean value of active demand and solar radiation is presented in Figure 7.16. Figure 7.17 presents the mean values of wind speed and wind direction. The angle is measured with respect to the axis defined by the fire and the conductor. Wind speed $V^{w,\mathrm{WT}}$ in Figure 7.17 refers to the height L^{WT} of WTs. The wind speed at L^l is given by [57]:

$$V^{w,l} = V^{w,\mathrm{WT}}\left(\frac{L^l}{L^{\mathrm{WT}}}\right)^{0.143} \tag{7.44}$$

The ambient temperature and the price of power exchange with the upstream system are presented in Figure 7.18. The price of buying and selling energy from and to the upstream system is considered equal and is adopted from [58].

7.6.2.2 Operation Against Approaching Wildfire

The proposed method is applied in order to optimize MG operation for enhancing its resilience against the approaching wildfire for the next 24 hours. Figure 7.19 shows the distance between the fire and line 1–2. It is observed that the fire approaches the line during the day, reaching zero distance in the last hours. Figure 7.20 presents the conductor heat gain rate from the fire. By comparing Figure 7.19 and Figure 7.20, it is observed that the conductor heat gain rate from the fire gets values greater than zero when the distance of the fire from the line is less than 300 m and increases exponentially, as the fire approaches the line. Figure 7.19 and Figure 7.20 also show the

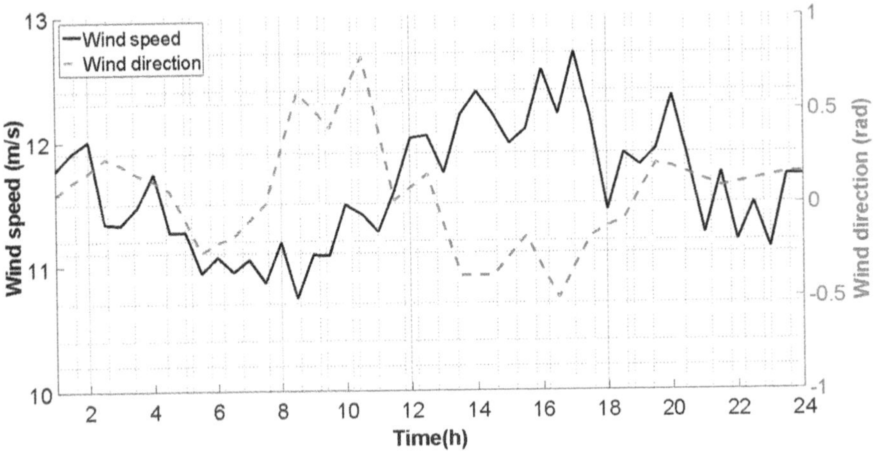

Figure 7.17 Mean value of wind speed at L^{WT} and wind direction. *Source:* [35] / IEEE.

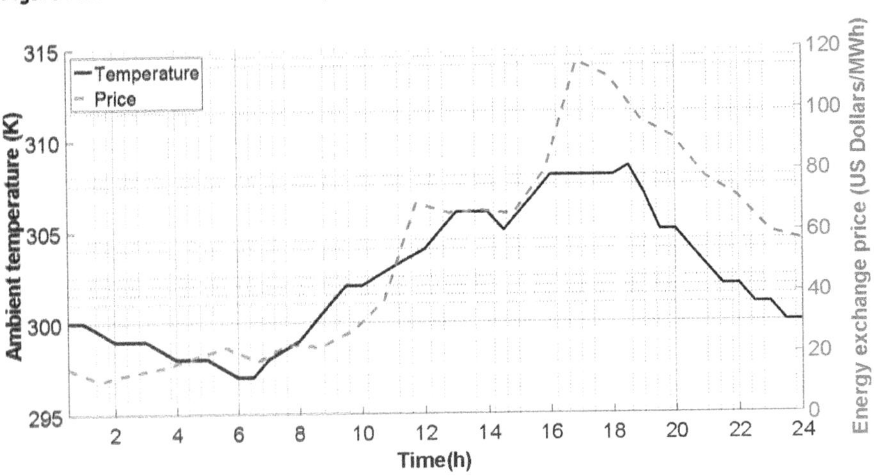

Figure 7.18 Ambient temperature and power exchange price. *Source:* [35] / IEEE.

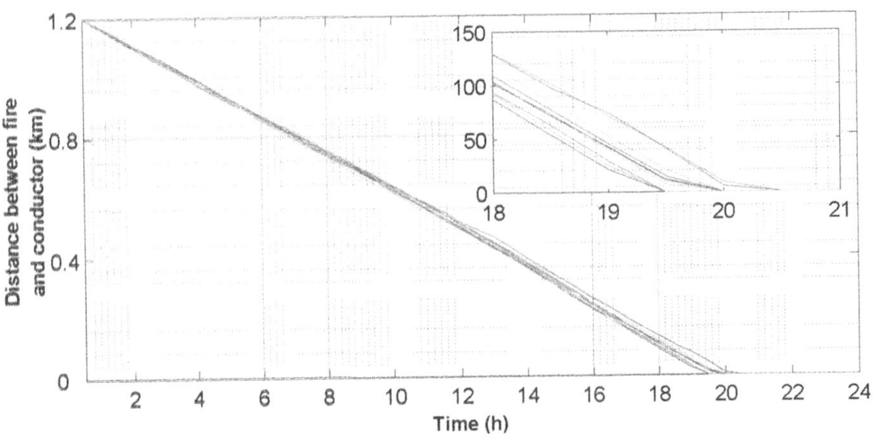

Figure 7.19 Distance between fire and line 1–2 in the reduced number of scenarios and magnified sub-figure. *Source:* [35] / IEEE.

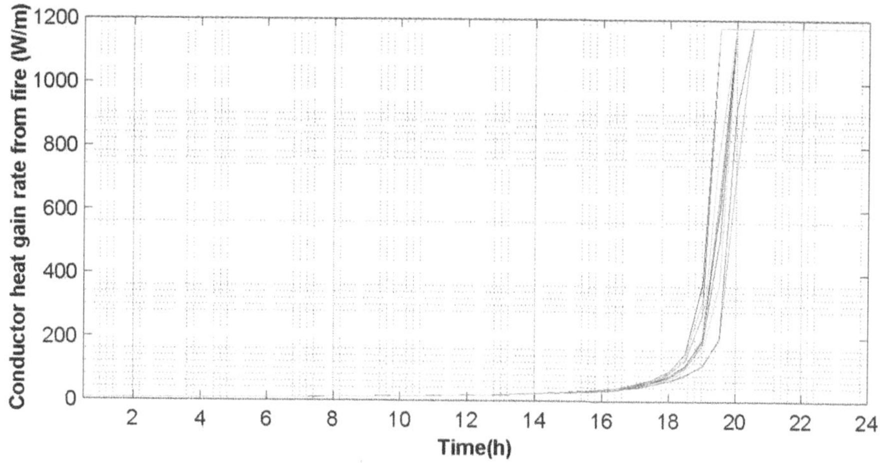

Figure 7.20 Conductor 1-2 heat gain rate from fire in the reduced number of scenarios. *Source:* [35] / IEEE.

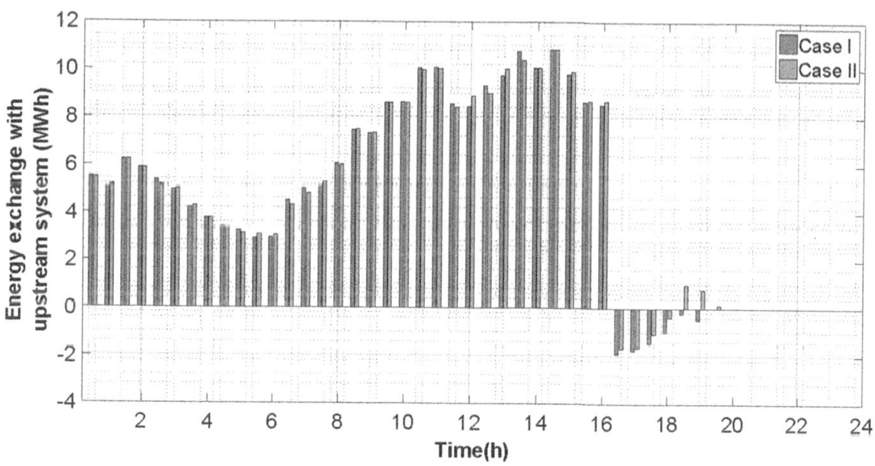

Figure 7.21 Expected energy exchange with the upstream system for Cases I and II. *Source:* [35] / IEEE.

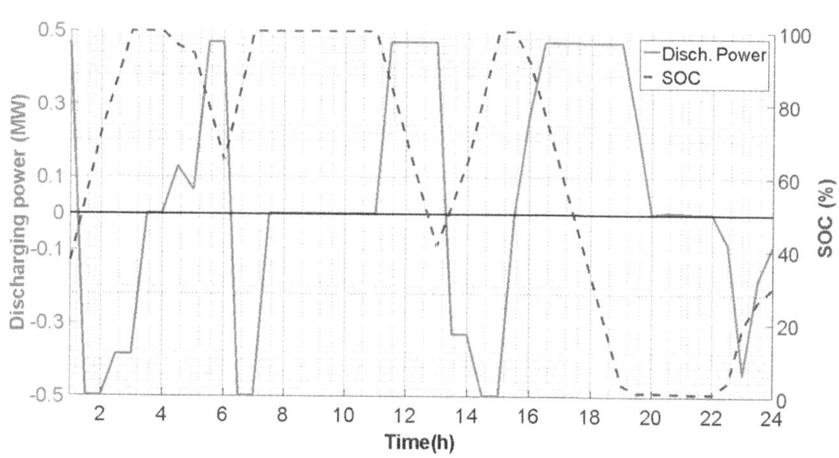

Figure 7.22 Expected discharging power and SOC of ESS at bus 19 for Case I. *Source:* [35] / IEEE.

added value of the stochastic analysis. In Figure 7.19, for example, it is observed that the wildfire reaches to a close distance from the line, (less than 15 m), in different time steps for each examined scenario. As a result, application of a deterministic optimization based on mean forecasted values, would not consider the line outage at an earlier, yet probable time step.

Two cases are considered in order to evaluate the effectiveness of the proposed approach. In *Case I*, the MG operation is optimized to minimize social cost without considering the wildfire progression. In *Case II*, the wildfire progression is considered, and the resilient operation of the MG is optimized.

The expected energy exchange with the upstream system for both cases is presented in Figure 7.21. It is observed that until 16:00 hours, the energy exchange is almost identical for both cases and energy is bought from the upstream system since the price of power bought is lower than the cost of the local MTs and the power from renewable generators does not cover the demand. After 16:00 hours and until 18:00, the power exchange price has high values and therefore energy is sold to the upstream system. For *Case I*, the exchanged energy is higher because, as shown in Figure 7.22, the ESS injects active power equal to its higher capacity in order to maximize the revenues from selling energy to the upstream system. For *Case II*, as shown in Figure 7.23, the discharging power of ESSs is lower since it is desired to maintain the SOC at the necessary level in order for the discharging power of ESSs to be used for minimizing load shedding when the connection with the upstream system will be disrupted due to the wildfire. After 18:00 hours and until 19:30, the power exchange price is lower but still higher than the cost of the local MTs. For *Case I*, the operator continues to sell energy to the upstream, while for *Case II*, energy is bought for charging the ESSs before line 1–2 is outaged due to the wildfire [35].

After this time, the power exchange with the upstream system drops to zero, since the line 1–2 is set out of order due to the violation of its maximum temperature. As a result of the ESS at bus 19 is at its lowest level. Meanwhile, for *Case II*, the ESSs are charged to the desired levels in order to serve the demand during the time the MG is disconnected from the upstream system and minimize the load shedding [35].

Figure 7.24 presents the expected amount of load that is shed to keep the balance between generation and demand, when line 1–2 is outaged. Until 19:00 hours, load is not shed in both cases. Between 19:30 and 22:30, an amount of load is shed due to the disconnection of line 1–2 and the inability of the DGs and ESSs to meet the demand. For *Case II*, the amount of load shedding is

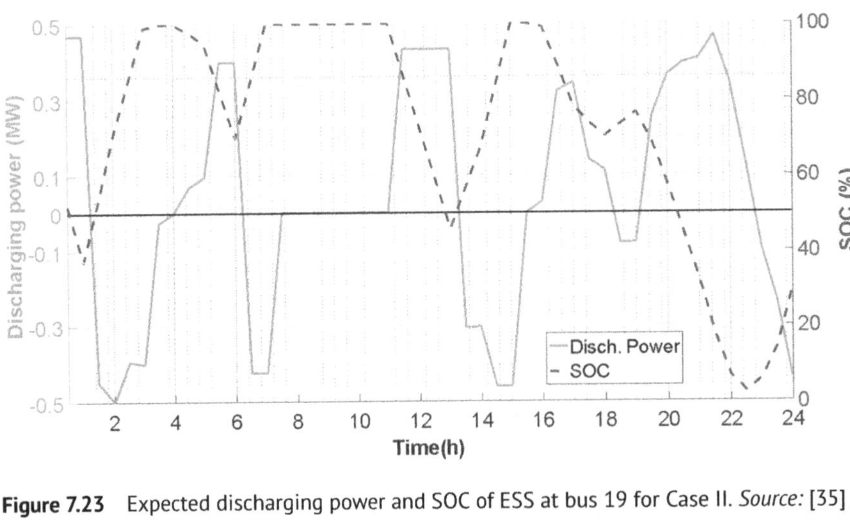

Figure 7.23 Expected discharging power and SOC of ESS at bus 19 for Case II. *Source:* [35] / IEEE.

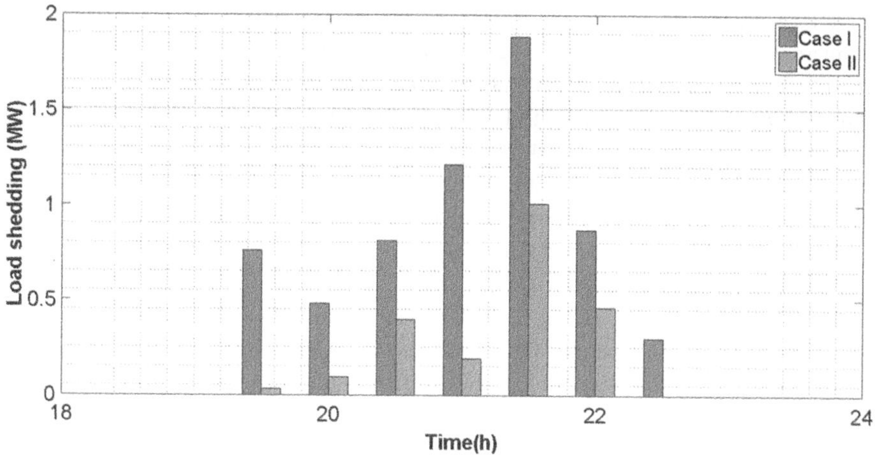

Figure 7.24 Expected load shedding for Cases I and II. *Source:* [35] / IEEE.

much lower due to the proper operation of the ESSs before the disconnection of line 1–2. After 23:00 hours, the demand is met by the DGs and load shedding is not recorded for both cases. Moreover, the ESSs are charged to reach the required levels [35].

References

1 G. Strbac, N. Hatziargyriou, J. P. Lopes, C. Moreira, A. Dimeas, and D. Papadaskalopoulos, "Microgrids: enhancing the resilience of the European megagrid," *IEEE Power Energy Mag.*, vol. 13, no. 3, pp. 35–43, 2015, doi: 10.1109/MPE.2015.2397336.

2 N. Hatziargyriou, Ed., *Microgrid: Architectures and Control.* Chichester, West Sussex, United Kingdom: Wiley, 2013.

3 IEEE standard for the specification of microgrid controllers, *IEEE Std 2030.7-2017*, pp. 1–43, April 23, 2018, doi: 10.1109/IEEESTD.2018.8340204.

4 CIGRE SC C6, Microgrids: engineering, economics, & experience. CIGRE SC C6, Tech. Brochure 635, 2015.

5 S. D. Manshadi and M. Khodayar, "Resilient operation of multiple energy carrier microgrids," in *2016 IEEE Power and Energy Society General Meeting (PESGM)*, Boston, MA, USA: IEEE, Jul. 2016, pp. 1–1. doi: 10.1109/PESGM.2016.7741950.

6 A. Vasilakis, I. Zafeiratou, D. T. Lagos, and N. D. Hatziargyriou, "The evolution of research in microgrids control," *IEEE Open Access J. Power Energy*, vol. 7, pp. 331–343, 2020, doi: 10.1109/OAJPE.2020.3030348.

7 D. Qiu, Y. Wang, J. Wang, N. Zhang, G. Strbac, and C. Kang, "Resilience-oriented coordination of networked microgrids: a Shapley Q-Value learning approach," *IEEE Trans. Power Syst.*, vol. 39, no. 2, pp. 3401–3416, 2024, doi: 10.1109/TPWRS.2023.3276827.

8 Y. Wang, D. Qiu, X. Sun, Z. Bie, and G. Strbac, "Coordinating multi-energy microgrids for integrated energy system resilience: a multi-task learning approach," *IEEE Trans. Sustain. Energy*, vol. 15, no. 2, pp. 920–937, 2024, doi: 10.1109/TSTE.2023.3317133.

9 D. Qiu, Y. Wang, T. Zhang, M. Sun, and G. Strbac, "Hierarchical multi-agent reinforcement learning for repair crews dispatch control towards multi-energy microgrid resilience," *Appl. Energy*, vol. 336, 120826, 2023, doi: 10.1016/j.apenergy.2023.120826.

10 L. Zhang, Z. Dong, X. Zhang, N. Zhang, and G. Strbac, "Resilience oriented planning for multi-energy microgrids in an Urban Area," in *2023 IEEE International Conference on Environment and Electrical Engineering and 2023 IEEE Industrial and Commercial Power Systems Europe (EEEIC/I&CPS Europe)*, Madrid, Spain: IEEE, Jun. 2023, pp. 1–6. doi: 10.1109/EEEIC/ICPSEurope57605.2023.10194676.

11 T. Chen, S. Bu, X. Liu, J. Kang, F. R. Yu, and Z. Han, "Peer-to-peer energy trading and energy conversion in interconnected multi-energy microgrids using multi-agent deep reinforcement learning," *IEEE Trans. Smart Grid*, vol. 13, no. 1, pp. 715–727, 2022, doi: 10.1109/TSG.2021.3124465.

12 X. Chen, G. Qu, Y. Tang, S. Low, and N. Li, "Reinforcement learning for selective key applications in power systems: Recent advances and future challenges," *IEEE Trans. Smart Grid*, vol. 13, no. 4, pp. 2935–2958, 2022, doi: 10.1109/TSG.2022.3154718.

13 S. Lei, C. Chen, H. Zhou, and Y. Hou, "Routing and scheduling of mobile power sources for distribution system resilience enhancement," *IEEE Trans. Smart Grid*, vol. 10, no. 5, pp. 5650–5662, 2019, doi: 10.1109/TSG.2018.2889347.

14 S. Lei, C. Chen, Y. Li, and Y. Hou, "Resilient disaster recovery logistics of distribution systems: Co-optimize service restoration with repair crew and mobile power source dispatch," *IEEE Trans. Smart Grid*, vol. 10, no. 6, pp. 6187–6202, 2019, doi: 10.1109/TSG.2019.2899353.

15 Y. Wang, D. Qiu, F. Teng, and G. Strbac, "Towards microgrid resilience enhancement via mobile power sources and repair crews: a multi-agent reinforcement learning approach," *IEEE Trans. Power Syst.*, vol. 39, no. 1, pp. 1329–1345, 2024, doi: 10.1109/TPWRS.2023.3240479.

16 J. Hu, S. You, M. Lind, and J. Ostergaard, "Coordinated charging of electric vehicles for congestion prevention in the distribution grid," *IEEE Trans. Smart Grid*, vol. 5, no. 2, pp. 703–711, 2014, doi: 10.1109/TSG.2013.2279007.

17 D. Qiu, Z. Dong, Y. Wang, N. Zhang, G. Strbac, and C. Kang, "Decarbonising the GB power system via numerous electric vehicle coordination," *IEEE Trans. Power Syst.*, vol. 39, no. 4, pp. 5880–5894, 2024, doi: 10.1109/TPWRS.2023.3342168.

18 E.-I. E. Stasinos, D. N. Trakas, and N. D. Hatziargyriou, "Microgrids for power system resilience enhancement," *iEnergy*, vol. 1, no. 2, pp. 158–169, 2022, doi: 10.23919/IEN.2022.0032.

19 P. Barter and T. Borer, "Case study: Microgrid at Princeton University," June 2015. [Online]. Available: https://www.csemag.com/articles/case-study-microgrid-at-princeton-university/.

20 K. Hirose,"Behavior of the Sendai Microgrid during and after the 311 Great East Japan Disaster," *Intelec 2013; 35th International Telecommunications Energy Conference, SMART POWER AND EFFICIENCY*, Hamburg, Germany, Oct. 13–17, 2013.

21 D. Carter *et al.*, "Demonstrating a secure, reliable, low-carbon community microgrid at the blue lake rancheria," Schatz Energy Research Center, CEC-500-2019-011, Jan. 2019. [Online]. Available: https://www.energy.ca.gov/sites/default/files/2021-05/CEC-500-2019-011.pdf.

22 P. Maloney, "Life won thanks to the blue lake rancheria microgrid," Microgrid Knowledge, Nov. 2019. [Online]. Available: https://www.microgridknowledge.com/editors-choice/article/11429301/life-won-thanks-to-the-blue-lake-rancheria-microgrid.

23 E. Wood, "Texas on the verge of an energy catastrophe: how microgrids are helping," Microgrid Knowledge, Feb. 2021. [Online]. Available: https://www.microgridknowledge.com/editors-choice/article/11428257/texas-on-the-verge-of-an-energy-catastrophe-how-microgrids-are-helping.

24 E. Wood, "Facing off climate disaster in Louisiana: the tale of two microgrid champions," Microgrid Knowledge, Oct. 2021. [Online]. Available: https://microgridknowledge.com/microgrids-footprint-project-powersecure-ida/.

25 Corporate Editorial Team, "Entergy system hurricane Ida Update – 9/2/21 @ 9 a.m," Sept. 2021 [Online]. Available: https://www.entergynewsroom.com/article/entergy-system-hurricane-ida-update-9-2-21-9-m/.

26 Paisley Dodds, Jessica Alexander,"Haiti earthquake aid slowed by gangs, blocked roads, and shipping delays," *The Haitian* Times, Sep. 10, 2021. [Online]. Available: https://haitiantimes.com/2021/09/10/haiti-earthquake-aid-slowed-by-gangs-blocked-roads-and-shipping-delays/.

27 L. Cohn, "Microgrids power through haiti earthquake," Microgrid Knowledge, Aug. 21. [Online]. Available: https://www.microgridknowledge.com/editors-choice/article/11427773/microgrids-power-through-haiti-earthquake.

28 N. Hatziargyriou, A. Dimeas, N. Vasilakis, D. Lagos, and A. Kontou, "The kythnos microgrid: a 20-year history," *IEEE Electrif. Mag.*, vol. 8, no. 4, pp. 46–54, 2020, doi: 10.1109/MELE.2020.3026439.

29 S. Tselepis, "12 years operation of the Gaidouromantra Microgrid in Kythnos island, COI-3869," *presented at the 5th International Conference on Integration of RES and DER*, Berli, Germany, Dec. 2012.

30 R. Moreno *et al.*, "Microgrids against wildfires: distributed energy resources enhance system resilience," *IEEE Power Energy Mag.*, vol. 20, no. 1, pp. 78–89, 2022, doi: 10.1109/MPE.2021.3122772.

31 R. Serrano *et al.*, "Fighting against wildfires in power systems: Lessons and resilient practices from the Chilean and Brazilian experiences," *IEEE Power Energy Mag.*, vol. 20, no. 1, pp. 38–51, 2022, doi: 10.1109/MPE.2021.3122731.

32 Y. Wang, A. O. Rousis, and G. Strbac, "On microgrids and resilience: a comprehensive review on modeling and operational strategies," *Renew. Sustain. Energy Rev.*, vol. 134, 110313, 2020, doi: 10.1016/j.rser.2020.110313.

33 B. Chen, J. Wang, X. Lu, C. Chen, and S. Zhao, "Networked microgrids for grid resilience, robustness, and efficiency: a review," *IEEE Trans. Smart Grid*, vol. 12, no. 1, pp. 18–32, 2021, doi: 10.1109/TSG.2020.3010570.

34 A. Hussain, V.-H. Bui, and H.-M. Kim, "Microgrids as a resilience resource and strategies used by microgrids for enhancing resilience," *Appl. Energy*, vol. 240, pp. 56–72, 2019, doi: 10.1016/j.apenergy.2019.02.055.

35 D. N. Trakas and N. D. Hatziargyriou, "Optimal distribution system operation for enhancing resilience against wildfires," *IEEE Trans. Power Syst.*, vol. 33, no. 2, pp. 2260–2271, 2018, doi: 10.1109/TPWRS.2017.2733224.

36 C. Chen, J. Wang, F. Qiu, and D. Zhao, "Resilient distribution system by microgrids formation after natural disasters," *IEEE Trans. Smart Grid*, vol. 7, no. 2, pp. 958–966, 2016, doi: 10.1109/TSG.2015.2429653.

37 K. Pang, C. Wang, N. D. Hatziargyriou, F. Wen, and Y. Xue, "Formulation of radiality constraints for optimal microgrid formation," *IEEE Trans. Power Syst.*, vol. 38, no. 6, pp. 5341–5355, 2023, doi: 10.1109/TPWRS.2022.3221048.

38 K. Pang, C. Wang, N. D. Hatziargyriou, and F. Wen, "Microgrid formation and real-time scheduling of active distribution networks considering source-load stochasticity," *IEEE Trans. Power Syst.*, vol. 39, no. 2, pp. 2801–2813, 2024, doi: 10.1109/TPWRS.2023.3276008.

39 R. H. Lasseter, Z. Chen, and D. Pattabiraman, "Grid-forming inverters: A critical asset for the power grid," *IEEE J. Emerg. Sel. Top. Power Electron*, vol. 8, no. 2, pp. 925–935, 2020, doi: 10.1109/JESTPE.2019.2959271.

40 L.-L. Li, S.-Y. Wen, M.-L. Tseng, and C.-S. Wang, "Renewable energy prediction: a novel short-term prediction model of photovoltaic output power," *J. Cleaner Prod.*, vol. 228, pp. 359–375, 2019, doi: 10.1016/j.jclepro.2019.04.331.

41 B. Chen, C. Chen, J. Wang, and K. L. Butler-Purry, "Multi-time step service restoration for advanced distribution systems and microgrids," *IEEE Trans. Smart Grid*, vol. 9, no. 6, pp. 6793–6805, 2018, doi: 10.1109/TSG.2017.2723798.

42 K. S. A. Sedzro, X. Shi, A. J. Lamadrid, and L. F. Zuluaga, "A heuristic approach to the post-disturbance and stochastic pre-disturbance microgrid formation problem," *IEEE Trans. Smart Grid*, vol. 10, no. 5, pp. 5574–5586, 2019, doi: 10.1109/TSG.2018.2887088.

43 M. Zhang and J. Chen, "Islanding and scheduling of power distribution systems with distributed generation," *IEEE Trans. Power Syst.*, vol. 30, no. 6, pp. 3120–3129, 2015, doi: 10.1109/TPWRS.2014.2382564.

44 H. Chavez, R. Baldick, and S. Sharma, "Governor rate-constrained OPF for primary frequency control adequacy," *IEEE Trans. Power Syst.*, vol. 29, no. 3, pp. 1473–1480, 2014, doi: 10.1109/TPWRS.2014.2298838.

45 C. L. Moreira, F. O. Resende, and J. A. P. Lopes, "Using low voltage microGrids for service restoration," *IEEE Trans. Power Syst.*, vol. 22, no. 1, pp. 395–403, 2007, doi: 10.1109/TPWRS.2006.888989.

46 D. Lagos, V. Papaspiliotopoulos, G. Korres, and N. Hatziargyriou, "Microgrid protection against internal faults: Challenges in islanded and interconnected operation," *IEEE Power Energy Mag.*, vol. 19, no. 3, pp. 20–35, 2021, doi: 10.1109/MPE.2021.3057950.

47 S. Boyd, "Distributed optimization and statistical learning via the alternating direction method of multipliers," *Found. Trends Mach. Learn.*, vol. 3, no. 1, pp. 1–122, 2010, doi: 10.1561/2200000016.

48 A. Alavian and M. C. Rotkowitz, "Improving ADMM-based optimization of mixed integer objectives," in *2017 51st Annual Conference on Information Sciences and Systems (CISS)*, Baltimore, MD, USA: IEEE, Mar. 2017, pp. 1–6. doi: 10.1109/CISS.2017.7926116.

49 F. Shen, Q. Wu, J. Zhao, W. Wei, N. D. Hatziargyriou, and F. Liu, "Distributed risk-limiting load restoration in unbalanced distribution systems with networked microgrids," *IEEE Trans. Smart Grid*, vol. 11, no. 6, pp. 4574–4586, 2020, doi: 10.1109/TSG.2020.2995099.

50 H. Heitsch and W. Römisch, "Scenario reduction algorithms in stochastic programming," *Comput. Optim. Appl.*, vol. 24, no. 2/3, pp. 187–206, 2003, doi: 10.1023/A:1021805924152.

51 P. A. Trodden, W. A. Bukhsh, A. Grothey, and K. I. M. McKinnon, "Optimization-based islanding of power networks using piecewise linear AC power flow," *IEEE Trans. Power Syst.*, vol. 29, no. 3, pp. 1212–1220, 2014, doi: 10.1109/TPWRS.2013.2291660.

52 C. Coffrin and P. Van Hentenryck, "A linear-programming approximation of AC power flows," *Inf. J. Comput.*, vol. 26, no. 4, pp. 718–734, 2014, doi: 10.1287/ijoc.2014.0594.

53 C. Wang and H. Z. Cheng, "Optimization of network configuration in large distribution systems using plant growth simulation algorithm," *IEEE Trans. Power Syst.*, vol. 23, no. 1, pp. 119–126, 2008, doi: 10.1109/TPWRS.2007.913293.

54 IEEE, *IEEE Standard for Calculating the Current-Temperature Relationship of Bare Overhead Conductors*, Nov. 1993, doi: 10.1109/IEEESTD.1993.120365.

55 M. Choobineh, B. Ansari, and S. Mohagheghi, "Vulnerability assessment of the power grid against progressing wildfires," *Fire Saf. J.*, vol. 73, pp. 20–28, 2015, doi: 10.1016/j.firesaf.2015.02.006.

56 M. Nick, O. Alizadeh-Mousavi, R. Cherkaoui, and M. Paolone, “Security constrained unit commitment with dynamic thermal line rating,” *IEEE Trans. Power Syst.*, vol. 31, no. 3, pp. 2014–2025, 2016, doi: 10.1109/TPWRS.2015.2445826.

57 D. M. Greenwood, G. L. Ingram, and P. C. Taylor, “Applying wind simulations for planning and operation of real-time thermal ratings,” *IEEE Trans. Smart Grid*, pp. 1–11, 2015, doi: 10.1109/TSG.2015.2488103.

58 A. Khodaei, “Resiliency-oriented microgrid optimal scheduling,” *IEEE Trans. Smart Grid*, vol. 5, no. 4, pp. 1584–1591, 2014, doi: 10.1109/TSG.2014.2311465.

8

Resilience of Low-Carbon Power Systems: Standards, Market, Regulatory, and Policy Aspects

8.1 Introduction

As widely discussed in this book, new methodologies and tools are needed, both in operation and in planning, to improve power system *resilience* to extreme, high-impact, low-probability (HILP)-type events. However, in spite of the exceptional amount of research and industry work that has been generated in the past few years, their *practical* implementation is still in its infancy. In particular, developing suitable policy and regulatory frameworks that can also adequately promote an efficient commercial environment is instrumental to support suitable planning, operational mechanisms, and solutions. Such frameworks are in fact not only based on technologies and engineering: they require an effective assessment and, when possible, quantification of the *full* costs that are associated with both the potential consequences of these events (weighted or not by their chance of occurrence) and the mechanisms and solutions needed to manage them [1]. What is more, the "new physics" [2] of a low-carbon grid dominated by renewables at different scales, low-carbon technologies embedded in distribution networks, and power electronics interfaces that challenge classical resource self-synchronization and damping, has direct impacts on the economics of the grid, given the manifold essential system services with differing characteristics. In this context, failing to properly link emerging technical requirements to the economics of the system may lead to the risk of creating market and commercial environments that are not fit for purpose to ensure reliability, not to mention resilience. It is therefore essential that suitable technical, commercial, regulatory, and policy measures be put in place in a coherent manner to ensure reliability and then, perhaps even more importantly, resilience, of low-carbon power systems. It is also important to highlight that such measures need to clearly address the role of *resource adequacy* and *generation resilience* too, besides *network resilience*, which has been the major focus of the previous chapters.

8.2 Weather and Low-Carbon Grid Reliability

While, as aforementioned, extreme events are increasingly affecting power systems worldwide, calling for new and effective ways to deal with HILP events, it is important to note that the situation is exacerbated by a transition to low-carbon grids. In fact, these are dominated by variable renewable energy (VRE) sources and distributed energy resources (DER) and are characterized by higher operational uncertainty. At the same time, renewable power plants are largely inverter-based resources (IBRs), whose power electronic decoupling of the resource output from grid synchronization introduces a number of stability challenges. As undermentioned, the operational challenges associated with low-carbon grids are material for both system reliability and resilience.

Fundamentals of Power System Resilience: Disruptions by Natural Causes, First Edition.
Mathaios Panteli, Rodrigo Moreno, Dimitris Trakas, Magnus Jamieson, Pierluigi Mancarella, Goran Strbac, and Nikos Hatziargyriou.

Historically, when looking at resource adequacy and general reliability studies for conventional systems based on thermal and hydro generators, focus has been on "discrete" outages and "credible" contingencies (with outages naturally assumed to be independent of each other) for generators as well as major transmission lines, for example, major tie-line interconnectors between synchronous areas. Classic reliability standards have thus been traditionally driven by discrete outage events, for example, modeled through parameters such as the forced outage rate (FOR). Overall, weather was only being considered mostly when modeling weather-sensitive (peak) demand and, in some cases, for weather-related network contingencies, with little or no consideration for the weather impact on the supply side, apart from the need to account for water availability in major hydro-dominated systems, such as Brazil or Norway. Historically, binomial distributions and their Gaussian approximations of the available conventional generation capacity have been considered a good first step in estimating different reliability parameters, including the required "capacity margins" of installed generation over the expected peak demand [3], and consequently, the potential tail risks too.

Low-carbon grids dominated by VRE, however, are much more heavily weather-dependent, which calls for a basically complete rethinking of the resource adequacy problem and, of course, system resilience. In particular, in a VRE-based system, the "available capacity" to meet demand at any point in time, and not only the peak, is now a function of the weather conditions and might significantly change on a sub-hourly, hourly, daily, seasonal, and yearly basis. In this context, traditional reliability proxies such as capacity margins may be misleading. Depending on the correlation between renewable generation output and demand, the potential tail risk of having very little renewable output available at peak times could now also be much more significant than in conventional generation systems. In fact, as supply resources might also largely be no longer "independent", given that the weather could create highly correlated renewable outputs, so limiting the "diversity" in often uncorrelated failures of conventional generators, the risk profile of a low-carbon grid could become much more heavily *tailed*. Weather could also play a further undesired effect when looking at tail risk due to the potential *anti-correlation* between available supply output from weather-based resources and weather-dependent demand, especially while electrification of heating and cooling makes demand even more weather-sensitive. Typical examples are extremely hot days with very high demand being driven by air conditioning, while the output of renewable plants is limited by loss of efficiency and derating of equipment. Other important examples are in the opposite direction too, that is, extremely cold days with very high demand driven by electrified heating, with generally lower efficiency of air-source heat pumps under such conditions, and low renewable output driven by incumbent atmospheric conditions, such as those uncommonly experienced in the UK[1].

8.3 Weather and Low-Carbon Grid Resilience

Overall, based on the previous discussions, it is evident that low-carbon grids that are naturally more weather-dependent on both the supply side and demand side are much more "fragile" [2] and may thus become intrinsically *less resilient*.

On the operational time scale, in particular, extreme weather may impact significantly system operation not only from a "normal" operation point of view (e.g. higher demand and lower renewable output, as in the examples mentioned for Australia and the UK), but also in terms of the

1 Anticyclonic weather in winter in the UK may often lead to extremely cold days with little wind output.

potential increase in risk of correlated outages and contingency events. For example, when the South Australia blackout of September 2017 event occurred, this was primarily due to issues in protection settings, which caused correlated tripping of wind farms subject to a large-front weather storm, while wind was the prevailing source of power in the region [4]. The system was also operating under a very low level of regional inertia, which led to a frequency collapse following an interconnector trip (see also Chapter 6). In a conventional system, weather events like this would not have likely led to a system collapse. However, in a weather-fueled, low-carbon environment, the correlation between widespread renewable output contingency risk and low-inertia operation has clearly increased system fragility and reduced resilience.

Another "textbook" example is the infamous Texas crisis of February 2021 [5], when tens of people died and thousands lost power and thermal comfort for several days following extremely cold weather across Texas and the nearby regions. In this case, the major cause of disruption was the correlation between extremely high demand and simultaneous unavailability of several gas generators due to the common-mode failure of gas supply through major pipelines, caused by the freezing conditions.

Challenges with operational resilience, of course, also reflect on planning aspects, as discussed in Chapter 5. In fact, while some operational emergency measures could effectively be put in place in anticipation of or in response to an unrolling extreme event, many of these measures would need to be enabled in the operational timescale, either in the form of resource investment or in the form of policy/regulatory setup, as it will be expanded on further below.

8.4 From Reliability to Resilience in Low-Carbon Grids: Need for New Security Standards

As aforementioned, low-carbon grids based on VRE, DER, and IBRs face much more significant operational challenges to maintain system security and more general reliability – historically associated with the system's ability to respond to credible contingencies and outages – and even more several challenges to provide resilience to rarer, more severe non-credible contingencies associated with HILP events. In fact, grids dominated by IBRs, VRE technologies, and DER exhibit increasing fragility of the system that makes cascading outages more likely [2]. As a consequence, *reliability* and *resilience* are concepts that somehow *merge* in low-carbon grids [6].

The fact that reliability and resilience become more intertwined is particularly evident when referring to the traditional $N-1$ or $N-2$ security criteria, based on which power systems have been historically operated to ensure customer supply even after the loss of one or two major components. Such security arrangements largely reflect the risks to the existing systems when they were developed, that is, when the generation mix was dominated by a limited number of large synchronous generation units. Most generators were located in hubs and in highly meshed parts of the grid to ensure the required security of supply. Given such structural arrangements, the main power system security risks would typically have to do with the sudden loss of specific generating units or major network elements. Normally, the outage causes would be independent as common-mode failures are extremely rare. Moreover, due to the way security standards were designed, such events would not normally lead to cascading loss of more and more components. Overall, then, contingency events could be specifically identified ("distinct") and "definable" in size, so that their impact could also be deterministically protected against with good approximation.[2]

---

2 Note that this does not mean that deterministic security criteria were most suitable or economically efficient even in a conventional environment – see [7] for further details.

However, extreme events, and particularly weather-related events as well as other natural hazards (e.g. earthquakes), can lead to multiple assets failing due to direct hazard impact. Such events may obviously be classified as N–X, with X (potentially much) greater than 2. Furthermore, extreme exogenous weather or other natural hazard stressors can lead to or exacerbate N–X contingency situations due to cascading resulting from natural system operation. Classical examples of cascading operation that have historically led to system blackouts correspond to rerouting of power flows after lines are lost due to faults or generally taken out of service by protection operation (sometimes due to "maloperation"), which overload more and more lines, causing thermal or voltage instability.

Besides the challenges introduced by extreme weather in a VRE-based supply system, the changing generation mix has introduced a completely new class of power system security risks. Security risks are in fact increasing due to what the Australian regulation has, for example, named *"indistinct" events*, following a review of the 2016 South Australia blackout [8]. These are essentially events that can affect multiple renewable generators as well as network assets in a certain affected area within seconds or minutes. However, due to the distributed nature of renewable generators, both large-scale, transmission-connected (e.g. wind farms) as well as embedded in the distribution network and down to the low voltage level (e.g. rooftop PV), the specific assets involved in the contingency event, and most importantly the size of its impact and its location, are not known before the event occurs. What is more, with a low-carbon grid featuring reduced operation, mothballing, or retirement of many large synchronous generating units traditionally providing essential system services (including inertia and system strength) coupled with the rapid deployment of VRE, DER, and IBRs, the grid becomes increasingly more fragile, with system stability becoming more precarious as both the volume of available grid services and the number of service providers becoming scarcer. Grid fragility makes cascading, again often in the form of "indistinct events", more likely to occur. Box 8.1 summarizes some examples of such emerging "indistinct" events, also schematically shown in Figure 8.1. Overall, higher weather dependency and grid fragility contribute to low-carbon systems naturally being less resilient and turning of events that in conventional systems would have been dealt with in a secure way into a threat to system resilience.

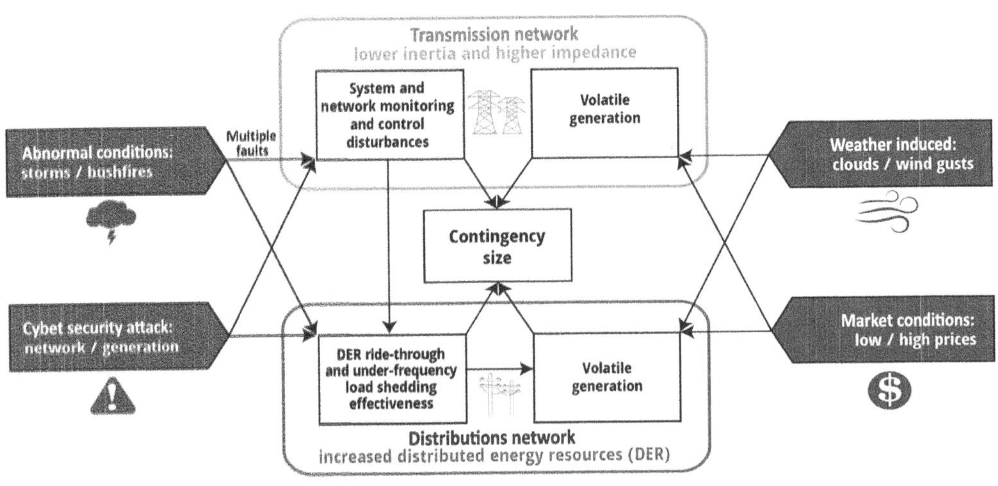

Figure 8.1 "Indistinct" events (as defined in the Australian regulatory framework) and threats to power system resilience. *Source:* [8] / IEEE.

Box 8.1 Examples of indistinct events that can challenge the resilient operation of low-carbon grids

Weather

There is an increasing risk associated with weather-based generation, such as fast-changing solar output due to cloud passage and formation or fast-changing wind output, for example, due to reaching of cut-out speed operation under storm conditions. Such generation changes could, in principle, be forecast and assessed probabilistically; however, there is huge uncertainty that emerges especially under extreme weather conditions, which makes even probabilistic tools less helpful. In addition, such weather events may affect a relatively large area and are by nature distributed, as also discussed before: they thus affect several units, and classical deterministic approaches may be difficult to apply. In several grids worldwide that are already experiencing very deep penetration of renewables, such as Australia, Great Britain, or Texas, the largest "credible" contingency could indeed be a change of outputs from several renewable units distributed across a certain area. One such example was seen in Western Australia on March 16, 2021, when, around 13:00 p.m., a large transient cloud formation over the city of Perth, which could not be forecasted by any of the three forecasters engaged by the system operator, led to a reduction of some 400 MW (about 15% of the incumbent demand of the Western Australian system) of rooftop PV output in a few minutes. This led the system frequency to plummet to 49.5 Hz, basically comparable to the impact of a large generator's trip [9].

System Response

The displacement of many conventional synchronous generation units from systems with deep penetration of renewables is leading to scarcity of a number of essential system services and variables that can support stable system operation, for example, inertia, system strength, fault level, and so forth [10]. As a consequence, the dynamic response of low-carbon grids to system disturbances may become much more uncertain and, overall, more sensitive – what we indicated as "grid fragility" [6]. Additionally, such system response uncertainty may be compounded by the "indistinct" response behavior of many DER, for example, due to sympathetic trips with transmission-level events. Hence, for example the size of a contingency due to a generator or interconnector trip may be significantly increased by further sympathetic loss of distributed generation. Typical examples of this kind may be seen in the load disconnection event of August 2019 in Great Britain [11], with large volumes of embedded generation tripping off at different times due to (mal)operation of protection relays on embedded generation. These relays responded to rate of change of frequency (ROCOF), vector shift, and under-frequency settings.

8.5 Toward a New Regulatory Framework for Resilience

Fundamentally, a new regulatory framework that fully incorporates resilience requires moving beyond *deterministic* security standards and beyond reliability metrics that are based on *expected values*.

When looking at *security*, it is not possible to economically run a power system to realize fully secure (and thus resilient too) operation against N–X contingencies that may be a direct consequence of extreme external (exogenous) events or internal (endogenous) system cascading due to grid fragility. Similarly, such a system cannot be economically built to be 100% *reliable* (and then resilient). New operational and planning approaches are therefore required, and these need to be facilitated by development of appropriate standards and regulations, within a more generally policy context, that move beyond traditional (deterministic) security criteria and established reliability standards (e.g. based on expected values). In this regard, while it is already challenging to technically understand and model system operation and planning in the presence of extreme events, it may be even more challenging to develop a suitable techno-economic regulatory framework that can deliver resilience.

However, various cases of such attempts already exist around the world. In fact, the development of regulatory frameworks in response to previous events is itself part of the "adaptation" phase of the *resilience trapezoid* (see Chapter 2). For instance, following the black system event in South Australia, an important regulatory development in Australia was the introduction of the concept of "indistinct event" aforementioned. It should be noted that these are not "formal nuances" but embody important practical consequences. In fact, in a traditional system, the system operator might only have the means to deal with system security in a deterministic way, and events that are of stochastic nature, such as basically all indistinct weather-driven events, might be difficult to act upon because of the lack of regulatory support and clear electricity rules in this regard. This was indeed the case of the South Australian blackout, whereby the system operator did not have operational tools that regulatorily would allow it to deal with stochastic weather impact, which was later resolved with the introduction of the "indistinct events" in the new security and resilience framework. In contrast, the regulatory introduction of "indistinct events" into the existing contingency management framework, for example, by *defining* a contingency event not only as "the failure or removal from operational service of plant", but also as "a sudden and unplanned change to the level of output, consumption or power flow of plant" [12], provides the system operator with much more flexibility to take decisions and schedule reserves and thus make the system more secure in the face of significant operational uncertainty.

From a more general *reliability* perspective, and including and particularly from the perspective of *resource adequacy* [13], there is a need for development of reliability standards (and more generally a reliability assessment regulatory framework), that go beyond "expected values" of typical reliability metrics (energy not supplied, loss of load probability, etc.). In practice, this might be achieved by incorporating some form of "tail risk" (see Chapter 5) in the analysis and building some form of risk aversion in a traditionally risk-neutral perspective. This was recognized in regulatory discussions in Australia [14], to deal not only with extreme events, but also, more generally, with a fast-changing risk profile due to the aforementioned low-carbon system operation challenges. Similarly, taking again as example the Texas event, the proper setup of some form of capacity mechanism and/or standards that enable adequate "weatherisation" of the available firm capacity requires a clear policy/regulatory framework that can capture the occurrence of "tail events", in this case driven by common-mode failures, and suitable ways to quantify, in terms of standards, the system's and individual resources' response to such events.

It should be noted that the same principles apply in the context of *DER*, particularly to highlight the role of DER in the provision of resilience. In fact, when looking at the regulation to incentivise adequate resource investment to enhance network resilience through non-network solutions (e.g. demand response and microgrids; see Chapter 7), there is a need to move beyond both deterministic security and expected value assessment based on general reliability metrics. This is essential in trying to capture the full value of DER, which are complementary to traditional network solutions but essentially also compete with them. While, of course, the specific outcomes may depend on the unique regulatory requirements in place in a given jurisdiction, in a traditional deterministic security context (such as *N*–1) that looks at expected events and relevant assessment, network reinforcement might just appear more appealing than DER schemes, including flexible demand disconnection and microgrids. In contrast, when extreme and rare events were to be factored in, even from an economic perspective, the value of flexibility of DER as opposed to network assets that might be relatively unutilized would fully emerge [15, 16] (see also Chapters 5 and 7).

8.6 Measuring Power System Resilience

A first essential question that has emerged in the last few years is how resilience should be assessed and measured for regulatory purposes, including, and particularly, through development of suitable standards and relevant procedures. This question stands on its own, basically independently from or, in any case, assuming the ability to model and assess power system resilience from a technical perspective, for example, based on the tools presented in the previous chapters. Of course, this question is also intrinsically related to more general techno-economic methodologies that would be required, including the development of suitable "resilience scenarios" to include in the resilience assessment and how to perform a cost-benefit analysis that looks at extreme events besides "normal" events.

It is natural to think that quantification of resilience, especially for regulatory and practical purposes, should be associated with "traditional" ways to quantify reliability and share some fundamental starting points, assumptions, and objectives. Resilience assessment could thus take the form of complementary or extended methodologies that have historically been adopted for reliability, while recognizing the specific features of resilience that have determined a need for moving beyond reliability from a technical perspective in the first place, as discussed throughout this book.

It is generally desirable that a reliable system, in the sense of its ability to respond to "credible" events, be resilient to some HILP events too, at least to some extent. In case, specific technical measures that would have been built to make the system reliable to credible events, for example some form of operating reserves or even load disconnection for frequency security purposes, could indeed also contribute to making the system resilient to extreme events. Indeed, security and reliability standards are generally already designed to deal with some extreme event and worst-case conditions (e.g. the mandatory requirements for black start services and resources). As a consequence, there is an expectation that a more reliable system, or one featuring tighter reliability standards (for instance, operated on the basis of an *N*–2 security standard as opposed to *N*–1), would also be more resilient. The opposite would also be true, having an expectation that a system that is more resilient to (at least some types of) extreme events would also *generally* be more reliable than a relatively less resilient system. Regulatory assessment methodologies, processes and standards should therefore be designed to avoid potential conflicts between the two concepts and practical implementation issues.

As a consequence, similar, or even the same, metrics could or should be used for reliability and resilience assessment. However, this requires a somehow appropriate definition of extreme event and, at least in the context of the current prevailing design of reliability standards and assessment methodologies, an estimate or at least an understanding of the potential probability of occurrence.

A natural candidate as a reliability assessment indicator that could also be extended to resilience is the *energy not supplied*, as it can indirectly capture a number of impact dimensions such as

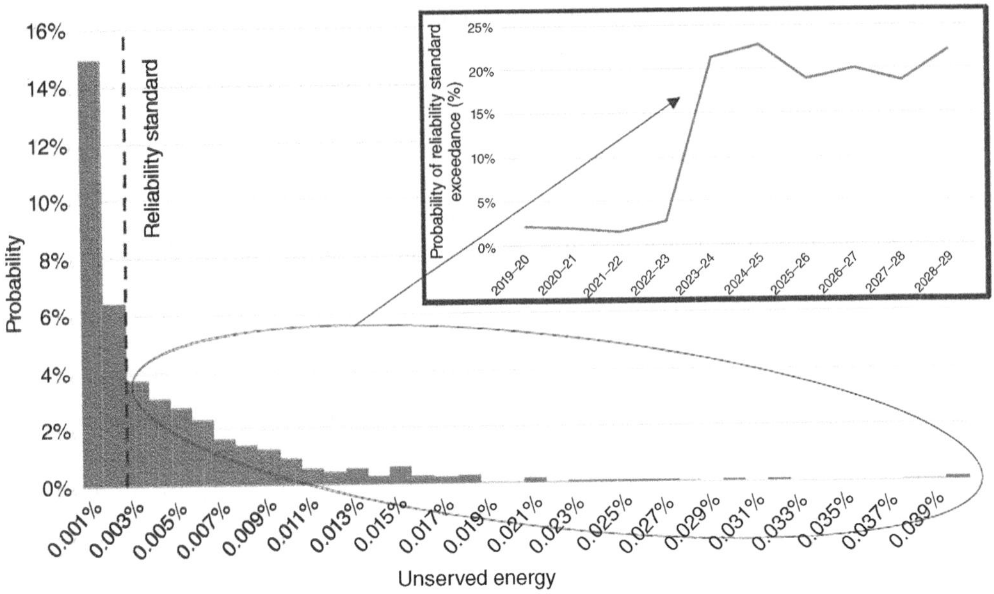

Figure 8.2 2022 projection of the distribution of annual unserved energy in New South Wales in 2023–2024 versus the incumbent reliability standard (defined as expected value). *Source:* Adapted from [14, 17].

magnitude, duration, and frequency, even if only in an integral form, especially if using the expected value of the general energy not supplied distribution as the reliability metric. However, there is a need to extend the metric to be able to capture more specifically the impact under HILP events: from the perspective of the *expected energy not supplied* (EENS) metric, many small events occurring in one year would be equivalent to one large, HILP event, without offering any specific insight on the need to protect the system against extreme events. A technically sound proposal would involve considering a *tail metric* of the energy not supplied, which would capture and somehow limit the impact that an extreme but rare event could have. This need was, for instance, captured by the Australian Energy Market Operator in [17]. This is illustrated in Figure 8.2, showing the projected distribution of energy not supplied compared to the reliability standard, highlighting the material risk of having a number of rare but high-impact events with extremely high levels of energy not supplied, even if the expected value of distribution were within the reliability standard (measured as EENS). The use of CVaR (see Chapter 5) as a tail metric that describes the energy not supplied risk of HILP events could thus be very suitable, and has, indeed, been considered as part of general regulatory discussion in Australia [14].

As resilience is also characterized by the evolution of the system before, during, and after the event, as expressed by the *resilience trapezoid* (see Chapter 2) and reported in Figure 8.3 for clarity, it is also natural to look for metrics and standards that deal with the system performance during a specific event and are inspired by the resilience trapezoid. The *area* of the trapezoid itself is a useful candidate, as such an "integral" metric synthesizes both impact and duration of an event and can be readily used in a broader reliability context. For instance, if we take the load not supplied as an indicator, the area of the trapezoid becomes the energy not supplied for a specific event under consideration. In practice, as we are considering HILP events that in the energy not supplied distribution (see Figure 8.2) would be in the tail, the area of the trapezoid would correspond to the CVaR of the energy not supplied, as previously discussed. As part of regulatory standards development,

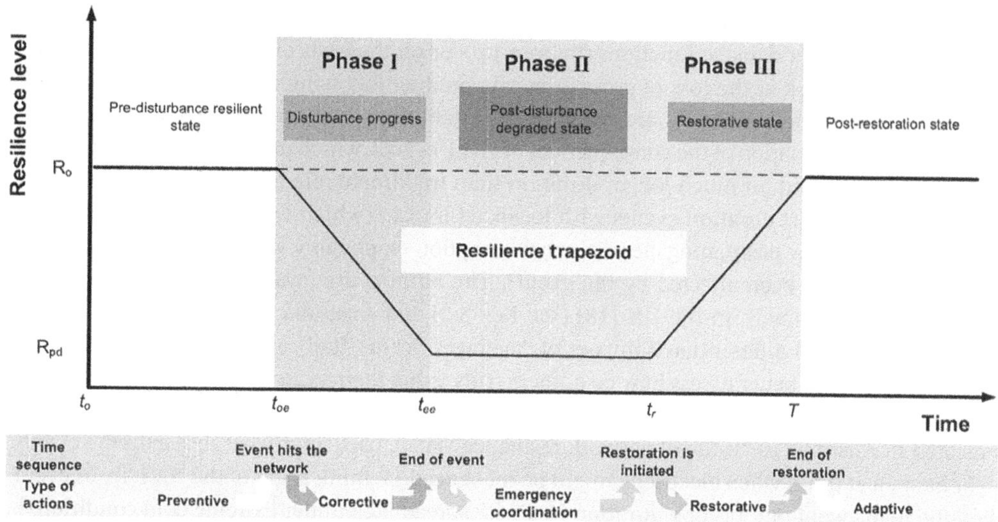

Figure 8.3 Resilience trapezoid.

then, limiting the CVaR of the energy not supplied[3] distribution, that is, the expected energy not supplied in the *tail* of the distribution, for example to be no more than a certain percentage of the expected energy not supplied for the *entire* distribution, would create incentives to limit the impact of HILP events.

It has also been discussed recently [13] that, especially in the context of weather-driven system operation, more than one metric might generally be required for reliability assessment. It is even more so for resilience, whereby there may be interest in constraining the value of specific dimensions of the impact of an extreme event as expressed by the trapezoid and the FLEP metric system (see Chapter 3), for example the magnitude of the impact or the duration of system degradation. This corresponds to creating standards that "bound" some feature of the trapezoid – for example, looking at impact duration by incentivizing the deployment of sufficient recovery resources.

The question, however, remains as to how to identify and model, in technical and then economic/regulatory terms, the relevant set of HILP events to be protected against and their impact on the grid. As previously discussed, in a decarbonized grid, the assessment of impact itself becomes more uncertain and difficult to assess, which compounds the difficulty in modeling the exogenous natural hazard itself. Regulatory processes around the world are now facing this challenge and are considering different approaches, also depending on the major events that could cause significant stress on the system (for instance, earthquakes in Chile, bushfires in Australia, and so on).

8.7 The Economics of Resilience: "Value of Customer Resilience" and Risk Aversion

Directly linked to the problem of measuring resilience for regulatory purposes is that of *valuing* resilience and integrating it in an appropriate cost-benefit analysis framework.

Approaches that may be used to assess the value of providing resilience are not dissimilar to what is done for reliability, essentially through metrics like the Value of Lost Load (VoLL). Such metrics

3 The same could be applied to other reliability indicators, such as loss of load probability.

can be estimated in different ways, for example starting from customer surveys and identification of the so-called customer damage functions (bottom-up), or on the basis of macro-economic studies that, for example, look at the loss of gross domestic product following major events (top-down). While this kind of methodologies are overall well established, one important issue is to determine whether the valuation of the consequences of HILP events, which may reach a geographically widespread impact and for much longer duration than traditional reliability events, may actually be different from short-duration events with localized impact (which would effectively allow for service substitution by postponing electricity consumption or, possibly, moving to a close location that would not have been affected by the event). The authors are aware of only a few projects around the world, namely, in the UK [18] (see Box 8.2) and Australia [19], pursuing an assessment of the potential differentiated impact of "average" versus "tail" events, and, possibly even more fundamentally, better assess how customers truly value their security of supply under different conditions, that is, the "value of customer reliability" versus the "value of customer resilience" (indicated in Australia as "value of network resilience" [19]), particularly because extreme events might happen at times when electricity might be most needed (think again of the 2021 Texas event when the lights went out for long duration in a widespread area under extreme cold conditions). The rollout of demand response programme facilitated by smart meters could provide insights into such "differentiated" levels of reliability/resilience "worth" of different customers. For example, customers buying some form of backup generation or storage may signal the desire for higher level of reliability and resilience, which could be proxied via the cost of these technologies.

Box 8.2 The "Value of Lost Load to Customers" Projects in Great Britain

1) A series of "Value of Lost Load to Customers" [18] projects was initiated in 2015 by Electricity North West (the local distribution company) in the Greater Manchester region of Great Britain. The key motivation was the observed and expected changes in electricity customer profiles, including the presence of more "electrified" customers with electric vehicles (EVs) and electric heat pumps (EHPs), besides rooftop PV and battery energy storage, and the increasing costs of outages due to an increasing number and magnitude of extreme weather events. Historically, as in other countries, *in Great Britain (GB) a single, uniform Value Of Lost Load (VoLL) has been used to evaluate the "dis-benefit" to customers of a supply interruption of average duration, and this figure is used to represent the economic measure of a supply interruption* [18]. Such VoLL is calculated as a weighted average, including different durations of interruption. However, the project, first of its kind, aimed at further delving into the factors that could differentiate the VoLL for different customers, including new, "electrified" customers (e.g. with EVs and EHPs) and customers that might be more subject to the impacts of extreme weather events such as rural customers that would be subject to bushfires, flooding, etc., and for whom the system recovery actions could be lengthier and more difficult. Different examples and findings from the project documentation point to the fact that customers with distributed energy resources, referred to as low-carbon technologies (LCTs), and electrified demand, as well as customers in areas potentially more prone to weather-driven interruptions, for example, in rural areas and especially if "fuel-poor" customers, might value reliability more than from traditional "average" assessments. Some important results could be summarized as follows: (i) Large-scale and lengthier interruptions have the potential to increase the perceived VoLL when compared to shorter, limited-scale interruptions. In particular, LCT users appear most sensitive to the longest interruption (3 days), and hence, arguably, with relatively larger-scale impact too, and this is consistent for EV and hybrid

vehicle as well as solar panel owners. (ii) The length of interruption was the biggest factor in determining the VoLL. (iii) The geographical scale of an interruption also affects its perceived impact, which may also depend on the specific consumer characteristics. For example, town/village-wide and region-wide interruptions exhibit a certain positive "multiplier effect" relative to interruptions that are limited to individual properties or streets, while the impact of a region-wide interruption (for example, involving the whole of the North West of England region) could be some 50% higher when compared to an interruption at the property level. In particular, EV/hybrid users place greater importance on a supply interruption to their property or in its immediate locality and potentially less on region-wide interruptions. This might reflect a perception of the availability of convenient local charging points and the fact that the majority of charging currently occurs at home. (iv) The frequency of an interruption might become an issue, but only when it becomes more frequent than once a year, after which the perceived VoLL increases.

In general, when coming to extreme events as opposed to more normal events, that is, when trying to assess the value that customers attach to resilience as opposed to reliability, it should be considered that customers may be much more *risk-averse* to extreme events, and this is especially true in rich countries that typically enjoy a very high level of "base" reliability. In rich countries or states like California, Australia, etc., the "social" perspective may prevail over techno-economic perspectives, and there is, in some way, a socio-economic rationale to provide hedge against extreme events. For example, load disconnection in rich suburbs of Australia on some of the hottest days of the decade, as in January 2019, caused outrage that was likely not related to practical, technical, or economic or even comfort-level impacts. Similarly, the South Australia blackout of September 2016 motivated security and resilience discussions that went well beyond the mere techno-economic impact and extended to the whole domain of energy policy to deal with renewable energy uptake and integration. The even more dramatic consequences of the 2021 Texas event led to interesting, and potentially surprising for many, developments in 2023 (see Box 8.3).

Box 8.3 Tangible Response of the Public to the Texas Blackout to Build Resources for Resilience

In April 2023, a $10 billion "Energy Insurance Program" had been discussed in Texas to improve the reliability of the Electric Reliability Council of Texas (ERCOT) grid. The greatest part of the programme would go to fund (under no little controversy) some 10 GW of (gas-fired) dispatchable power plants. While approved by the Senate, the bill did not pass in the House of Representatives. Then, in November 2023, the public approved, with great majority, a constitutional amendment for a $10 billion low-interest loan in the form of a "Texas Energy Fund" that would be administered by the Public Utility Commission of Texas (PUCT). The largest part of the fund aimed to support the construction of new or retrofitting (with weatherization) of existing dispatchable (essentially gas-fired) power plants and relevant infrastructure, while some 20% would be used for development of back-up resources and microgrids in particularly critical areas.

Besides the specific details, what is interesting to think about in the 2023 debate for, effectively, funding resources for resilience, is that the public clearly demonstrated risk aversion and the desire to "buy an insurance product" to avoid similar events in the future. While this is, of course, only anectodical, as aforementioned, there have been similar situations in other jurisdictions. In fact, recent research [20] points to the fact that consumers' "estimate", or rather "perception", of the VoLL heavily depends on the "perception" of the existing reliability level. The numerical value of VoLL may (and generally will) change with time, depending on the perceived reliability level at the time of the survey being performed to assess consumer preferences. In particular, if reliability levels are very high, as in most developed countries, normally, there will be no action taken by consumers to prepare for and respond to interruptions. Implicitly, they will value their reliability less. In contrast, a low incumbent reliability level will foster awareness and possibly consumers' investment in local back-up supply as generation or storage. Consequently, the perceived value attached to reliability will increase, and many will even go for local back-up (whose investment, per se, could represent a reasonable proxy for the equivalent VoLL). The public vote in Texas can be seen as reflecting these principles.

8.8 The Economics of Resilience: Cost–Benefit Analysis and Risk Aversion

The risk-aversion paradigm in dealing with extreme events, for which there seems to be anecdotical evidence from the consumer perspective, also appears naturally aligned with the general perspective of system planners, regulators and policy makers that aim at hedging against events that could put at significant risk social (and political) welfare and stability.

When coming to developing means to enhance system resilience on the basis of a cost–benefit analysis (CBA), both the value of customer reliability/resilience and risk aversion principles should be applied. In particular, classical CBA methodologies that look at expected (technical and economic) impacts and system performance values may not be appropriate as they do not capture risk explicitly and are fundamentally risk-neutral. As discussed for example in [21], this might stop several important developments to enhance system resilience, including through new roles played by emerging technologies such as microgrids [15, 22].

A more resilience-oriented methodology that could adopt risk-averse models such as those based on CVaR could, in contrast, be more appropriate. A hybrid approach using composite metrics, for example based on a linear combination of expected and CVaR values of a relevant economic indicator, could also be put forward [23]. As a consequence, the planning options resulting from more resilience-oriented methodologies could differ significantly from more conventional, reliability-oriented ones, with a general expectation that both system investment and operation might become more "conservative", that is, expensive, than when applying for example classical security criteria (see for instance the resilience-constrained operation of interconnected regions discussed in Chapter 6). This is effectively the resilience "premium" to pay to hedge against extreme events that would not normally be considered, which is what again we might have witnessed in the public response to the Texas event (see Box 8.3). How to agree on mature, robust decision-making and CBA resilience-oriented regulatory approaches is an open debate that also requires significant research that could involve areas such as insurance economics that specialise in dealing with valuation of rare and extreme events (see for example [24]).

8.9 The Economics of Resilience: Cost-Effectiveness Analysis

An alternative but somehow similar view of the problem has also been discussed in [25], in this case in the context of the challenges associated with the huge uncertainty in estimating the VoLL for different customer groups and circumstances already previously discussed. The alternative view provided is that "cost-effectiveness" analysis (CEA) could replace a more traditional CBA. CEA is a common tool used by policy makers in those cases where benefits or costs may be hard to evaluate. For example, for climate change, a CEA framework entails agreeing on a target temperature rise and then developing an economic analysis to identify the least-cost path to achieve that target given the relevant system constraints. Interestingly, Gorman's paper also notices how, given the 2021 blackouts in California and Texas and the resulting political backlash, using the current CBA frameworks to balance system costs and power outage costs through the VoLL remains highly contentious.

In the context of power system resilience, such a CEA framework would be equivalent to identifying different technical indicators (e.g. and in particular, energy not supplied, etc.) and relevant metrics (e.g. expected values, tail values, etc.) and establishing targets of "acceptable" values for such metrics – thus not dissimilar to what has largely been done for many years, and as is still common practice in many countries worldwide. Such technical heuristics would then be used in making investment and operational decisions to *minimize system costs while achieving the selected targets*. Of course, deciding the "acceptable" level of such metrics remains a difficult and important task, which may, however, be supported by relevant technical and techno-economic studies and expert considerations. A version of this way of thinking could also be recognized in the form of introducing "budget constraints" to provide a predefined level of resilience-oriented planning interventions, as for example shown in [23]. In general, this goes back to the issue of defining the right level of risk aversion to make optimal decisions. Hence, for example *risk aversion* could be defined as the "*weight*" used to balance tail versus average indicators in the objective function used in a planning problem (see, for instance, [23]).

8.10 Regulation, Policy, and Markets: Who Should Provide Resilience?

It is natural to think that to large extent governments are responsible to "provide resilience", especially in the context of resilience to extreme events being seen as "public good" that ensure public safety and social and political stability (thinking for example of the dramatic consequences of prolonged blackouts, both in terms of economic impact but also social unrest). This has indeed happened historically, with governments prompting the development of major initiatives (again, the "adaptation" phase of the resilience trapezoid) following extreme events, like the creation of NERC in the US in 1968 following major blackouts. With this perspective, planners, policy makers, and regulators take a *centralized* view of the resilience problem and the means to solve it, potentially by *mandating* specific interventions and requirements. In this case, the resilience enhancement problem could be "simply" delegated to engineers who could optimise for social welfare (in the public good assumption of resilience) by suitable methodologies and metrics and propose what resources to build and actions to undertake and then mandate it. For example, this could take place through grid code requirements and performance standards to be met under extreme conditions, for instance, by defining minimum strength levels of transmission tower subject to storms, minimum vegetation clearance levels for transmission corridors to prevent bushfires, and so forth – all being part of network and system planning *before* any disaster took place. However, the essential question here would still be, for governments and regulators, to establish how much should be spent (and who should pay – see later on this) to build a resilience insurance premium (presumably on top of the "normal" requirements for power system resilience) and hedge against extreme events. Likely, estimates of this kind could be based on historical data, although with the uncertainty arising with climate change using the past as the predictor of the future results increasingly challenging.

On another outlook, *could also market "deliver resilience"*, and if yes, to what extent and how? This would involve a more *decentralized* approach whereby policy and regulation should facilitate market design that can incentivise different stakeholders to react to appropriate incentives, for example consumers to reveal their level of reliability and resilience worth and market participants and investors their risk attitude. The question then arises as to how different interests and risk attitudes could be aligned, given that, for example consumers and regulators would naturally be more risk-averse than market investors. It should also be highlighted that with DER, electrification of other sectors, and the digitalization of the energy system, new technologies are likely to emerge that could provide operational resilience *in addition to* their day-to-day objectives (see [6]; Chapter 7). In particular, different forms of responsive demand could significantly improve the resilience profile of a system by providing load disconnection availability and potentially preventing cascading failures when the system is most needed.

While to a large extent, DER and demand-side flexibility could *naturally* provide some resilience, as it is likely that periods of system stress and scarcity pricing would generate high market prices to which demand would respond, a more structured approach to market design (and more generally a CBA or CEA model) to deal with extreme events is desirable. This is particularly important also to ensure some form of "fairness" across different market participants and, in particular, end-users, also considering the extreme social conditions associated with natural hazards, under which it may be (socially and politically) unacceptable to sustain extremely high prices – as for instance witnessed during the energy crisis in Australia (which could be seen as a "market resilience" issue) in June 2022, when the market had to be suspended due to sustained periods of excessively high prices [26]. Even from a technical perspective, actually, what might be perceived as fair provision of similar levels of reliability to different customers when dealing with N–1 or N–2 "credible" events could turn into very unfair, non-uniform impacts in the case of extreme events leading to N–k outages [7]. Again, probabilistic approaches, and particularly those which could incorporate some form of tail metrics and risk constraint, such as based on CVaR, are likely to be more suitable to provide more uniform outcomes. However, the situation is even more complex when looking into the economics and how different customers, also with different socio-economic backgrounds, could experience very different services under extreme event scenarios. Taking again the example of Texas in 2021, many customers were fully interrupted (with dramatic consequences including loss of life) while some parts of the grid remained totally unaffected. Under those conditions, there are several questions as to whether priority curtailment services or mandatory rationing schemes could/would have helped, whether the system operator could have done something different with a larger operational toolset or different market services at disposal, and to what extent economics should still play a predominant role under such dire circumstances or suspending markets could rather be more appropriate. All in all, economic incentives should be developed as much as possible in *preventive* mode – for example, in the case of the Texas event, by creating appropriate weatherization incentives that could enhance plant resilience.

Can markets alone provide resilience? The likely pragmatic reality is that, as hybrid portfolios of generation, storage, demand-side resources, and network and non-network solutions may be needed, and some of these technologies may compete against each other and some might be regulated, some form of regulatory coordination would probably be needed in any case. Also, even with efficient pricing that could overcome the potential issues aforementioned, concerns would still remain regarding the performance of market-driven investments and appropriate creation of incentives for investors. In fact, especially in light of climate change, the probability distribution functions of rare events are unknown and non-stationary, and private-led investment portfolios meant to hedge these risks would be difficult to justify. This problem is exacerbated by the risk aversion of self-interested investors, who require more confidence about the revenue streams associated with their investments and thus ensure "bankability" of their projects. Also, unfortunately, market operation under extreme events could still be subject to gaming opportunities, and investors might act strategically to not fully provide a robust system design and indeed preserve high prices in times of scarcity conditions. Overall, we believe that it is essential to have regulatory mechanisms and supports that could complement market developments.

8.11 Regulation, Policy and Markets: Who Should Pay for Resilience?

Whether resilience is provided through response to market incentives, centralized or decentralized planning and investment, etc., the other important question to address is who should pay for resilience and how resilience costs and benefits should be properly allocated and reflected, including the "fairness" aspect aforementioned.

Some form of stakeholder engagement and awareness may be essential here, for example by incentivizing network businesses to follow cost allocation methodologies based on "beneficiaries-pay" method. This approach would also help address the issue of the available budget to enhance resilience, as the appropriate beneficiaries would self-regulate and pay based on their marginal return and risks, thus possibly also revealing their risk attitude. On the other hand, resilience costs socialization would disintegrate these kinds of incentives, while on the other hand, at least in some cases and hopefully, improve fairness. Hence, developing suitable methodologies for appropriate cost allocation is crucial to boost resilience enhancement plans, especially in the attempt to relatively decentralize the planning and decision-making processes and in the presence of relatively little mandatory requirements.

Currently, post-extreme-event costs are generally dealt with through insurance mechanisms and Government support as part of their disaster management investment schemes. However, there also exist examples of very high penalties paid by utilities that, by not meeting the rules and standards, might have contributed to blackouts. Clear allocation of responsibilities and wielding financial penalties based on the outcomes of specific, *individual* extreme events, also depending on the impact on customers and the levels by which rules were not followed, could indeed be quite an effective way to regulate resilience, as for the cases of the Arizona–Southern California blackout in September 2011 and the UK blackout in August 2019. However, it is still open how some form of resilience services and standards should be defined, quantified, valued, and in case allocated to different actors (e.g. distributors, system operators, utilities, etc.), and whether associated penalties and incentives applied to the relevant stakeholders could ensure in the best/easiest/most practical way an optimal (e.g. "least-cost" or "least-regret") system-level provision of resilience in a decentralized form and create an appropriate environment to identify and value the risk attitude of different stakeholders.

8.12 Resilience, Decarbonization Policies, and Digitalization Paradigm

As highlighted in previous sections and chapters, both research and real-world developments are increasingly demonstrating how DER in general and microgrids in particular are able to contribute to providing system security, reliability, and resilience [8, 15, 21, 22]. In this regard, for example, the concept of resilience is now included in the UK regulatory framework that, for the first time, requires distribution network operators to develop plans for enhancing resilience of supply in district systems in case of bad weather events or driven infrastructure outages at the national level [27].

However, focusing solely on the power sector and neglecting interdependencies among multiple energy sectors would limit the resilience of future energy supply infrastructures. Furthermore, while energy decarbonization pathways and system resilience are generally considered in isolation, to achieve a resilient carbon-free energy supply cost-effectively, it is crucial to link resilience and future energy system decarbonization strategies. In this context, as also discussed in Chapter 7, Multi-Energy MicroGrids (MEMGs) [28, 29] with advanced data-driven smart control strategies could enhance system resilience while pursuing energy sector decarbonization. Furthermore, MEMGs could also provide black-start services to support the restoration of energy supply at the national level. However, overall this topic requires that fundamentally novel policy and regulatory frameworks are developed to provide appropriate incentives for development and implementation of MEMGs that would radically enhance resilience of supply of essential demand and support cost-effective transition to low/zero carbon whole-energy future.

Further links between climate policy and energy decarbonization objectives on the one hand and cost-effective delivery of infrastructure resilience on the other come from the significant growth in controllable DER associated with electrification of the heat sector and, perhaps even more, of the transport sector and the relevant transition to digitalized energy paradigms. In fact, flexibility and controllability enabled by new types of DER and digitalization will eventually enable us to differentiate between essential and non-essential demand, which may fundamentally change the way we operate and design energy systems and networks. This is because, as previously discussed, electricity supply interruptions have financial and social impacts on customers that vary by season, time of day, customer type, and customer load. Understanding the VoLL is important for energy suppliers and network operators when determining cost-effective investment and operation strategies for future energy systems. And this will become increasingly important as customers become more reliant on electricity in the low-carbon future. At present, potential overloads of energy infrastructure would be managed by demand disconnections, with some consumers being completely disconnected and some consumers fully supplied. However, the roll-out of smart metering and advanced demand control concepts (including application of artificial intelligence, energy Internet, etc.), will provide a unique opportunity for smarter management by switching off non-essential loads when system is stressed while keeping supply of essential loads, as the VoLL is radically different for essential and non-essential demand. This would result in a significant enhancement of the reliability of supply delivered by the available energy networks, as more consumers will have their essential load supplied under extreme conditions. Generally, essential demand would include medical services, emergency lighting, data center servers, traffic lights, and underground transport to save passengers. On the other side, non-essential/non-critical loads would be disconnected during major outages as these are not fundamental to the end users, such as tumble-driers, washing machines for domestic consumers, and printers, office lighting, and desk fans, for commercial demand. To the extent that flexible electrified heating could be intelligently managed without affecting the users' comfort level [30], some heating demand may also be considered non-essential, and the same can be applied said for electric vehicles, especially in the context of automated demand response [31].

Hence, the transition to electrification for whole-system decarbonization, which will bring along very significant amount of flexible and to large extent non-essential demand, and to digitalized energy paradigm, which will bring along relevant DER communication and control infrastructure, platform and algorithms, will significantly enhance resilience of supply through ability to differentiate between essential and non-essential demand when managing the system under extreme conditions in which the access to electricity supply resources is massively reduced [32]. This will also need the development of appropriate regulatory and policy strategies and incentives to support the enhancement of resilience of supply through smart control of end-use demand, which will provide cost-effective transition to resilient low-carbon energy future. In particular, in order to make use of electric buses, taxes and private vehicles that could radically enhance resilience of supply in district areas (see Chapter 7), appropriate policy strategy will need to be developed.

8.13 A Regulatory Perspective on Resilience: Final Considerations

The "new physics" of low-carbon grids, with more variable and uncertain weather-fuelled supply, a more weather-sensitive demand due to electrification of heating and transport, and pervasive IBRs *in lieu* of synchronously coupled machines, calls for new arrangements to ensure security, reliability and resilience in both operation and planning. In this context, regulation for future low-carbon grids needs to develop efficient mechanisms to value how new resources and methodologies could provide operational and planning flexibility and resilience, the latter in case with risk awareness considerations. While it is unlikely that markets alone would be able to deliver on system-level resilience, there are opportunities for a hybrid approach to technical standards, markets, and regulation to be jointly developed. For this to happen, there are different ingredients for a successful overarching regulatory framework. Such a framework, in particular, should aim to: discriminate resilience events from reliability ones while maintaining a uniform techno-economic treatment; recognize some form of risk awareness and risk aversion in decision making; identify robust methodologies to assess the impact of resilience events for customers; efficiently incorporate resilience into cost benefit analysis adopted for reliability decisions or in case introduce relevant CEA methodologies; attempt to allocate resilience costs (and benefits) in a "fair" way across consumers and more generally network users; determine the most suitable mechanisms to provide resilience (e.g. technical standards versus mandates versus market approaches); coordinate (regulated) network and (market-based) nonnetwork investments contributing to system resilience; create suitable incentives for market stakeholders, especially in the context of DER, to invest in and provide resilience resources; operate across multi-energy systems and infrastructures, especially in the context of electrification of heating and transport and the associated energy system digitalization that enable the distinction and practical implementation of essential and non-essential demand services; and finally align whole-system decarbonization energy policy and regulatory objective and pathways with resilience enhancement strategies. These are obviously non-trivial tasks, and we expect that significant resources, in terms of both research and practical developments from industry, regulatory, and institutional bodies around the world, will be needed in the coming years to identify and then build an optimal array of relevant standards, regulatory and market instruments.

References

1 F. Billimoria, P. Mancarella, and R. Poudineh, "Market and regulatory frameworks for operational security in decarbonizing electricity systems: from physics to economics," *Oxf. Open Energy*, vol. 1, 2022, doi: 10.1093/ooenergy/oiac007.

2 P. Mancarella and F. Billimoria, "The fragile grid: The physics and economics of security services in low-carbon power systems," *IEEE Power Energy Mag.*, vol. 19, no. 2, pp. 79–88, 2021, doi: 10.1109/MPE.2020.3043570.

3 R. Billinton and R. N. Allan, *Reliability Evaluation of Power Systems*. New York, NY: Springer, 1996.

4 AEMO, Integrated final black system incident report, 2017.

5 Wikipedia, (2021). Texas power crisis, https://en.wikipedia.org/wiki/2021_Texas_power_crisis.

6 P. Mancarella, "Electricity grid fragility and resilience in a future net-zero carbon economy," *Oxf. Energy Forum – Electr. Netw. Net-Zero-Carbon Econ.*, vol. 124, pp. 41–45, 2020.

7 G. Strbac, D. Kirschen, and R. Moreno, "Reliability standards for the operation and planning of future electricity networks," *Found. Trends Electr. Energy Syst.*, vol. 1, no. 3, pp. 143–219, 2016.

8 J. J. Eggleston, C. Zuur, and P. Mancarella, "From security to resilience: technical and regulatory options to manage extreme events in low-carbon grids," *IEEE Power Energy Mag.*, vol. 19, pp. 67–75, 2021.

9 J. Susanto, Operational impacts of DER integration in the SWIS, April 2021, 10.13140/RG.2.2.16261.88806.

10 L. Meegahapola, P. Mancarella, D. Flynn, and R. Moreno, "Power system stability in the transition to a low carbon grid: A techno-economic perspective on challenges and opportunities," *WIREs Energy Environ.*, vol. 10, e399, 2021.

11 C. MacIver, K. Bell, and M. Nedd, "An analysis of the August 9th 2019 GB transmission system frequency incident," *Electr. Power Syst. Res.*, vol. 199, 107444, 2021.

12 Australian Energy Market Commission (AEMC), National Electricity Amendment (Enhancing operational resilience in relation to indistinct events) Rule 2022, 2022.

13 Energy Systems Integration Group (ESIG), Redefining resource adequacy for modern power systems, 2021.

14 Australian Energy Market Commission (AEMC), 2022 Reliability Standard and Settings Review, 2022.

15 R. Moreno *et al.*, "Microgrids against wildfires: distributed energy resources enhance system resilience," *IEEE Power Energy Mag.*, vol. 20, pp. 78–89, 2022.

16 P. Apablaza, S. Püschel-Løvengreen, R. Moreno, S. Mhanna, and P. Mancarella, "Assessing the impact of DER on the expansion of low-carbon power systems under deep uncertainty," *Electr. Power Syst. Res.*, vol. 235, 110824, 2024, doi: 10.1016/j.epsr.2024.110824.

17 Australian Energy Market Operator (AEMO), 2019 NEM electricity statement of opportunities, 2019.

18 Electricity North West, (2021). Value of lost load to customer, https://www.enwl.co.uk/voll.

19 Australia Energy Regulator (AER), Value of network resilience 2024, https://www.aer.gov.au/industry/registers/resources/reviews/value-network-resilience-2024.

20 M. Over, E. Heylen, S. Proost, G. Deconinck, and D. Van Hertem, "How detailed value of lost load data impact power system reliability decisions," *Energy Policy*, vol. 132, pp. 1064–1075, 2019.

21 R. Moreno *et al.*, "From reliability to resilience: Planning the grid against the extremes," *IEEE Power Energy Mag.*, vol. 18, no. 4, pp. 41–53, 2020, doi: 10.1109/MPE.2020.2985439.

22 Y. Zhou, M. Panteli, R. Moreno, and P. Mancarella, "System-level assessment of reliability and resilience provision from microgrids," *Appl. Energy*, vol. 230, pp. 374–392, 2018.

23 T. Lagos *et al.*, "Identifying optimal portfolios of resilient network investments against natural hazards, with applications to earthquakes," *IEEE Trans. Power Syst.*, vol. 35, no. 2, pp. 1411–1421, 2020.

24 F. Billimoria, F. Fele, I. Savelli, T. Morstyn, and M. McCulloch, "An insurance paradigm for improving power system resilience via distributed investment," *IEEE Trans. Energy Mark. Policy Regul.*, vol. 1, no. 4, pp. 499–511, 2023.

25 W. Gorman, "The quest to quantify the value of lost load: A critical review of the economics of power outages," *Electr. J.*, vol. 35, 107187, 2022.

26 Australian Energy Regulator (AER), June 2022 market events report, 2022.

27 OFGEM, RIIO-ED2 business plan guidance, 2021, https://www.ofgem.gov.uk/publications/riio-ed2-business-plan-guidance.

28 E. A. Martinez-Cesena, N. Good, A. Syrri, and P. Mancarella, "Techno-economic and business case assessment of multi-energy microgrids with co-optimization of energy, reserve and reliability services," *Appl. Energy*, vol. 210, pp. 896–913, 2018.

29 Y. Wang, D. Qiu, X. Sun, Z. Bie, and G. Strbac, "Coordinating multi-energy microgrids for integrated energy system resilience: a multi-task learning approach," *IEEE Trans. Sustain. Energy*, vol. 15, pp. 920–937, 2023.

30 N. Good, E. Karangelos, A. Navarro-Espinosa, and P. Mancarella, "Optimization under uncertainty of thermal storage based flexible demand response with quantification of residential users' discomfort," *IEEE Trans. Smart Grid*, vol. 6, no. 5, pp. 2333–2342, 2015.

31 S. Altaher, P. Mancarella, and J. Mutale, "Automated demand response from home energy management system under dynamic pricing and power and comfort constraints," *IEEE Trans. Smart Grid*, vol. 6, no. 4, pp. 1874–1883, 2015.

32 Imperial College London, Review of distribution network security standards, Report to Energy Networks Association, 2015.

Index

Fundamentals of Power System Resilience: Disruptions by Natural Causes, First Edition.
Mathaios Panteli, Rodrigo Moreno, Dimitris Trakas, Magnus Jamieson, Pierluigi Mancarella, Goran Strbac, and Nikos Hatziargyriou.

Printed and bound by CPI Group (UK) Ltd, Croydon, CR0 4YY
17/09/2026
14964257-0001